Lecture Notes in Computer Science 7371

Commenced Publication in 1973
Founding and Former Series Editors:
Gerhard Goos, Juris Hartmanis, and Jan van Leeuwen

W0246084

Editorial Board

David Hutchison
  Lancaster University, UK
Takeo Kanade
  Carnegie Mellon University, Pittsburgh, PA, USA
Josef Kittler
  University of Surrey, Guildford, UK
Jon M. Kleinberg
  Cornell University, Ithaca, NY, USA
Alfred Kobsa
  University of California, Irvine, CA, USA
Friedemann Mattern
  ETH Zurich, Switzerland
John C. Mitchell
  Stanford University, CA, USA
Moni Naor
  Weizmann Institute of Science, Rehovot, Israel
Oscar Nierstrasz
  University of Bern, Switzerland
C. Pandu Rangan
  Indian Institute of Technology, Madras, India
Bernhard Steffen
  TU Dortmund University, Germany
Madhu Sudan
  Microsoft Research, Cambridge, MA, USA
Demetri Terzopoulos
  University of California, Los Angeles, CA, USA
Doug Tygar
  University of California, Berkeley, CA, USA
Gerhard Weikum
  Max Planck Institute for Informatics, Saarbruecken, Germany

Nora Cuppens-Boulahia
Frédéric Cuppens
Joaquin Garcia-Alfaro (Eds.)

Data and Applications Security and Privacy XXVI

26th Annual IFIP WG 11.3 Conference, DBSec 2012
Paris, France, July 11-13, 2012
Proceedings

 Springer

Volume Editors

Nora Cuppens-Boulahia
Frédéric Cuppens
Joaquin Garcia-Alfaro
Télécom Bretagne, Campus de Rennes 2
rue de la Châtaigneraie
35512 Cesson Sévigné Cedex, France
E-mail: {nora.cuppens, frederic.cuppens, joaquin.garcia} @telecom-bretagne.eu

ISSN 0302-9743   e-ISSN 1611-3349
ISBN 978-3-642-31539-8   e-ISBN 978-3-642-31540-4
DOI 10.1007/978-3-642-31540-4
Springer Heidelberg Dordrecht London New York

Library of Congress Control Number: 2012940756

CR Subject Classification (1998): C.2.0, K.6.5, C.2, D.4.6, E.3, H.4, C.3, H.2.7-8,
E.1

LNCS Sublibrary: SL 3 – Information Systems and Application, incl. Internet/Web
and HCI

Typesetting: Camera-ready by author, data conversion by Scientific Publishing Services, Chennai, India

Printed on acid-free paper

Springer is part of Springer Science+Business Media (www.springer.com)

Preface

This volume contains the papers presented at the 26th Annual WG 11.3 Conference on Data and Applications Security and Privacy (DBSec 2012). The conference, hosted for the first time in Paris, France, July 11–13, 2012, offered outstanding research contributions to the field of security and privacy in Internet-related applications, cloud computing and information systems.

In response to the call for papers, 49 papers were submitted to the conference. These papers were evaluated on the basis of their significance, novelty and technical quality. Each paper was reviewed by at least three members of the Program Committee. The Program Committee meeting was held electronically with intensive discussion over a period of one week. Of the papers submitted, 17 full papers and 6 short papers were accepted for presentation at the conference. The conference program also included two invited talks by Patrick McDaniel (Pennsylvania State University) and Leon van der Torre (University of Luxembourg).

Several trends in computer security have become prominent since the beginning of the new century and are considered in the program. These include the proliferation of intrusions that exploit new vulnerabilities, the emergence of new security threats against security and privacy, the need to adapt existing approaches and models to handle these threats and the necessity to design new security mechanisms for cloud computing infrastructure. Reflecting these trends, the conference includes sessions on security and privacy models, privacy-preserving technologies, secure data management, smart card, intrusion, malware, probabilistic attacks and cloud computing security.

The success of this conference was the result of the effort of many people. We would especially like to thank Joaquin Garcia-Alfaro (Publication Chair), Said Oulmakhzoune (Web Chair), Ghislaine Le Gall (Local Arrangements Chair) and Artur Hecker (Sponsor Chair). We also thank EADS/Cassidian and the Institut Mines Télécom for their financial support.

We gratefully acknowledge all authors who submitted papers for their efforts in continually enhancing the standards of this conference. It is also our pleasure to thank the members of the Program Committee and the external reviewers for their work and support.

Last but not least, thanks to all the attendees. We hope you will enjoy reading the proceedings.

July 2012

David Sadek
Frédéric Cuppens
Nora Cuppens-Boulahia

Organization

Executive Committee

General Chair

David Sadek   Institut Mines-Télécom, France

Program Chair

Nora Cuppens-Boulahia   Télécom Bretagne, France

Program Co-chair

Frédéric Cuppens   Télécom Bretagne, France

Publication Chair

Joaquin Garcia-Alfaro   Télécom Bretagne, France

Web Chair

Said Oulmakhzoune   Télécom Bretagne, France

Local Arrangements Chair

Ghislaine Le Gall   Télécom Bretagne, France

Sponsor Chair

Artur Hecker   Télécom Bretagne, France

IFIP WG 11.3 Chair

Vijay Atluri   Rutgers University, USA

Program Committee

Kamel Adi	Université du Québec en Outaouais, Canada
Gail-Joon Ahn	Arizona State University, USA
Claudio Agostino Ardagna	Università degli Studi di Milano, Italy
Vijay Atluri	Rutgers University, USA
Joachim Biskup	Technische Universität Dortmund, Germany
Marina Blanton	University of Notre Dame, USA
David Chadwick	University of Kent, UK

Additional Reviewers

Massimiliano Albanese
Damià Castellà-Martínez
Tom Chothia
William Fitzgerald
Xingze He
Deguang Kong
Meixing Le
Min Li
Giovanni Livraga
Santi Martinez-Rodriguez
Takashi Nishide
Adam O'Neill
Ruben Rios
Georgios Spathoulas
Xiaoyan Sun
Emre Uzun
Duminda Wijesekara
Jia Xu
Lei Zhang
Yufeng Zhen

Sergiu Bursuc
Ramaswamy Chandramouli
Nicholas Farnan
Nurit Gal-Oz
Masoud Koleini
Kostas Lambrinoudakis
Younho Lee
Jia Liu
Luigi Logrippo
Dieudonne Mulamba
David Nuñez
Thao Pham
Jordi Soria-Comas
Chunhua Su
Isamu Teranishi
Guan Wang
Lei Xu
Shengzhi Zhang
Yulong Zhang

Table of Contents

Privacy-Preserving Technologies

Data Management

Intrusion and Malware

Probabilistic Attacks and Protection (Short Papers)

Logics for Security and Privacy

Leendert van der Torre

Computer Science and Communication, University of Luxembourg, Luxembourg

Abstract. In this presentation I first review new developments of deontic logic in computer science, then I discuss the use of dynamic epistemic deontic logic to reason about privacy policies, and finally I discuss the use of modal logic for access control. This presentation is based on joint work with Guillaume Aucher, Guido Boella, Jan Broersen, Dov Gabbay and Valerio Genovese.

1 Introduction

In the past two decades, a number of logics and formal frameworks have been proposed to model and analyse interconnected systems from the security point of view. Recently, the increasing need to cope with distributed and complex scenarios forced researchers in formal security to employ non-classical logics to reason about these systems. I believe that logicians have a lot to benefit from specifying and reasoning about real-world scenarios as well as researchers in security can apply recent advances in non-classical logics to improve their formalisms.

2 Deontic Logic in Computer Science [3]

Over the past two decades, research in deontic logic has changed due to the participation of computer science. Broersen and van der Torre [3] discuss many traditional and new questions, centered around ten problems of deontic logic and normative reasoning in computer science. Five of these problems were discussed as philosophical problems in deontic logic by Hansen, Pigozzi and van der Torre [11], and five problems are addressed in particular in computer science.

Problem 1 - In what sense are obligations different from norms? Traditionally, people wondered whether there can be a deontic logic, given that norms do not have truth values. Nowadays, many people identify logic with reasoning, and the question is how *norms* and *obligations* are related. Instead of saying that a set of norms is consistent, two sets of norms are logically equivalent, a norm is implied by a set of norms, we have to define when a normative system is coherent, two normative systems are equivalent, or a norm is redundant in a normative system. Moreover, a new meta theory has to be developed, and relevant meta theoretic properties have to be identified.

Problem 2 - How to reason about contrary to duty norms? A difference between norms and other kinds of constraints is that norms can be *violated*, and the most discussed challenge to normative reasoning is the formalization of the *contrary-to-duty* paradoxes such as the Chisholm and Forrester paradoxes. These paradoxes receive less attention

N. Cuppens-Boulahia et al. (Eds.): DBSec 2012, LNCS 7371, pp. 1–7, 2012.

nowadays, also because they are not confined to contrary-to-duty reasoning but also contain other challenges such as according to duty reasoning associated with deontic detachment, and reasoning about time and action. But the challenge to reason about and recovering from violations is alive and kicking.

Problem 3 - How do norms change? Though *norm change* has been discussed since the early eighties, only during the last decade it has become one of the most discussed challenges. For example, researchers in normative multiagent systems identified that it is essential for a normative system application in computer science not only that norms can be violated, but in addition that norms can be changed by the agents in the system. Moreover, belief merging and its relation to judgment aggregation and social choice is emerging only recently.

Problem 4 - What is the role of time in deontic reasoning? Norms and *time* have been intimately related from the start of deontic logic, but it seems that most problems discussed in the area are not restricted to the deontic setting, but problems about temporal reasoning in general. Also in computer science and artificial intelligence, issues like deadlines where addressed in planning before they were addressed in deontic logic. For practical problems, for example in computer science, we now know that temporal references are the most elusive part of norms. However, it seems that little progress is made in understanding the challenges in the role of time in deontic logic.

Problem 5 - How to relate various kinds of permissions? In a sense, the relation between obligation and *permission* is the oldest problem in deontic logic, since Von Wright wrote his seminal paper in 1951 after he observed a similarity between the relation between necessity and possibility on the one hand, and obligation and permission on the other hand. The general opinion is that there are several kinds of permission, and it is not so easy to disentangle them. However, since permission plays a much less central role than obligation, it has received also less attention. By itself the notion of permission is also simpler than the notion of obligation, because permissions cannot be violated. The main challenge is the interaction between permission and obligation. The main interest nowadays seems to be in related legal concepts like rights and authorizations.

Problem 6 - What is the role of action in deontic reasoning? Von Wright considered his deontic *action* logic as his main contribution to the field of normative reasoning, and the first work of significance in the area was the use of dynamic deontic logic to model obligations on actions. Moreover, this is the rst problem where the agents subject to the norms come to the forefront, raising the questions how agents make decisions based on norms, or how norms are interpreted. Nevertheless, it seems that only few challenges have emerged.

Problem 7 - What is the role of constitutive norms? Constitutive norms have been used to dene meaning postulates and intermediate concepts, to define the creation of social reality using counts-as conditionals, to dene legal and institutional powers of agents, to dene the way normative systems can change, to define the interpretation of norms, and so on. However, their logical analysis has not achieved much attention. It may be expected, however, that more attention will be given to them in the future. They play a central role in many applications, for example in legal texts, there are often (much) more constitutive norms than regulative norms.

Problem 8 - How do norms influence, solve, or control games? One of our favorite challenges is to understand the relation between norms and *games*. On the one hand, it is now common to see norms as a mechanism to influence, solve, or control the interaction among agents, in particular in the area of multiagent systems. Thus, norms are useful tools in a wider context. Moreover, many problems of normative reasoning, such as norm creation, norm acceptance and norm compliance can be viewed as games, and existing game theoretic theories apply in the normative context. On the other hand, games may be seen as the foundation of deontic logic itself, defining norms as descriptions of violation or norm creation games.

Problem 9 - How do we check norm compliance? If you want to make money with deontic logic or normative reasoning, there is only one candidate: the challenge of norm *compliance*, i.e. the development of tools for automated checking of compliance to formalized sets of rules, laws and policies.

Problem 10 - How do norms interact with other modalities? How to represent and reason about boid agents and knowledge-based obligations? Traditionally norms and obligations have been studied by themselves, but nowadays the focus is on the interaction between them and *other modalities*. Some obligations hold only if you know something, and there are obligations and permissions about what you know or belief. For example, privacy policies are often expressed in what knowledge may be disclosed to who. In decision making in normative settings, there may be a trade off between fulfilling your obligations or your desires, and it may depend on your personality how you resolve such conflicts. Some interactions, such as between obligations and intentions, have hardly been studied thus far.

Finally, Broersen and van der Torre note that deontic logic has inherited from its philosophical origins the emphasis on conceptual and semantic issues, and only a few questions have actually addressed *computational* issues. This in contrast to, for example, decision theory, game theory and social choice, where new interdisciplinary disciplines of computational decision theory, computational game theory, and computational social choice have emerged over the past years. For further information on deontic logic in computer science, see:

http://www.deonticlogic.org

3   Dynamic Epistemic Deontic Logic for Privacy Compliance [1]

In general, privacy policies can be defined either in terms of permitted and forbidden *knowledge*, or in terms of permitted and forbidden *actions*. For example, it may be forbidden to know the medical data of a person, or it may be forbidden to disclose these data. Both of these approaches have their advantages and disadvantages. Implementing a privacy policy based on permitted and forbidden *actions* is relatively easy, since we can add a filter on the system checking the outgoing messages. Such a filter is an example of a security monitor. If the system attempts to send a forbidden message, then the security monitor blocks the sending of that message. However, the price to pay for this relatively straightforward implementation is that it is difficult to determine privacy policies using permitted and forbidden actions only, in the sense that it is difficult to decide which actions are permitted or forbidden so that a piece of information is not disclose. For example, it is a well known database problem that you may be able to find out my identity without asking for it explicitly, for example by asking a very detailed question (all the people who are born in Amsterdam on September 11 1986, who drive a blue Mercedes, and who are married to a person from Paris on November 9, 2009), or by combining a number of queries on a medical database [12]. Aucher, Boella and van der Torre [1] are therefore interested in privacy policies expressed in terms of permitted and forbidden knowledge.

Expressing a privacy policy in terms of permitted and forbidden knowledge is relatively easy, since it lists the situations which should not occur. These situations are typically determined by the fact that it may not be permitted to know some sensitive information. In many cases it is more efficient or natural to specify that a given piece of information may not be known, than explicitly forbidding the different ways of communicating it. The policies are more declarative, more concise and therefore easier to understand by the user. They may also cover unforeseen sequences of actions leading to forbidden situation. However, implementing a privacy policy based on permitted and forbidden knowledge is relatively difficult, since the system has to reason about the relation between permitted knowledge and actions. The challenge is that the exchange of messages changes the knowledge, and the security monitor therefore needs to reason about these changes.

To express privacy policies in terms of permitted and forbidden knowledge, we use modal logic, since both knowledge and obligations (and permissions) are traditionally and naturally modeled in branches of modal logic called epistemic and deontic logic respectively. Cuppens introduced in 1993 a modal logic for a logical formalization of secrecy [4], and together with Demolombe he developed a logic for reasoning about confidentiality [5] and a modal logical framework for security policies [6]. The logic models the knowledge of the users of the system, and allows the security monitor to reason about them. It expresses formulas such as 'the user knows the address of someone', and epistemic norms, i.e. norms regulating what is permitted to know. The security monitor is able to foresee the inferences that the users can do by combining their knowledge. For example, if the user knows street name, number, town and state of a person, then he knows his address. Moreover, since privacy policies are specified in terms of knowledge that the recipient of information is permitted/forbidden to have, we can represent violations. This is an advantage over privacy policy languages modeling norms as strict constraints that cannot be violated, because in some situations it is necessary to cope with violations. These violations can be due for example to occasional and unintentional disclosures, or to the creation of new more restrictive norms.

The main task of a security monitor reasoning about a situation given a privacy policy is to check compliance – regardless of whether these policies are expressed in terms of permitted and forbidden actions or permitted and forbidden knowledge. In our approach, to check compliance one has therefore to be able to derive the permitted, obligatory and forbidden actions in a given context, just like a decision maker needs to know whether his alternative actions do not violate norms and may therefore be subject to sanctions. In this paper, we further distinguish between regulatory compliance and behavioural compliance. Regulatory compliance checks whether the permissions and obligations set up by the security monitor of an organization (e.g., company, web-service ...) are compliant with respect to the privacy policies set up by the law/policy makers. Behavioural compliance checks whether these very obligations and permissions are indeed enforced in the system by the security monitor of the organization.

Despite its strengths, the Cuppens-Demolombe logic cannot express whether the situation is (regulatory or behaviourally) compliant with respect to a privacy policy. The problem is that the logic can define privacy policies in terms of the permitted and forbidden knowledge of the resulting epistemic state of the recipient of information, but it cannot derive the permitted messages nor the obligatory messages by combining and reasoning on this knowledge. Our modal logic addresses these problems and extends the Cuppens-Demolombe logic with dynamic update operators inspired from the ones of dynamic epistemic logic [13]. These dynamic operators model both the dynamics of knowledge and of privacy policies. They can add or remove norms from the policy, and we add constants expressing whether the system is regulatorily and behaviourally compliant with a policy, i.e., there is no violation.

Aucher, Boella and van der Torre [1] discuss the following scenario of privacy policies. They consider a single agent (Sender) communicating information from a knowledge base to another agent (Recipient), with the effect that the Recipient knows the information. The Sender is subject to privacy policies which restrict the messages he is permitted to send to the Recipient. The Sender is therefore a security monitor. They illustrate the distinction between norms of transmission of information and epistemic norms with an example:

Example 1. Consider a Sender s, e.g., a web server, which is subject to a privacy regulation: he should not communicate the address a of a person to the Recipient r. We could write this as a norm of transmission of information, regulating the sending of a message: $\neg P_s(Send\ a)$, which denotes the denial that the Sender sends message a. Instead, in an epistemic norm perspective, this prohibition can be derived from the prohibition for the Sender that the Recipient comes to know the address: $K_r a$. This is expressed by a deontic operator indexed by the Sender and having as content the ideal knowledge K_r of the Recipient: $\neg P_s K_r a$.

This distinction is bridged by modelling sending actions performed by the Sender which update the knowledge of the Recipient.

Example 2. The action of sending the message, $[Send\ a]$, expresses that the Sender sends to the Recipient the address a. The result of this action is that the Recipient knows a: $K_r a$. Since $K_r a$ is not permitted by the epistemic norm $\neg P_s K_r a$, the Sender during his decision process derives that also the action $[Send\ a]$ is not permitted: $\neg P_s(Send\ a)$. Analogously, all other possible actions leading to the forbidden epistemic state $K_r a$, if any, are prohibited too. For example, if the address is composed by street m, number n and town t such that $(m \wedge n \wedge t) \leftrightarrow a$, then the sequence of messages $[Send\ m][Send\ n][Send\ t]$ leads to the forbidden epistemic state $K_r a$.

4   Modal Logic for Access Control [2]

Boella *et al.* [2] study access control policies based on the says operator by introducing a logical framework called Fibred Security Language (FSL) which is able to deal with features like joint responsibility between sets of principals and to identify them by means of first-order formulas. FSL is based on a multimodal logic methodology. They first discuss the main contributions from the expressiveness point of view, they give semantics for the language (both for classical and intuitionistic fragment), they then prove that in order to express well-known properties like speaks-for or hand-off, defined in terms of says, they do not need second-order logic (unlike previous approaches) but a decidable fragment of first-order logic suffices. They propose a model-driven study of the says axiomatization by constraining the Kripke models in order to respect desirable security properties, they study how existing access control logics can be translated into FSL and they give completeness for the logic.

Genovese *et al.* [10] study the applicability of constructive conditional logics as a general framework to define decision procedures in access control logics. They formalize the assertion A says ϕ, whose intended meaning is that principal A says that ϕ, as a conditional implication. They introduce Cond$_{ACL}$, which is a conservative extension of the logic ICL recently introduced by Garg and Abadi. They identify the conditional axioms needed to capture the basic properties of the "says" operator and to provide a proper definition of boolean principals. They provide a Kripke model semantics for the logic and they prove that the axiomatization is sound and complete with respect to the semantics. Moreover, they define a sound, complete, cut-free and terminating sequent calculus for Cond$_{ACL}$, which allows them to prove that the logic is decidable. They argue for the generality of our approach by presenting canonical properties of some further well known access control axioms. The identification of canonical properties provides the possibility to craft access control logics that adopt any combination of axioms for which canonical properties exist.

Genovese and Garg [9] present a new modal access control logic ACL$^+$ to specify, reason about and enforce access control policies. The logic includes new modalities for permission, control, and ratification to overcome some limits of current access control logics. They present a Hilbert-style proof system for ACL$^+$ and a sound and complete Kripke semantics for it. They exploit Kripke semantics to define Seq-ACL$^+$: a sound, complete, cut-free and terminating calculus for ACL$^+$, proving that ACL$^+$ is decidable. They point at a Prolog implementation of Seq-ACL$^+$ and discuss possible extensions of ACL$^+$ with axioms for subordination between principals.

The same authors [8,7] introduce also labeled sequent calculi for access control logics.

References

1. Aucher, G., Boella, G., van der Torre, L.: A dynamic logic for privacy compliance. Artif. Intell. Law 19(2-3), 187–231 (2011)
2. Boella, G., Gabbay, D.M., Genovese, V., van der Torre, L.: Fibred security language. Studia Logica 92(3), 395–436 (2009)

3. Broersen, J., van der Torre, L.: Ten problems of deontic logic and normative reasoning in computer science. In: ESSLLI 2010/2011 Lecture Notes in Logic and Computation (2012)
4. Cuppens, F.: A logical formalization of secrecy. In: IEEE Computer Security Foundations Workshop CSFW 1993. IEEE Computer Society, Los Alamitos (1993)
5. Cuppens, F., Demolombe, R.: A deontic logic for reasoning about confidentiality. In: Deontic Logic, Agency and Normative Systems, Third International Workshop on Deontic Logic in Computer Science, DEON 1996. Springer, Berlin (1996)
6. Cuppens, F., Demolombe, R.: A Modal Logical Framework for Security Policies. In: Raś, Z.W., Skowron, A. (eds.) ISMIS 1997. LNCS, vol. 1325, pp. 579–589. Springer, Heidelberg (1997)
7. Garg, D., Genovese, V., Negri, S.: Countermodels from sequent calculi in multi-modal logics. In: 27th Annual ACM/IEEE Symposium on Logics in Computer Science - LICS 2012 (2012)
8. Genovese, V., Garg, D., Rispoli, D.: Labeled sequent calculi for access control logics: Countermodels, saturation and abduction. In: 25th IEEE Computer Security Foundations Symposium - CSF 2012 (2012)
9. Genovese, V., Garg, D.: New Modalities for Access Control Logics: Permission, Control and Ratification. In: Meadows, C., Fernández-Gago, C. (eds.) STM 2011. LNCS, vol. 7170, pp. 56–71. Springer, Heidelberg (2012)
10. Genovese, V., Giordano, L., Gliozzi, V., Pozzato, G.L.: A Conditional Constructive Logic for Access Control and Its Sequent Calculus. In: Brünnler, K., Metcalfe, G. (eds.) TABLEAUX 2011. LNCS, vol. 6793, pp. 164–179. Springer, Heidelberg (2011)
11. Hansen, J., Pigozzi, G., van der Torre, L.W.N.: Ten philosophical problems in deontic logic. In: Boella, G., van der Torre, L.W.N., Verhagen, H. (eds.) Normative Multi-agent Systems. Dagstuhl Seminar Proceedings, vol. 07122, Internationales Begegnungs-und Forschungszentrum für Informatik (IBFI), Schloss Dagstuhl, Germany (2007)
12. Sweeney, L.: k-anonymity: a model for protecting privacy. International Journal of Uncertainty, Fuzziness and Knowledge-Based Systems 10(5), 557–570 (2002)
13. van Ditmarsch, H., van der Hoek, W., Kooi, B.: Dynamic Epistemic Logic. Synthese library, vol. 337. Springer, Berlin (2007)

A User-to-User Relationship-Based Access Control Model for Online Social Networks⋆

Yuan Cheng, Jaehong Park, and Ravi Sandhu

Institute for Cyber Security, University of Texas at San Antonio
ycheng@cs.utsa.edu, {jae.park,ravi.sandhu}@utsa.edu

Abstract. Users and resources in online social networks (OSNs) are interconnected via various types of relationships. In particular, user-to-user relationships form the basis of the OSN structure, and play a significant role in specifying and enforcing access control. Individual users and the OSN provider should be allowed to specify which access can be granted in terms of existing relationships. We propose a novel user-to-user relationship-based access control (UURAC) model for OSN systems that utilizes regular expression notation for such policy specification. We develop a path checking algorithm to determine whether the required relationship path between users for a given access request exists, and provide proofs of correctness and complexity analysis for this algorithm.

Keywords: Access Control, Security, Social Networks.

1 Introduction

Access control in OSNs presents several unique characteristics different from traditional access control. In mandatory and role-based access control, a system-wide access control policy is typically specified by the security administrator. In discretionary access control, the resource owner defines access control policy. However, in OSN systems, users may want to regulate access to their resources and activities related to themselves, thus access in OSNs is subject to user-specified policies. Other than the resource owner, some related users (e.g., user tagged in a photo owned by another user, parent of a user) may also expect some control on how the resource or user can be exposed. To prevent users from accessing unwanted or inappropriate contents, user-specified policies that regulate how a user accesses information need to be considered in authorization as well. Thus, the system needs to collect these individualized partial policies, from both the accessing users and the target users, along with the system-specified policies and fuse them for the overall control decision.

In OSN, access to resources is typically controlled based on the relationships between the accessing user and the controlling user of the target found on the social graph. This type of relationship-based access control [10] takes into account the existence of a particular relationship or a particular sequence of relationships

---

⋆ This work is supported by grants from the US National Science Foundation.

N. Cuppens-Boulahia et al. (Eds.): DBSec 2012, LNCS 7371, pp. 8–24, 2012.

between users and expresses access control policies in terms of such user-to-user (U2U) relationships.

Facebook-like systems allow users to specify access control policy to related resources based on topology of the social graph, by choosing options such as "public", "private", "friend" or "friend of friend". Circles in Google+ allow users to create customized relationships. In recent years, researchers have proposed more advanced relationship-based access control models, such as [1–9, 11]. Policies in [1–6, 8, 9] can be composed of multiple types of relationships. [4–6] also adopt the depth and the trust value of relationship to control the spread of information. Although only having the "friend" relationship type, [7] provides additional topology-based policies, such as known quantity, common friends and stranger of more than k distance. While these works have their own advantages, one of the common drawbacks they share is that they do not allow different relationship types and multiple possible types on each hop.

In this paper, we propose a novel user-to-user relationship-based access control (UURAC) model and a regular expression-based policy specification language which enable more sophisticated and fine-grained access control in OSNs. To the best of our knowledge, this is the first relationship-based access control model for OSNs with such capability.

2   Motivation and Related Work

This section discusses characteristics of access control in OSNs, related works, our approach and shows our contributions.

2.1   Characteristics of Access Control for OSNs

Below, we discuss some essential characteristics [13, 14] that need to be supported in access control solutions for OSN systems.

Policy Individualization. OSN users may want to express their own preferences on how their own or related contents should be exposed. A system-wide access control policy such as we find in mandatory and role-based access control, does not meet this need. Access control in OSNs further differs from discretionary access control in that users other than the resource owner are also allowed to configure the policies of the related resource. In addition, users who are related to the accessing user, e.g. parent to child, may want to control the accessing user's actions. Therefore, the OSN system needs to collectively utilize these individualized policies from users related to the accessing user or the target, along with the system-specified policies for control decisions.

User and Resource as a Target. Unlike traditional user access where the access is against target resource, activities such as poking and friend recommendations are performed against other users. User as a target is particularly crucial for access control in OSNs since policies for users can specify rules for incoming actions as well as outgoing actions.

User Policies for Outgoing and Incoming Actions. Notification of a particular friends' activities could be bothersome and a user may want to block it. This type of policy is captured as incoming action policy. Also, a user may want to control her own or other users' activities. For example, a user may restrict her own access from any violent contents or a parent may not want her child to invite her coworker as a friend. This type of policy is captured as an outgoing action policy. In OSN, it is necessary to support policies for both types of actions.

Necessity for Relationship-Based Access Control. Access control in OSNs is mainly based on relationships among users and resources. For example, only Alice's direct friends can access her blogs, or only user who owns the photo or tagged users can modify the caption of the photo. Depth is another significant parameter, since people tend to share resources with closer users (e.g., "friend", or "friend of friend").

2.2   Prior Access Control Models for OSNs

Fong et al [7] developed a formal algebraic model for access control in Facebook-like systems. This model generalizes the Facebook-style access control mechanism into two stages: reaching the search listing of the resource owner and accessing the resource. The model formalizes policies for accessing resources as well as policies for search, traversal and communications. The policy vocabulary supports expressing arbitrary topology-based properties, such as "k common friends" and "k clique", which are beyond what Facebook offers.

In [8], Fong proposed a formal model for social computing applications, in which authorization decisions are based on the user-to-user relationships. This model employs a modal logic language for policy specification. Fong et al extended the policy language and formally characterized its expressiveness power [9]. In contrast to [7], this model allows multiple relationship types and directional relationships. Relationships and authorizations are articulated in access contexts and context hierarchy to support sharing of relationships among contexts. Bruns et al [1] later improved [8, 9] by using hybrid logic to enable better efficiency in policy evaluation and greater flexibility of atomic formulas.

Carminati et al [4–6] proposed a series of access control solutions for OSNs where the access rules are specified by the users at their discretion. The access requirements that the accessing user must satisfy are specified as type, depth, and trust metrics of the user-to-user relationships between the accessing user and the resource owner. The system features a centralized certificate authority that asserts the validity of the relationship path, while access control enforcement is carried out on decentralized user side.

In [2, 3], an access control model for OSNs is proposed by Carminati et al by utilizing semantic web technologies. Unlike many other works, this model exhibits different relationships between users and resources. It defines three kinds of access policies with the Web Ontology Language (OWL) and the Semantic Web Rule Language (SWRL), namely authorization, administration and filtering policies. Similar to [2, 3], Masoumzadeh et al [12] proposed ontology-based social network access control. Their model captures delegation of authority and empowers both users and the system to express finer-grained access control policies.

2.3 Comparison of Access Control Models for OSNs

The first four columns of Table 1 summarize the salient characteristics of the models discussed above. The fifth column gives these characteristics for the new UURAC model to be defined in this paper.

Table 1. Comparison of Access Control Models for OSNs

	Fong [7]	Fong [8, 9]	Carminati [6]	Carminati [2, 3]	UURAC
Relationship Category					
Multiple Relationship Types		✓	✓	✓	✓
Directional Relationship		✓	✓		✓
U2U Relationship	✓	✓	✓	✓	✓
U2R Relationship				✓	
Model Characteristics					
Policy Individualization	✓	✓	✓	✓	✓
User & Resource as a Target				(partial)	✓
Outgoing/Incoming Action Policy				(partial)	✓
Relationship Composition					
Relationship Depth	0 to 2	0 to n	1 to n	1 to n	0 to n
Relationship Composition	f, f of f	exact type sequence	path of same type	exact type sequence	path pattern of different types

All the models deal only with U2U relationships, except [2, 3] also recognize U2R (user-to-resource) relationships explicitly. U2R relationships can be captured implicitly via U2U with the last hop being U2R. Nevertheless, we believe that explicit treatment of U2R and R2R (resource-to-resource) relationships is important but leave it for future work.

2.4 Our Contributions

This paper develops a novel UURAC model for OSNs, using regular expression notation. UURAC supports policy individualization, user and resource as a target, distinction of user policies for outgoing and incoming actions, and relationship-based access control. It incorporates greater generality of path patterns in its policy specifications than prior models, including the incorporation of inverse relationships. We also provide an effective path checking algorithm for access control policy evaluation, along with proofs of correctness and complexity analysis.

3   UURAC Model Foundation

In this section, we develop the foundation of UURAC including basic notations, access control model components and social graph model.

3.1   Basic Notations

We write Σ to denote the set of relationship type specifiers, where $\Sigma = \{\sigma_1,\sigma_2,\ldots,\sigma_n,\sigma_1^{-1},\sigma_2^{-1},\ldots,\sigma_n^{-1}\}$. Each relationship type specifier σ is represented by a character recognizable by regular expression parser. Given a relationship type $\sigma_i \in \Sigma$, the inverse of the relationship is $\sigma_i^{-1} \in \Sigma$.

We differentiate the active and passive forms of an action, denoted *action* and *action*$^{-1}$, respectively. If Alice pokes Bob, the action is *poke* from Alice's viewpoint, whereas it is *poke*$^{-1}$ from Bob's viewpoint.

3.2   Access Control Model Components

The model comprises five categories of components as shown in Figure 1.

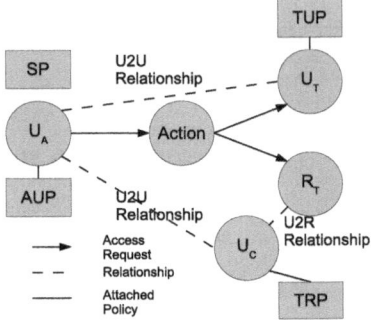

Accessing User (u_a) represents a human being who performs activities. An accessing user carries access control policies and U2U relationships with other users.

Each **Action** is an abstract function initiated by accessing user against target. Given an action, we say it is *action* for the accessing user, but *action*$^{-1}$ for the recipient user or resource.

Target is the recipient of an action. It can be either *target user* (u_t) or *target resource*

Fig. 1. Model Components

(r_t). Target user has her own policies and U2U relationship information, both of which are used for authorization decisions. Target resource has U2R relationship (i.e., ownership) with *controlling users* (u_c). An accessing user must have the required U2U relationships with the controlling user in order to access the target resource.

Access Request denotes an accessing user's request of a certain type of action against a target. It is modeled as a tuple $< u_a, action, target >$, where $u_a \in U$ is the accessing user, *target* is the user or resource that u_a tries to access, whereas *action* $\in Act$ specifies from a finite set of supported functions in the system the type of access the user wants to have with *target*. If u_a requests to interact with another user, *target* $= u_t$, where $u_t \in U$ is the target user. If u_a tries to access a resource owned by another user u_c, *target* is resource $r_t \in R$ where R is a finite set of resources in OSN.

Policy defines the rules according to which authorization is regulated. As shown in Figure 2, policies can be categorized into user-specified and system-specified policies, with respect to who defines the policies. System-specified policies (SP) are system-wide general rules enforced by the

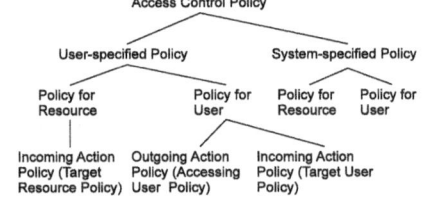

Fig. 2. Access Control Policy Taxonomy

OSN system; while user-specified policies are applied to specific users and re-sources. Both user- and system-specified policies include policies for resources and policies for users. Policies for resources are used to control who can access a resource, while policies for users regulate how users can behave regard-ing an action. User-specified policies for a resource are called *target resource policies* (*TRP*), which are policies for *incoming actions*. User-specified policies for users can be further divided into *accessing user policies* (*AUP*) and *target user policies* (*TUP*), which correspond to user's outgoing and incoming access (see examples in Section 2.1), respectively. *Accessing user policies*, also called *outgoing action policies*, are associated with the accessing user and regulate this user's outbound access. *Target user policies*, also called *incoming action policies*, control how other users can access the target user. Note that system-specified policies do not have separate policies for incoming and outgoing actions, since the accessor and target are explicitly identified.

3.3 Modeling Social Graph

As shown in Figure 3, an OSN forms a directed labeled simple graph[1] with nodes (or vertices) representing users and edges representing user-to-user relationships. We assume every user owns a finite set of resources and specifies access control policies for the resources and activities related to her. If an accessing user has the U2U relationship required in the policy, the accessing user will be granted permission to perform the requested action against the corresponding resource or user.

We model the social graph of an OSN as a triple $G = <U, E, \Sigma>$:

- U is a finite set of registered users in the system, represented as nodes (or vertices) on the graph. We use the terms user and node interchangeably from now on.
- $\Sigma = \{\sigma_1, \sigma_2, .., \sigma_n \ \sigma_1^{-1}, \sigma_2^{-1}, .., \sigma_n^{-1}\}$ denotes a finite set of relationship types, where each type specifier σ denotes a relationship type supported in the system.
- $E \subseteq U \times U \times \Sigma$, denoting social graph edges, is a set of existing user relationships.

Since not all the U2U relationships in OSNs are mutual, we define the relation-ships E in the system as directed. For ev-ery $\sigma_i \in \Sigma$, there is $\sigma_i^{-1} \in \Sigma$ representing the inverse of relationship type σ_i. We do not explicitly show the inverse relation-ships on the social graph, but assume the original relationship and its inverse twin always exist simultaneously. Given a user $u \in U$, a user $v \in U$ and a relationship

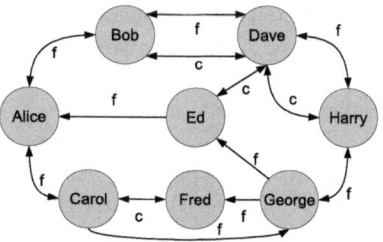

Fig. 3. A Sample Social Graph

type $\sigma \in \Sigma$, a relationship (u, v, σ) expresses that there exists a relationship of type σ starting from user u and terminating at v. It always has an equivalent form (v, u, σ^{-1}). $G = <U, E, \Sigma>$ is required to be a simple graph.

---

[1] A simple graph has no loops (i.e., edges which start and end on the same vertex) and no more than one edge of a given type between any two different vertices.

4 UURAC Policy Specifications

This section defines a regular expression based policy specification language, to represent various patterns of multiple relationship types.

4.1 Path Expression Based Policy

The user relationship path in access control policies is represented by regular expressions. The formulas are based on a set Σ of relationship type specifiers. Each specification in this language describes a pattern of required relationship types between the accessing user and the target/controlling user. We use three kinds of wildcard notations that represent different occurrences of relationship types: asterisk (*) for 0 or more, plus (+) for 1 or more and question mark(?) for 0 or 1.

4.2 Graph Rule Specification and Grammar

An access control *policy* consists of a requested action, optional target resource and a required *graph rule*. In particular, *graph rule* is defined as (*start, path rule*), where *start* denotes the starting node of relationship path evaluation, whereas *path rule* represents a collection of *path specs*. Each path spec consists of a pair (*path, hopcount*), where *path* is a sequence of characters, denoting the pattern of relationship path between two users that must be satisfied, while *hopcount* limits the maximum number of edges on the path.

Typically, a user can specify one piece of policy for each action regarding a user or a resource in the system, and the *path rule* in the policy is composed of one or more *path specs*. Policies defined by different users for the same action against same target are considered as separate policies. Multiple *path specs* can be connected by disjunction or conjunction. For instance, a path rule ($f*$, 3) ∨ ($\Sigma*$, 5) ∨ (fc, 2), where f is friend and c is coworker, contains disjunction of three different pieces of path specs, of which one must be satisfied in order to grant access. Note that, there might be a case where only users who do not have particular types of relationships with the target are allowed to access. To allow such negative relationship-based access control, a boolean negation operator over *path specs* is allowed, which implies the non-existence of the specified pair of relationship type pattern *path* and hopcount limit *hopcount* following ¬. For example, ¬ ($fc+$, 5) means the involved users should not have relationship of pattern $fc+$ within depth of 5 in order to get access.

Each graph rule usually specifies a starting node, the required types of relationships between the starting node and the evaluating node, and the hopcount limit of such relationship path. A grammar describing the syntax of such policy language is defined in Table 2. Here, *GraphRule* stands for the graph rule to be evaluated. *StartingNode* can be either the accessing user u_a, the target user u_t or the controlling user u_c, denoting the given node from which the required relationship path begins. *Path* represents a sequence of type specifiers from the starting node to the evaluating node. *Path* will typically be non-empty. If *path* is

Table 2. Grammar for graph rules

$GraphRule ::= \text{“(”} < StartingNode > \text{“,”} < PathRule > \text{“)”}$
$PathRule ::= < PathSpecExp > \mid < PathSpecExp >< Connective >< PathRule >$
$Connective ::= \vee \mid \wedge$
$PathSpecExp ::= < PathSpec > \mid \neg < PathSpec >$
$PathSpec ::= \text{“(”} < Path > \text{“,”} < HopCount > \text{“)”} \mid \text{“(”} < EmptySet > \text{“,”} < Hopcount > \text{“)”}$
$HopCount ::= < Number >$
$Path ::= < TypeExp > \mid < TypeExp >< Path >$
$EmptySet ::= \emptyset$
$TypeExp ::= < TypeSpecifier > \mid < TypeSpecifier >< Wildcard >$
$StartingNode ::= u_a \mid u_t \mid u_c$
$TypeSpecifier ::= \sigma_1 \mid \sigma_2 \mid .. \mid \sigma_n \mid \sigma_1^{-1} \mid \sigma_2^{-1} \mid .. \mid \sigma_n^{-1} \mid \Sigma$ where $\Sigma = \{\sigma_1, \sigma_2, .., \sigma_n, \sigma_1^{-1}, \sigma_2^{-1}, .., \sigma_n^{-1}\}$
$Wildcard ::= \text{“ * ”} \mid \text{“?”} \mid \text{“ + ”}$
$Number ::= [0-9]+$

empty and *hopcount* $= 0$ we assign the special meaning of "only me". *Wildcard* captures the three wildcard characters, which facilitate specifying more powerful and expressive path expressions. Given a graph rule from the access control policy, this grammar specifies how to parse the expression and to extract the containing path pattern and hopcount from the expression.

4.3   User- and System-Specified Policy Specifications

User-specified policies specify how individual users want their resources or services related to them to be released to other users in the system. These policies are specific to actions against a particular resource or user. System-specified policies allow the system to specify access control on users and resources. Different from user policies, the statements in system policies are not specific to particular accessing user or target, but rather focus on the entire set of users or resources (see Table 3).

Table 3. Access Control Policy Representations

Accessing User Policy	$< action, (start, path\ rule)>$
Target User Policy	$< action^{-1}, (start, path\ rule)>$
Target Resource Policy	$< action^{-1}, r_t, (start, path\ rule)>$
System Policy for User	$< action, (start, path\ rule)>$
System Policy for Resource	$< action, r.type, (start, path\ rule)>$

In *accessing user policy*, *action* denotes the requested action, whereas (*start, path rule*) expresses the graph rule. Similarly, *action*$^{-1}$ in *target user policy* and *target resource policy* is the passive form of the corresponding *action* applied to target user. Target resource policy contains an extra parameter r_t, representing the resource to be accessed.

This paper considers only U2U relationships in policy specification. In general, there could be one or more controlling users who have certain types of U2R relationships with the resource and possess policies for the corresponding target resource. For simplicity, we assume the only such U2R relationship is ownership. To access the resource, the accessing user must have the required relationships with the controlling user. The policies associated with the controlling users are defined on the basis of per action per resource. For instance, when querying *read* access request on r_t, *owner*(r_t) returns the list of users who have ownership with r_t. Access to r_t is under the authority of all the controlling users who have *read* policies for r_t. Note that in this paper we are not introducing the policy administration model, so who can specify the policy is not discussed.

System-specified policies do not differentiate the active and passive forms of an action. *System policy for users* carries the same format as accessing user policy does. However, when specifying *system policy for resources*, one system-wide policy for one type of access to all resources may not be fine-grained and flexible enough. Sometimes we need to refine the scope of the resources that applied to the policies in terms of resource types *r.type*. Examples of resource type *r.type* are photo, blog post, status update, etc. Thus, $<read, photo, (u_c, f*, 4)>$ is a system policy applied to all *read* access to photos in the system.

4.4 Access Evaluation Procedure

Algorithm 1. $AccessEvaluation(u_a, action, target)$

1: (Policy Collecting Phase)
2: **if** $target = u_t$ **then**
3:    $AUP \leftarrow u_a$'s policy for $action$, $TUP \leftarrow u_t$'s policy for $action^{-1}$, $SP \leftarrow$ system's policy for $action$
4: **else**
5:    $u_c \leftarrow owner(r_t)$, $AUP \leftarrow u_a$'s policy for $action$, $TRP \leftarrow u_c$'s policy for $action^{-1}$ on r_t, $SP \leftarrow$ system's policy for $action, r.type$
6: (Policy Evaluation Phase)
7: **for all** policies in AUP, TUP/TRP and SP **do**
8:    Extract graph rules (*start, path rule*) from policies
9:    **for all** graph rules extracted **do**
10:       Determine the starting node, specified by *start*, where the path evaluation starts
11:       Determine the evaluating node which is the other user involved in access
12:       Extract path rules *path rules* from graph rules
13:       Extract each path spec *path, hopcount* from path rules
14:       Path-check each path spec using Algorithm 2
15:       Evaluate a combined result based on conjunctive or disjunctive connectives between path specs
16:    Compose the final result from the result of each policy

---

Algorithm 1 specifies how the access evaluation procedure works. When an accessing user u_a requests an *action* against a target user u_t, the system will look up u_a's *action* policy, u_t's $action^{-1}$ policy and the system-specified policy corresponding to *action*. When u_a requests an *action* against a resource r_t, the system will first find out the controlling user u_c via $owner(r_t)$ and retrieve all the corresponding policies. Although each user can only specify one policy per action per target, there might be multiple users specifying policies for the same pair of action and target. Multiple policies might be collected in each of the three policy sets: AUP, TUP/TRP and SP.

Example Given the following policies and social graph in Figure 3:

- Alice's policy P_{Alice}: $< poke, (u_a, (f*, 3))> < poke^{-1}, (u_t, (f, 1))> < read, (u_a, (\Sigma*, 5))> < read^{-1}, file1, (u_c, (cf*, 4))>$
- Harry's policy P_{Harry}: $< poke, (u_a, (cf*, 5) \lor (f*, 5))> < poke^{-1}, (u_t, (f*, 2))> < read^{-1}, file2, (u_c, \neg(p+, 2)>$
- System's policy P_{Sys}: $< poke, (u_a, (\Sigma*, 5))> < read, photo, (u_a, (\Sigma*, 5))>$

When Alice requests to poke Harry, the system will look up the following policies: $< poke, (u_a, (f*, 3))>$ from P_{Alice}, $< poke^{-1}, (u_t, (f*, 2))>$ from P_{Harry}, and $< poke, (u_a, (\Sigma*, 5))>$ from P_{Sys}. When Alice requests to read photo $file2$ owned by Harry, the policies $< read, (u_a, (\Sigma*, 5))>$ from P_{Alice}, $< read^{-1}, file2, (u_c, \neg(p+, 2)>$ from P_{Harry}, and $< read, photo, (u_a, (\Sigma*, 5))>$ from P_{Sys} will be used for authorization.

For all the policies in the policy sets, the algorithm first extracts the graph rule $(start, path\ rule)$ from each policy. Once the graph rule is extracted, the system can determine where the path checking evaluation starts (using $start$), and then extracts every path spec $path, hopcount$ (from $path\ rules$). Then, it runs a path-checking algorithm (see the next section) for each path spec. The path-checking algorithm returns a boolean result for each path spec. To get the evaluation result of a particular policy, we combine the results of all path specs in the policy using conjunction, disjunction and negation. At last, the final evaluation result for the access request is made by composing all the evaluation results of the policies in the chosen policy sets.

4.5   Discussion

The existence of multi-user policies can result in decision conflicts. To resolve this, we can adopt a disjunctive, conjunctive, or prioritized approach. When a disjunctive approach is enabled, the satisfaction of any corresponding policy is sufficient for granting the requested access. In a conjunctive approach, the requirements of every involved policy should be satisfied in order that the access request would be granted. In a prioritized approach, if, for example, parents' policies get a priority over children's policies, the parents' policies overrule children's policies. While policy conflicts are inevitable in the proposed model, we do not discuss this issue in further detail here. For simplicity we assume system level policies are available to resolve conflicts in user-specified authorization policies and do not consider user-specified conflict resolution policies.

One observation from user-specified policies is that $action$ policy starts from u_a whereas $action^{-1}$ policy starts from u_t. This is because at the time of policy configuration, users are not aware of who are the other participants in the action hence cannot specify graph rule starting from the other side. When $hopcount = 0$ and $path$ equals to empty, it has special meaning of "only me". For instance, $< poke, (u_a, (\emptyset, 0))>$ says that u_a can only poke herself, and $< poke^{-1}, (u_t, (\emptyset, 0))>$ specifies u_t can only be poked by herself. The above two policies give a complementary expressive power that the regular policies do not cover, since regular policies are simply based on existing paths and limited hopcount.

As mentioned earlier, the social graph is modeled as a simple graph. Further we only allow simple path with no repeating nodes. Avoiding repeating nodes on the relationship path prevents unnecessary iterations among nodes that have been visited already and unnecessary hops on these repeating segments. On the other hand, this "no-repeating" could be quite useful when a user wants to expose her resource to farther users without granting access to nearer users. For example, in a professional OSN system such as LinkedIn, a user may want to promote her resume to users outside her current company, but does not want her coworkers to know about it. Note that the two distinct paths denoted by $(fffc)$ and (fc) may co-exist between a pair of users. The path specs $fffc \wedge \neg fc$ allows the coworkers of the user's distant friends to see the resume, while the coworkers of the user's direct friends (fc) are not authorized.

In general, conventional OSNs are susceptible to the multiple-persona problem, where users can always create a second persona to get default permissions. In a default-denial system, a new persona initially has no permission to access others, thus allowing multiple new personas from the same user is safe to the existing users. Our approach follows the default-denial design, which means if there is no explicit positive authorization policy specified, there is no access permitted at all. Based on the default-denial assumption, negative authorizations in our policy specifications are mainly used to further refine permissions allowed by the positive authorizations specified (e.g., $f * c \wedge \neg fc$). A single negative authorization without any positive authorization has the same effect as there is no policy specified at all. Nonetheless it is possible for the coworker of a direct friend to have a second persona that meets the criteria for coworker of a distant friend and thereby acquires access to the resume. Without strong identities we can only provide persona level control in such policies.

5   Path Checking Algorithm

In this section, we present the algorithms for determining if there exists a qualified path between two involved users in an access request.

As mentioned, in order to grant access, relationships between the accessor and the target/controlling user must satisfy the graph rules specified in access control policies regarding the given request. We formulate the problem as follows: given a social graph G, an access request $< u_a, action, target >$ and an access policy, the system decision module explores the graph and verifies the existence of a path between u_a and $target$ (or u_c of $target$) matching the graph rule $< start, path$ $rule >$.

Path checking is performed by Algorithm 2, which takes as input the social graph G, the path pattern $path$ and the hopcount limit $hopcount$ specified by $path$ $spec$ in the policy, the starting node s specified by $start$ and the evaluating node t which is the other user involved, and returns a boolean value as output. Note that $path$ is non-empty, so this algorithm only copes with cases where $hopcount \neq 0$. The starting node s and the evaluating node t can be either the accessing user or the target/controlling user, depending on the given policy. The algorithm starts by constructing a DFA (deterministic finite automata) from the regular expression $path$. The REtoDFA() function receives $path$ as input, and converts it to a NFA (non-deterministic finite automata) then to a DFA, by using the well-known Thompson's Algorithm [16] and Subset Construction Algorithm (also known as Büchi's Algorithm) [15], respectively.

The algorithm uses a depth-first search (DFS) to traverse the graph, because it requires only one running DFA and, correspondingly, one pair of variables

Algorithm 2. $PathChecker(G, path, hopcount, s, t)$

1: $DFA \leftarrow REtoDFA(path)$; $currentPath \leftarrow NIL$; $d \leftarrow 0$
2: $stateHistory \leftarrow$ DFA starts at the initial state
3: **if** $hopcount \neq 0$ **then**
4:      **return** DFST(s)

Algorithm 3. $DFST(u)$

1: **if** $d + 1 > hopcount$ **then**
2:      **return** FALSE
3: **else**
4:      **for all** (v, σ) where (u, v, σ) in G **do**
5:          **switch**
6:          **case 1** $v \in currentPath$
7:          break
8:          **case 2** $v \notin currentPath$ and $v = t$ and DFA with transition σ is at accepting state
9:          $d \leftarrow d + 1$; $currentPath \leftarrow currentPath.(u, v, \sigma)$
10:          $currentState \leftarrow$ DFA takes transition σ
11:          $stateHistory \leftarrow stateHistory.(currentState)$
12:          **return** TRUE
13:          **case 3** $v \notin currentPath$ and $v = t$ and transition σ is valid for DFA but DFA with transition σ is not at accepting state
14:          break
15:          **case 4** $v \notin currentPath$ and $v \neq t$ and transition σ is invalid for DFA
16:          break
17:          **case 5** $v \notin currentPath$ and $v \neq t$ and transition σ is valid for DFA
18:          $d \leftarrow d + 1$; $currentPath \leftarrow currentPath.(u, v, \sigma)$
19:          $currentState \leftarrow$ DFA takes transition σ
20:          $stateHistory \leftarrow stateHistory.(currentState)$
21:          **if** (DFST(v)) **then**
22:              **return** TRUE
23:          **else**
24:              break
25:      **if** $d = 0$ **then**
26:          **return** FALSE
27:      **else**
28:          $d \leftarrow d - 1$; $currentPath \leftarrow currentPath \backslash (u, v, \sigma)$
29:          $previousState \leftarrow$ last element in $stateHistory$
30:          DFA backs off the last taken transiton σ to $previousState$
31:          $stateHistory \leftarrow stateHistory \backslash (previousState)$
32:          **return** FALSE

keeping the current status and the history of exploration in a DFS traversal. Whereas, a breadth-first search (BFS) traversal has to maintain multiple DFAs and multiple variables simultaneously and switch between these DFAs back and forth constantly, which makes the costs of memory space and I/O operations proportional to the number of nodes visited during exploration. Note that DFS could avoid a target node for a longer time, even if the node is close to the starting node. If the hopcount is unlimited, a DFS traversal may pursue lengthy useless exploration. However, activities in OSN typically occur among people with close relationships. Hence, DFS with limited hopcount fits our model.

The variable $currentPath$, initialized as NIL, holds the sequence of the traversed edges between the starting node and the current node. Variable $stateHistory$, initialized as the initial DFA state, keeps the history of DFA states during algorithm execution. The main procedure starts by setting d to 0 and launches the DFS traversal function $DFST()$, given in Algorithm 3, from the starting node s.

Given a node u, if $d + 1$ does not exceed the hopcount limit, it indicates that traversing one step further from u is allowed. Otherwise, the algorithm returns false and goes back to the previous node. If further traversal is allowed, then the algorithm picks up an edge (u, v, σ) from the list of the incident edges leaving u. If (u, v, σ) is unvisited, we get the node v on the opposite side of the edge (u, v, σ). Now we have five different cases. If v is on $currentPath$, we will never visit v again, because doing so creates a cycle on the path. Rather, the algorithm breaks out of for loop, and finds the next unchecked edges of u. When v is not on $currentPath$ and v is the target node t and DFA taking transition σ reaches an accepting state, we find a path between s and t matching the pattern $Path$. We increment d by one, concatenate edge (u, v, σ) to $currentPath$, and save the current DFA state to history. If v is the target node but DFA with transition σ is not at an accepting state, then the path from s to v does not match the pattern. When v is not on $currentPath$ and is not the target node, there are two cases depending on whether the transition type σ is a valid transition for DFA. If it is not, we break out of for loop and continue to check the next unchecked edge of u. Otherwise, the algorithm increments d by one, concatenates e to $currentPath$, moves DFA to the next state via transition type σ, updates the DFA state history, and repeatedly executes $DFST()$ from node v. If the recursive function call discovers a matching path, the previous call also returns true. Otherwise, it checks next edge of node u.

After all the outgoing edges of u have been checked, the algorithm has to step back to the previous node of u and reset all variables to the previous values. But if $d = 0$, all the outgoing edges of the starting node are checked, thus the whole execution completes without a matching path.

In Figure 3, suppose user Harry owns a resource r_t and expresses the target resource policy as $(read^{-1}, r_t, (f * cf*, 3))$, where $read$ is the permitted action, $(f * cf*, 3)$ is the path pattern and hopcount limit. Path pattern $f * cf*$ means the accessing user and Harry must be either a pair of coworkers (c) or direct or indirect friend (f) of a pair of coworkers. Hopcount 3 constrains the distance between the two users to be within three hops. Figure 4 shows the DFA accepting the path pattern $f * cf*$. If Alice requests read access to the resource owned by Harry, the algorithm starts exploration from node H (Harry) by checking all the edges leaving H. If it picks the edge (H, D, f) or (H, D, c) first, it will eventually find out that there exists a satisfiable path $(H, D, f), (D, E, c), (E, A, f)$ or $(H, D, c), (D, B, f), (B, A, f)$ that also moves the DFA from the starting state π_0 to the accepting state π_3 in three hops. $(H, G, f), (G, F, f), (F, C, c), (C, A, f)$ also matches the path pattern, but it is invalid because it takes four hops to reach node A.

Suppose Harry specifies a target user policy for him as $(poke^{-1}, (f+, 2))$. This implies only his friends or indirect friends can poke him. Then, Bob, Dave, Ed, Fred and George can poke Harry because the paths between Harry and them contain relationship f and are within depth of two. Carol and Harry do not have friend relationship with Harry, while Alice is too far away from Harry.

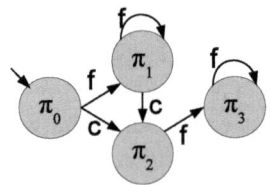

Fig. 4. DFA for $f * cf*$

6 Conclusions and Future Work

We proposed a UURAC model and a regular expression based policy specification language. We provided a DFS-based path checking algorithm and established its correctness and complexity. Correctness of the algorithm is proved by induction on hopcount. Due to the sparseness nature of social graph, given the constraints on relationship types and hopcount limit in policy, the complexity of the algorithm can be dramatically reduced. Proofs of correctness and complexity are given in appendix.

While this work only uses user-to-user relationships for authorization, we plan to extend our model to exploit user-to-resource and resource-to-resource relationships. To improve the expressiveness of the model, we also plan to incorporate some predicate expressions for attribute-based control and filtering users and relationships. Another future direction is to capture some unconventional relationships in OSNs, such as temporary relationships (i.e., vicinity) and one-to-many relationships (i.e., network, group). Last but not least, we will be working on implementing our approach into a prototype and doing some experiments to analyze the approach.

References

1. Bruns, G., Fong, P.W., Siahaan, I., Huth, M.: Relationship-based access control: its expression and enforcement through hybrid logic. In: ACM CODASPY (2012)
2. Carminati, B., Ferrari, E., Heatherly, R., Kantarcioglu, M., Thuraisingham, B.: A semantic web based framework for social network access control. In: ACM SACMAT (2009)
3. Carminati, B., Ferrari, E., Heatherly, R., Kantarcioglu, M., Thuraisingham, B.: Semantic web-based social network access control. Computers and Security 30(2-3) (2011); Special Issue on Access Control Methods and Technologies
4. Carminati, B., Ferrari, E., Perego, A.: Rule-Based Access Control for Social Networks. In: Meersman, R., Tari, Z., Herrero, P. (eds.) OTM 2006 Workshops, Part II. LNCS, vol. 4278, pp. 1734–1744. Springer, Heidelberg (2006)
5. Carminati, B., Ferrari, E., Perego, A.: A decentralized security framework for web-based social networks. Int. Journal of Info. Security and Privacy 2(4) (2008)
6. Carminati, B., Ferrari, E., Perego, A.: Enforcing access control in web-based social networks. ACM Trans. Inf. Syst. Secur. 13(1) (2009)
7. Fong, P.W.L., Anwar, M., Zhao, Z.: A Privacy Preservation Model for Facebook-Style Social Network Systems. In: Backes, M., Ning, P. (eds.) ESORICS 2009. LNCS, vol. 5789, pp. 303–320. Springer, Heidelberg (2009)
8. Fong, P.W.: Relationship-based access control: protection model and policy language. In: ACM CODASPY (2011)
9. Fong, P.W., Siahaan, I.: Relationship-based access control policies and their policy languages. In: ACM SACMAT (2011)
10. Gates, C.E.: Access control requirements for web 2.0 security and privacy. In: Proc. of Workshop on Web 2.0 Security and Privacy, W2SP 2007 (2007)
11. Kruk, S.R., Grzonkowski, S., Gzella, A., Woroniecki, T., Choi, H.-C.: D-FOAF: Distributed Identity Management with Access Rights Delegation. In: Mizoguchi, R., Shi, Z.-Z., Giunchiglia, F. (eds.) ASWC 2006. LNCS, vol. 4185, pp. 140–154. Springer, Heidelberg (2006)

12. Masoumzadeh, A., Joshi, J.: Osnac: An ontology-based access control model for social networking systems. In: IEEE Social Computing, SocialCom (2010)
13. Park, J., Sandhu, R., Cheng, Y.: Acon: Activity-centric access control for social computing. In: Int. Conf. on Availability, Reliability and Security, ARES (2011)
14. Park, J., Sandhu, R., Cheng, Y.: A user-activity-centric framework for access control in online social networks. IEEE Internet Computing 15(5) (September-October 2011)
15. Rabin, M.O., Scott, D.: Finite automata and their decision problems. IBM J. Res. Dev. 3 (April 1959)
16. Thompson, K.: Programming techniques: Regular expression search algorithm. Commun. ACM 11 (June 1968)

A   Proof of Correctness

Theorem 1. *Algorithm 2 will halt with true or false.*

Proof. Base case ($Hopcount = 1$): d is initially set to 0. Each outgoing edge from the starting node s will be examined once and only once. If taking an edge reaches the target node t and its type matches the language $Path$ denotes (case 2), the algorithm returns true. If the edge type matches the prefix of an expression in $L(Path)$ (lines 17-24), d increments to 1 followed by a recursive call to $DFST()$. The second level call will return false, since incremented d has exceeded $Hopcount$. In all other cases, the examined edge is discarded and d remains the same. Eventually, if a matching edge is not found, the algorithm will go through every outgoing edge from s and exit with false thereafter (lines 25-26).

Induction step: Assume when $Hopcount = k$ ($k \geq 1$), Theorem 1 is true. When $Hopcount$ is $k+1$, all the $(k+1)$th level recursive calls will examine every outgoing edge from the $(k+1)$th node on $currentPath$. If visiting an edge reaches t and the updated $currentPath$ matches $L(Path)$, the $(k+1)$th level call returns true and exits to the previous level, making all of the previous level calls all the way back to the first level exit with true as well. If an edge falls into case 5, d is incremented to $k+2$ and a $(k+2)$th level recursive call invokes, which will halt with false and return to the $(k+1)$th level as d has exceeded $Hopcount$. After all edges are examined without returning true, the algorithm will exit with false to the previous level. In the kth level, when $Hopcount = k+1$, edges without taking a recursive call are treated the same as they are when $Hopcount = k$. Since when $Hopcount = k$ the theorem holds, the algorithm will terminate with true or false when $Hopcount = k+1$ as well.

Lemma 1. *At the start and end of each $DFST()$ call, the DFA corresponding to Path is at currentState reachable from the starting state π_0 by transitions corresponding to the sequence of symbols in currentPath.*

Proof. The proof is straightforward. New edge is added to $currentPath$ only when it reaches the target node (lines 8-12) or it may possibly lead to the target node by taking a recursive $DFST()$ call (lines 17-24). In both cases the DFA starting from π_0 will move to $currentState$ by taking the transition regarding the edge. Removing the last edge on $currentPath$ after all edges leaving the current node are checked always accompanies one step back-off of the DFA to its previous state (lines 28-32), which can eventually take the DFA all the way back to the starting state π_0.

Theorem 2. *If Algorithm 2 returns true, currentPath gives a simple path of length less than or equal to Hopcount and the string described by currentPath belongs to the language described by Path (L(Path)). If Algorithm 2 returns false, there is no simple path p of length less than or equal to Hopcount such that the string representing p belongs to L(Path).*

Proof. Base case ($Hopcount = 1$): At first, $d = 0$, $currentPath = $ NIL, and the DFA is at the starting state π_0. When $d = 0$, case 1 requires that the edge being checked is a self loop which is not allowed in a simple graph. $DFST()$ only returns true in case 2, where edge (s, t, σ) to be added to $currentPath$ finds the target node t in one hop. The transition σ moves the DFA to an accepting state. Case 5 cannot return true, because incrementing d by one will exceed $Hopcount$ in the recursive $DFST()$ run. When DFST() exits with true, due to Lemma 1, $currentPath$, which is (s, t, σ), can move the DFA from π_0 to an accepting state π_1, implying that $\sigma \in L(Path)$. If the first $DFST()$ call returns false (lines 29-30), the algorithm has searched all the edges leaving node s. However, these examined edges either do not match the pattern specified by $L(Path)$ (case 2 and 3), or may possibly match $L(Path)$ but require more than one hop (case 5). Hence, Theorem 2 is true when $Hopcount = 1$.

Induction step: Assume when $Hopcount = k$ ($k \geq 1$), Theorem 2 is true. For the same G, $Path$, s and t, executions of $DFST()$ when $Hopcount = k$ and $k + 1$ only differ after invoking the recursive $DFST()$ call in case 5. If an edge being checked can make the algorithm return true when $Hopcount = k$, $currentPath$ is a string of length $\leq k$ which is in $L(Path)$. When $Hopcount$ is $k + 1$, the same $currentPath$ gives the same string and is of length $< k + 1$, thus making the function exit with true as well. The only difference between $Hopcount = k$ and $Hopcount = k + 1$ is that adding edges that lie in case 5 to $currentPath$ and incrementing d by one may not exceed the larger $Hopcount$ during the recursive call. If taking one of these edges leads to the target node and its corresponding transition moves the DFA to an accepting state, the algorithm will return true. The new $currentPath$ gives a simple path of length $k + 1$ that connects node s and t. The algorithm only returns true in these two scenarios. In both scenarios, based on Lemma 1, the DFA can reach an accepting state by taking the transitions corresponding to $currentPath$, so the string corresponding to $currentPath$ is in $L(Path)$. If the algorithm returns false when $Hopcount = k$, there is no simple path p of length $\leq k$, where the string of symbols in p is in $L(Path)$. When $Hopcount$ is $k + 1$, given the same G, such a path still does not exist. By taking a recursive $DFST()$ call in case 5, the algorithm will go through all 5 cases again to check all the edges leaving the new node. If the recursive call returns false, it means there is no simple path of length $k + 1$

with its string of symbols in $L(Path)$. Combining the results from all $k+1$ level recursive calls, there exists no simple path of length $\leq k+1$ with its string of symbols in $L(Path)$. Hence, Theorem 2 is true when $Hopcount = k+1$.

B Complexity

In this algorithm, every possible path from s to t will be visited at most once until it fails to reach t, while every outgoing edge of a visited node may be checked multiple times during the search. In the extreme case, where every relationship type is acceptable and the graph is a complete directed graph, the overall complexity would be $O(|V|^{Hopcount})$. However, users in OSNs usually connect with a small group of users directly, thus the social graph is actually very sparse. We define the maximum and minimum out-degree of node on the graph as $dmax$ and $dmin$, respectively. Then, the time complexity can be bounded between $O(dmin^{Hopcount})$ and $O(dmax^{Hopcount})$. Given the constraints on the relationship types and hopcount limit in the policies, the size of graph to be explored can be dramatically reduced. The recursive DFST() call terminates as soon as either a matching path is found or the hopcount limit is reached.

Automated and Efficient Analysis of Role-Based Access Control with Attributes

Alessandro Armando[1,2] and Silvio Ranise[2]

[1] DIST, Università degli Studi di Genova, Italia
[2] Security and Trust Unit, FBK-Irst, Trento, Italia

Abstract. We consider an extension of the Role-Based Access Control model in which rules assign users to roles based on attributes. We consider an open (allow-by-default) policy approach in which rules can assign users negated roles thus preventing access to the permissions associated to the role. The problems of detecting redundancies and inconsistencies are formally stated. By expressing the conditions on the attributes in the rules with formulae of theories that can be efficiently decided by Satisfiability Modulo Theories (SMT) solvers, we characterize the decidability and complexity of the problems of detecting redundancies and inconsistencies. The proof of the result is constructive and based on an algorithm that repeatedly solves SMT problems. An experimental evaluation with synthetic benchmark problems shows the practical viability of our technique.

1 Introduction

Role Based Access Control (RBAC) [27] is one of the most widely adopted model for information security. It regulates access by assigning users to roles which, in turn, are granted permissions to perform certain operations. Despite several advantages (e.g., reduction of the complexity of security administration), it has been observed that RBAC suffers some inflexibility in adapting to rapidly evolving domains in particular when there is a need to take into account dynamic attributes to determine the permissions of a user. The problem is the explosion in the number of roles that should be considered in order to specify the various sets of permissions that a user may acquire depending on the values of the dynamic attributes.

To overcome this problem, [16] proposes to add attributes to RBAC and overviews several approaches to do this. Among these, the one called *Dynamic Roles (DRs)* determines the user's role depending on his/her attributes. The interest of DRs is that they retain the structure of the RBAC model while providing additional flexibility to cope with attributes. A similar approach have already been investigated in [2,3] and problems concerning redundancies in the rules that assign users to roles and possible inconsistencies with the standard RBAC role hierarchy are studied. Furthermore, in [4], negative roles are introduced to widen the scope of applicability of the RBAC model to those situations in which an open (allow-by-default) policy approach is adopted (i.e. access is

N. Cuppens-Boulahia et al. (Eds.): DBSec 2012, LNCS 7371, pp. 25–40, 2012.
© IFIP International Federation for Information Processing 2012

denied if there exists a corresponding negative authorization and is permitted otherwise [26]). In this context, conflicts may arise among rules that assign users to a role and, at the same time, to the negated role.

Clearly, simply authoring a set of rules that assign users to roles is not sufficient: an organization must also be able to analyse it in order to avoid the kind of problems sketched above. Testing, while useful, is not exhaustive and, as organizations grow, their rule set can become very large, which places particular burden on the quality of testing. Designers of RBAC policies with DRs would thus benefit from complementing testing with more exhaustive and automatic, formal analysis techniques. In this respect, the paper makes three contributions.

First, we give an abstract definition for policies with DRs that naturally extends the RBAC model (Section 2). Negative authorization (via negated roles) can be easily accommodated in the proposed framework (Section 2.3).

Second, we provide a rule-based characterization of the association between users' attributes and roles (Section 2) that allows us to formally state two crucial problems for the design of RBAC policies with DRs: detection of redundancies (Section 2.1) and detection of conflicts (Section 2.3).

Third (Section 4.1), we show how conditions on the users' attributes of authorization rules can be expressed by theories, whose satisfiability problems can be efficiently decided by Satisfiability Modulo Theories (SMT) solvers (Section 3) and characterize the decidability and the complexity of the problems of detecting redundancies and conflicts under natural assumptions on the theories of the attributes (Sections 4.2 and 4.3). An experimental evaluation with a prototype implementation of the algorithm (Figure 1) used for the decidability results on a synthetic benchmark confirms the viability of the approach (Section 4.4).

Related work and conclusions are also discussed (Section 5). Proofs can be found in the extended version of this paper available at http://st.fbk.eu/SilvioRanise#Papers.

2   RBAC with Dynamic Roles

Preliminarily, we recall some basic notions concerning the RBAC model [27]. RBAC regulates access through roles. Roles in a set R associate permissions in a set P to users in a set U by using the following two relations: $UA \subseteq U \times R$ and $PA \subseteq R \times P$. Roles are structured hierarchically so as to permit permission inheritance. Formally, a role hierarchy is a partial order RH on R, where $(r_1, r_2) \in RH$ means that r_1 is *more senior than* r_2 for $r_1, r_2 \in R$. A user u is an *explicit member* of role r when $(u, r) \in UA$, u is an *implicit member* of r if there exists $r' \in R$ such that $(r', r) \in RH$ and $(u, r') \in UA$, and u is a *member* of role r if he/she is either an implicit or explicit member of r. Given UA and PA, a user u *has permission* p if there exists a role $r \in R$ such that $(p, r) \in PA$ and u is a member of r. A *RBAC policy* is a tuple (U, R, P, UA, PA, RH).

When extending RBAC with dynamic roles, the relation UA is not given explicitly but it is defined in terms of pairs (attribute name, attribute value). A *RBAC policy with Dynamic Roles (RBAC-DR)* is a tuple $(U, R, P, \underline{a}, \underline{D_a}, \underline{s_a}, AR,$

PA, RH) where U, R, P, PA, and RH are as in the RBAC model above, $\underline{a}$ is a (finite) sequence of *attributes*, $\underline{D}_{\underline{a}}$ is a (finite) sequence of domains associated to $\underline{a}$, $\underline{s}_{\underline{a}}$ is a sequence of *user-attribute mappings associated to $\underline{a}$* (the sequences $\underline{a}$, $\underline{D}_{\underline{a}}$, and $\underline{s}_{\underline{a}}$ have equal length) such that for each a in $\underline{a}$ (written $a \in \underline{a}$) s_a is a function from U to 2^{D_a}, and AR is the *attribute-role relation* that contains tuples of the form (C, r) where C is a set of pairs (a, e_a) with $e_a \in D_a$ for $a \in \underline{a}$ and $r \in R$. (In the following, the sub-script of $\underline{D}_{\underline{a}}$ and $\underline{s}_{\underline{a}}$ will be dropped to simplify the notation.) Given a RBAC-DR policy $(U, R, P, \underline{a}, \underline{D}, \underline{s}, AR, PA, RH)$, a user $u \in U$ *is an explicit member of role $r \in R$ under the user-attribute mapping $\underline{s}$* iff there exist a pair $(C, r) \in AR$ and a set $S_u \subseteq C$ such that $e_a \in s_a(u)$ for every $(a, e_a) \in S_u$. The notions of "being an implicit member of a role" and "having a permission" are defined as those of RBAC policies above.

2.1   Rule-Based Authorization Rules

The most difficult part in the design of a RBAC-DR policy is the definition of the attribute-role relation. To do this, following [2], we use authorization rules that associate a role to a user provided that his/her attribute values satisfy certain conditions. Let $\underline{a} = a_1, ..., a_n$ be a sequence of attributes, a pair $(\underline{b}, C)$ is a *condition C on a subsequence* $\underline{b} = b_1, ..., b_m$ of $\underline{a}$ (i.e. for each b_j there exists a_k such that $b_j = a_k$ for $j = 1, ..., m$, $k = 1, ..., n$, and $m \leq n$), also written as $C(\underline{b})$, where C is a sub-set of $D_{b_1} \times \cdots \times D_{b_m}$. A *(rule-based) authorization rule* is a pair $(C(\underline{b}), r)$, also written as $C(\underline{b}) \rightsquigarrow r$, such that $C(\underline{b})$ is a condition and r is a role in R. The *attribute-role relation AR associated to the set AU of authorization rules* is $\{(\{(b_1, e_1), ..., (b_m, e_m)\}, r) \mid (C(b_1, ..., b_m) \rightsquigarrow r) \in AU$ and $(e_1, ..., e_m) \in C\}$. By abuse of notation, a tuple $(U, R, P, \underline{a}, \underline{D}, \underline{s}, AU, PA, RH)$, where AU is a finite set of authorization rules and all the other components are as in a RBAC-DR policy, will also be called a RBAC-DR policy. Given a RBAC-DR policy $(U, R, P, \underline{a}, \underline{D}, \underline{s}, AU, PA, RH)$, a user u *satisfies the condition $C(b_1, ..., b_m)$ of an authorization rule in AU under the user-attribute mapping $\underline{s}$*, in symbols $u, \underline{s} \vdash C(b_1, ..., b_m)$, iff $(e_1, ...e_m) \in C$ and $e_j \in s_{b_j}(u)$ for $j = 1, ..., m$. The *implicit user-role relation $IUA \subseteq U \times R$* is defined as $IUA := \{(u, r) \mid$ there exists $(C \rightsquigarrow r) \in AU$ such that $u, \underline{s} \vdash C\}$. A user $u \in U$ is a member of role $r \in R$ under the user-attribute mapping $\underline{s}$ (via the notion of attribute-role relation associated to the set AU of authorization rules) iff $(u, r) \in IUA$.

For effectiveness, we assume the computability of (a) the user-attribute assignments and of (b) the membership to the conditions of the authorization rules as well as to (c) the relations RH and PA. Requirement (a) is reasonable as the notion of user-attribute assignment is an abstraction of the mechanism that associates users with attributes (e.g., an LDAP). Requirement (b) is necessary for the effectiveness of the satisfaction relation $\vdash$. Concerning (c), we observe that since the sets R and P are finite in several applications, then membership to RH and PA is obviously computable.

Example 1. We consider a refined version of the example in [2] for an on-line entertainment store streaming movies to users. The store needs to enforce an access control policy that is based on the user's age and the country where the user lives that may have different regulations for considering someone as an adult, a teen, or a child. E.g., for the legislation of Italy and France, one is considered adult when he/she is 18 or older while in Japan the adult age is 20.

For simplicity, we assume there are only three users Alice, Bob, and Charlie. Alice is 12 years old and lives in Italy, Bob is 39 and lives in Japan, and Charlie is 17 and lives in France. We leave unspecified the permissions, the role hierarchy, and the role-permission assignment as we want to focus on the user-role assignment based on the age and country attributes. In the corresponding RBAC-DR policy $(U, R, P, \underline{a}, \underline{D}, \underline{s}, AU, PA, RH)$, we have that $U := \{Alice, Bob, Charlie\}$, $R := \{Adult, Teen, Child\}$, $\underline{a} = age, country$, $\underline{D} = \mathbb{N}, WC$ where WC is an enumerated set containing all the countries in the world, $\underline{s} = s_{age}, s_{country}$ where $s_{age} = \{Alice \mapsto 12, Bob \mapsto 39, Charlie \mapsto 17\}$, $s_{country} = \{Alice \mapsto Italy, Bob \mapsto Japan, Charlie \mapsto France\}$, and AU contains, among others, the following authorization rules:

$$\rho_1 : \{(a,c) \in \mathbb{N} \times WC | a \geq 20 \text{ and } c \in \{Japan, Indonesia\}\} \rightsquigarrow Adult,$$
$$\rho_2 : \{(a,c) \in \mathbb{N} \times WC | a \geq 18 \text{ and } c \in \{Italy, France, ...\}\} \rightsquigarrow Adult,$$
$$\rho_3 : \{(a,c) \in \mathbb{N} \times WC | 13 \leq a \leq 17 \text{ and } c \in \{Italy, France, ...\}\} \rightsquigarrow Teen, \text{ and}$$
$$\rho_4 : \{(a,c) \in \mathbb{N} \times WC | a \leq 12 \text{ and } c \in \{Italy, France, ...\}\} \rightsquigarrow Child$$

that correspond to the policy informally described above (a and c abbreviate *age* and *country*, respectively). It is not difficult to see that $Bob, \underline{s} \vdash \rho_1$ but $Bob, \underline{s} \nvdash \rho_2$, that $Charlie, \underline{s} \vdash \rho_3$ but $Charlie, \underline{s} \nvdash \rho_i$ for $i = 1, 2, 4$, and that $Alice, \underline{s} \vdash \rho_4$ but $Alice, \underline{s} \nvdash \rho_i$ for $i = 1, 2, 3$. As a consequence, we have that, e.g., $(Alice, Child) \in IUA$, $(Bob, Adult) \in IUA$, and $(Charlie, Teen) \in IUA$. □

2.2   Redundancies in RBAC-DR Policies

As observed in [3], for RBAC-DR policies, besides the usual role hierarchy RH, it is possible to define another hierarchy IRH induced by the authorization rules in AU. We follow [3] and since IRH is defined for every possible user-attribute mapping, we introduce the notion of *RBAC-DR family of policies* as a tuple $(R, P, \underline{a}, \underline{D}, AU, PA, RH)$ where its components are the same as those of a RBAC-DR policy; in other words, a RBAC-DR family of policies is a RBAC-DR policy where the users and the user-attribute assignment are omitted. For every pair (ρ_1, ρ_2) of authorization rules in AU, ρ_1 is *more senior than* ρ_2 (in symbols, $\rho_1 \sqsupseteq \rho_2$) iff for every user u and every user-attribute assignment $\underline{s}$ if $u, \underline{s} \vdash C_1$ then also $u, \underline{s} \vdash C_2$, where C_i is the condition of ρ_i for $i = 1, 2$, i.e. when the set of users satisfying C_1 is a subset of that satisfying C_2 under every possible user-attribute mapping $\underline{s}$. Then, we define $(r_1, r_2) \in IRH$, i.e. the pair (r_1, r_2) is in the *induced role hierarchy IRH*, iff for each rule $(C_1 \rightsquigarrow r_1) \in AU$ there exists a rule $(C_2 \rightsquigarrow r_2) \in AU$ such that $(C_1 \rightsquigarrow r_1) \sqsupseteq (C_2 \rightsquigarrow r_2)$.

The key difference between RH and IRH is that the former is designed so that seniority among roles reflects inheritance of permissions associated to them whereas the latter characterizes seniority according to the sets of users associated to them by the set AU of authorization rules. It may happen that the induced role hierarchy IRH is such that both $(r, r') \in IRH$ and $(r', r) \in IRH$ for two distinct roles $r, r' \in R$, i.e. the set of users associated to r and that associated to r' are identical. This is so because IRH is a quasi-order (i.e. it is reflexive and transitive) and not a partial order as RH (that is also anti-symmetric). This implies a redundancy in the definition of IRH since the two roles r_1 and r_2 can be considered equivalent as both $(r_1, r_2) \in IRH$ and $(r_2, r_1) \in IRH$. This does not necessarily imply a problem in the authorization rules but it is desirable that designers become aware of all the redundancies in IRH (see [3] for more on this issue). Notice that once the "more senior than" relation among the authorization rules in AU is known, detecting redundancies in IRH becomes obvious. As a consequence, we define the *problem of computing the "more senior than relation" over a set AU of authorization rules (MS-AU problem, for short)* as follows: given a RBAC-DR family $(R, P, \underline{a}, \underline{D}, AU, PA, RH)$ of policies, the MS-AU problem amounts to checking whether $\rho_1 \sqsupseteq \rho_2$ for every pair ρ_1 and ρ_2 of authorization rules in AU. Notice that this problem is stated for a RBAC-DR family of policies and thus must be solved regardless of the set of users and the user-attribute mapping.

Solving the MS-AU problem allows one to eliminate the redundancies in IRH. Formally, this amounts to first turning $\sqsupseteq$ into a partial order in such a way that also IRH is so. This is a crucial pre-requisite to apply the techniques in [3] to detect (and eliminate) any "disagreement" between RH and IRH, i.e. to identify those situations in which the security policies encoded by the authorization rules and the business practices captured by RH do not match. Although we do not explore this problem further, we observe that the automated technique to solve MS-AU problems in Section 4.2 facilitates the application of the techniques in [3].

2.3 RBAC-DR with Negative Authorization

In [4], negative authorizations are added to the RBAC-DR model by allowing negated roles in authorization rules. The intuition is that an authorization rule with a negative role r denies access to the permissions associated to r via the permission-assignment relation. This characterization of negative roles allows for the adoption of the so-called *open* (allow-by-default) *policy* semantics [13,26] in an extension of the RBAC model, that is based on positive permissions conferring the ability to do something on the holders of the permissions [27]. This is crucial to widen the scope of applicability of the RBAC model extended with dynamic roles to applications where the set of users is not known a priori, e.g., web-services. Unfortunately, as pointed out in [4], the addition of negative authorizations introduces conflicts that need to be detected and then resolved.

Formally, we define the set $SR = \{+, -\} \times R$ of *signed roles* and write the pairs $(+, r)$ and $(-, r)$ as $+r$ and $-r$, respectively, for a role $r \in R$. We extend the definition of authorization rule as follows: a pair $C \leadsto sr$ is a *signed authorization rule* where C is a condition and $sr \in SR$. Intuitively, a user u satisfying the condition C of a signed authorization rule of the form $C \leadsto -r$ under a certain user-attribute mapping is forbidden to be assigned the role r. A *RBAC-DR policy with negative roles* (RBAC-NDR) is a tuple $(U, R, P, \underline{a}, \underline{D}, \underline{s}, AU, PA, RH)$ where $U, R, P, \underline{a}, \underline{D}, \underline{s}, PA$ and RH are as in a RBAC-DR policy and AU is a

set of signed authorization rules. A RBAC-NDR family of policies is the tuple
$(R, P, \underline{a}, \underline{D}, AU, PA, RH)$

To understand the kind of conflicts that may arise from the use of signed
authorization rules, consider $\rho_1 := (C_1 \rightsquigarrow +r)$ and $\rho_2 := (C_2 \rightsquigarrow -r)$ such that
a user u satisfies both C_1 and C_2 under a certain user-attribute mapping. At
this point, we are faced with the problem of deciding whether the user u should
be assigned or not the role r. As discussed in [4], there are several strategies to
resolve the problem, e.g., by adopting a "deny-override" strategy where a user u
can be assigned a role r provided that u satisfies the condition C of a rule of the
form $C \rightsquigarrow +r$ in AU and there is no rule in AU of the form $C' \rightsquigarrow -r$ such that
u satisfies the condition C' that forbids the assignment of r to u. However, this
or alternative conflict elimination strategies may be unsatisfactory for certain
applications; see again [4] for details. Furthermore, certain conflicts may arise
from errors in the design of the authorization rules and must be identified in order
to correct them. As a consequence, we define the *problem of detecting conflicts in
the set AU of signed authorization rules* (*CD-SAU problem*, for short) as follows:
given a RBAC-NDR family $(R, P, \underline{a}, \underline{D}, AU, PA, RH)$ of policies, the CD-SAU
problem amounts to checking whether for each pair $(C \rightsquigarrow +r, C' \rightsquigarrow -r)$ in AU,
there is no user u satisfying both C and C' under a user-attribute mapping $\underline{s}$.
As for the MS-AU problem, also the CD-SAU problem is stated for a family
of policies and thus must be solved for any set of users and any user-attribute
mapping.

3   Satisfiability Modulo Theories Solving

We assume the usual syntactic (e.g., signature, variable, term, atom, literal) and
semantic (e.g., structure, truth) notions of many-sorted first-order logic with
equality (see, e.g., [10]). A *constraint* (*clause*, resp.) is a conjunction (disjunction,
resp.) of literals (i.e. atoms or their negations) and a *quantifier-free formula* is
an arbitrary Boolean combination of atoms.

According to [24], a *theory* T is a pair $(\Sigma, \mathcal{C})$, where Σ is a signature and $\mathcal{C}$ is
a class of Σ-structures; the structures in $\mathcal{C}$ are the *models* of T. Given a theory
$T = (\Sigma, \mathcal{C})$, a quantifier-free Σ-formula $\phi(\underline{v})$—i.e. a quantifier-free formula built
out of the symbols in Σ and the variables in the sequence $\underline{v}$—is T-*satisfiable*
if there exists a Σ-structure $\mathcal{M}$ in $\mathcal{C}$ and a valuation μ mapping the variables
in $\underline{v}$ to values of the domain of $\mathcal{M}$ such that $\phi(\underline{v})$ is true in $\mathcal{M}$ (in symbols,
$\mathcal{M}, \mu \models \phi$); it is T-*valid* (in symbols, $T \models \varphi$) if $\mathcal{M}, \mu \models \varphi(\underline{v})$ for every $\mathcal{M} \in \mathcal{C}$
and every valuation μ. The quantifier-free formula $\varphi(\underline{v})$ is T-valid iff its negation
$\neg\varphi(\underline{v})$ is T-unsatisfiable. For example, if φ is the implication $\varphi_1 \rightarrow \varphi_2$, then its
T-validity is equivalent to the T-unsatisfiability of the conjunction of φ_1 with
the negation of φ_2 ($\varphi_1 \wedge \neg\varphi_2$).

The *satisfiability modulo the theory T (SMT(T)) problem* amounts to establishing the T-satisfiability of quantifier-free Σ-formulae. Many state-of-the-art
SMT solvers, in their simplest form, tackle the SMT(T) as follows. Initially,
each atom occurring in the input quantifier-free formula φ is considered simply

as a propositional letter (in other words, the theory T is forgotten). Then, the "propositional abstraction" of φ is sent to a SAT solver. If this reports propositional unsatisfiability, then φ is also T-unsatisfiable. Otherwise, an assignment of truth values to the atoms in φ is returned that makes its propositional abstraction true. Such an assignment can be seen as a conjunction of literals and checked by a specialized *theory solver* for T that can only deal with constraints (i.e. conjunctions of literals). If the theory solver establishes the T-satisfiability of the constraint then φ is also T-satisfiable. Otherwise, the theory solver computes a clause that, once added to the SAT solver, precludes the assignment of truth values that has been considered. The SAT solver is then started again. This process is repeated until the theory solver reports the T-satisfiability of one assignment or all the assignments returned by the SAT solver are found T-unsatisfiable so that also φ is reported to be T-unsatisfiable. Various refinements of this basic schema have been proposed for efficiency, the interested reader is pointed to, e.g., [28,9].

The SMT(T) problem is NP-hard as it subsumes the SAT problem. To evaluate the additional complexity due to the theory T, we focus on theory solvers and consider the complexity of checking the T-satisfiability of constraints, called the *constraint T-satisfiability problem* for some theories that have been found useful for the declarative specifications of authorization policies (see, e.g., [18]). The theory $\mathcal{EUF}$ of *equality with uninterpreted function symbols* is the theory interpreting the symbol $=$ as an equivalence relation that is also a congruence. The constraint $\mathcal{EUF}$-satisfiability problem is decidable and polynomial. The theory $\mathcal{ED}(\{v_1, ..., v_n\}, S)$ of the *enumerated data-type S with values $\{v_1, ..., v_n\}$* (for $n \geq 1$) is the theory whose signature consists of the sort S, the constant symbols $v_1, ..., v_n$ of sort S, and its class of models contains all the structures whose domain contains exactly n distinct elements. The constraint $\mathcal{ED}$-satisfiability problem is decidable and NP-complete. Linear Arithmetic on the Rationals (Integers, resp.) $\mathcal{LAR}$ ($\mathcal{LAI}$, resp.) is the theory whose class of models is the singleton containing the usual structure $\mathbb{R}$ of the Reals ($\mathbb{Z}$ of the integers, resp.) and whose atoms are equalities, disequalities, and inequalities of linear polynomials whose coefficients are integers and the variables take values over the Reals (integers, resp.). The constraint satisfiability problems for $\mathcal{LAR}$ and $\mathcal{LAI}$ are decidable, the former is polynomial and the latter is NP-complete. The theory $\mathcal{RDL}$ of *real difference logic* is the sub-theory of $\mathcal{LAR}$ whose atoms are written as $x - y \bowtie c$ where $\bowtie$ is an arithmetic operator, x and y are variables, and $c \in \mathbb{R}$. The theory $\mathcal{IDL}$ of *integer difference logic* is defined as $\mathcal{RDL}$ but it is a sub-theory of $\mathcal{LAI}$. Both the constraint $\mathcal{RDL}$- and $\mathcal{IDL}$-satisfiability problems are decidable and polynomial. For more details about these and other theories, see, e.g., [28,9].

Many applications require to reason about conjunctions of constraints coming from several theories $T_1, ..., T_n$. In this situation, it is easy to build a theory solver by reusing those available for the theories $T_1, ..., T_n$ as follows. Let $\alpha = \alpha_1(\underline{v}_1) \wedge \cdots \wedge \alpha_n(\underline{v}_n)$ be a *composite constraint*, i.e. α_i is a literal over the signature of T_i for $i = 1, ..., n$ whose variables $\underline{v}_i$ are disjoint with those of the other constraints. Then, α is satisfiable iff each α_i is T_i-satisfiable and it is unsatisfiable otherwise, i.e. there exists $i \in \{1, ..., n\}$ such that α_i is T_i-unsatisfiable. Thus, if the constraint T_i-satisfiability problem is decidable and polynomial for each $i = 1, ..., n$, then it is also decidable and polynomial to check the satisfiability of composite constraints.

4   Solving the MS-AU and CD-SAU Problems

We explain how RBAC-DR and RBAC-NDR policies can be specified by using theories and how this allows us to prove the decidability of the MS-AU and CD-SAU problems. The proof is constructive and reduces both problems to a sequence of SMT problems that can be efficiently solved by invoking state-of-the-art SMT solvers.

4.1   Specifying RBAC-DR and RBAC-NDR Policies with Theories

The idea is to use quantifier-free formulae of a suitable theory to specify the conditions of (signed) authorization rules. Formally, we assume the availability of a *theory* $T_A = (\Sigma_A, \mathcal{C}_A)$ *of attributes* and define a *RBAC-DR policy with background theory* T_A *of attributes* as a tuple $(U, R, P, \underline{a}, \underline{s}, AU_{QF}, PA, RH)$ where U, R, P, PA, and RH are as specified above, and the following conditions hold:
(**C1**) each $a \in \underline{a}$ is a first-order variable of sort S_a in Σ_A and each $s_a \in \underline{s}$ is a mapping $s_a : U \to 2^{S_a^{\mathcal{M}_A}}$ for some $\mathcal{M}_A \in \mathcal{C}_A$ (where $\sigma^{\mathcal{M}_A}$ is the interpretation in $\mathcal{M}_A$ of the symbol $\sigma \in \Sigma$; e.g., $S_a^{\mathcal{M}_A}$ is a subset of the domain of $\mathcal{M}_A$), (**C2**) for every $e \in S^{\mathcal{M}_A}$ there exists a constant c of sort S_a in Σ_A such that $c^{\mathcal{M}_A} = e$ for some $\mathcal{M}_A \in \mathcal{C}_A$,[1] and (**C3**) the set AU_{QF} contains finitely many pairs of the form $\varphi(\underline{b}) \rightsquigarrow r$, called *syntactic authorization rules*, where φ is a quantifier-free formula built out of the symbols of Σ_A and the variables in $\underline{b}$, that is a sub-sequence of $\underline{a}$.

The crucial observation is that each quantifier-free formula $\varphi(\underline{b})$ in a syntactic authorization rule $\varphi(\underline{b}) \rightsquigarrow r$ in AU_{QF} defines a condition $C(\underline{b})$, in the sense of Section 2.2, as follows: $(e_1, ..., e_m) \in C(\underline{b})$ iff there exists $\mathcal{M}_A \in \mathcal{C}_A$ and a valuation μ such that $\mu(b_j) = e_j \in S_{b_j}^{\mathcal{M}_A}$ for $j = 1, ..., m$ and $\mathcal{M}_A, \mu \models \varphi(\underline{b})$. In the following, the condition $C(\underline{b})$ associated to the quantifier-free formula $\varphi(\underline{b})$ is written as $[[\varphi(\underline{b})]]$. Thus, a RBAC-DR policy $(U, R, P, \underline{a}, \underline{s}, AU_{QF}, PA, RH)$ with background theory $T_A = (\Sigma_A, \mathcal{C}_A)$ of the attributes defines the following RBAC-DR policy: $(U, R, P, \underline{a}, \underline{D}, \underline{s}, AU, PA, RH)$ where $D_a = S_a^{\mathcal{M}_A}$ for each $a \in \underline{a}$ and $AU = \{[[\varphi(\underline{b})]] \rightsquigarrow r \mid (\varphi(\underline{b}) \rightsquigarrow r) \in AU_{QF}\}$ for some $\mathcal{M}_A \in \mathcal{C}_A$. Similar definitions can be given for RBAC-NDR policies and RBAC-DR (RBAC-NDR) family of policies with background theory T_A of attributes.

---

[1] This condition is not restrictive as it is always possible to add a constant c_e for each element e in the domain whose interpretation is e itself. The addition of these constants does not change the set of satisfiable quantifier-free formulae (see, e.g., [10]).

Example 2. For Example 1, we consider composite constraints from the enumerated data-type theory $\mathcal{ED}(\{\mathit{Italy}, \mathit{France}, \mathit{Japan}, \mathit{Indonesia}, ...\}, C)$ for the set WC of world countries and the theory $\mathcal{IDL}$ for the constraints on the age (see Section 3). The attribute *country* is also seen as a variable of sort C and the attribute *age* as a variable of sort $\mathbb{Z}$. As a consequence, the syntactic authorization rules corresponding to the authorization rules in Example 1 are:

$\rho_1 : -age \leq -20 \wedge (\mathit{country} = \mathit{Japan} \vee \mathit{country} = \mathit{Indonesia}) \rightsquigarrow \mathit{Adult},$
$\rho_2 : -age < -18 \wedge (\mathit{country} = \mathit{Italy} \vee \mathit{country} = \mathit{France} \vee \cdots) \rightsquigarrow \mathit{Adult},$
$\rho_3 : -age \leq -13 \wedge age \leq 17 \wedge (\mathit{country} = \mathit{Italy} \vee \mathit{country} = \mathit{France} \vee \cdots) \rightsquigarrow \mathit{Teen},$
$\rho_4 : age \leq 12 \wedge (\mathit{country} = \mathit{Italy} \vee \mathit{country} = \mathit{France} \vee \cdots) \rightsquigarrow \mathit{Child}$

Notice that $\mathit{country} \in \{c_1, ..., c_n\}$ has been translated as $\bigvee_{i=1}^{n} \mathit{country} = c_i$. □

By generalizing the observations in the example, it is not difficult to see that the language to express authorization rules introduced in [2] can be translated to syntactic authorization rules provided that a suitable background theory of attributes is used.

Proposition 1. *Given a RBAC-DR policy* $(U, R, P, \underline{a}, \underline{s}, AU_{QF}, PA, RH)$ *with background theory* $T_A = (\Sigma_A, \mathcal{C}_A)$ *of the attributes. If the SMT(T_A) problem is decidable, then it is decidable to check if a user* $u \in U$ *is a member of a role* $r \in R$.

An obvious corollary of this proposition and the fact that the membership to RH and PA is computable is that checking whether a user is an implicit member of a role or he/she has a permission are also decidable.

   In the following, without loss of generality, we assume that **(A1)** any condition of an authorization rule is T_A-satisfiable but not T_A-valid (i.e. its negation is also T_A-satisfiable) and **(A2)** the conditions of the syntactic authorization rules in AU_{QF} are constraints and not arbitrary Boolean combinations of atoms. The two situations ruled out by assumption **(A1)** can be automatically identified under the assumption that the SMT(T_A) problem is decidable and can be safely discarded as uninteresting: those that are T_A-satisfiable never assigns a user to a role while those that are T_A-valid assigns any user to a role (this kind of problems have been considered in the context of a similar rule-based specification framework for access control policies [25]). To see why assumption **(A2)** is without loss of generality, consider a rule $\varphi \rightsquigarrow r$ in AU_{QF} where φ is a quantifier-free formula (and $r \in R$). Since it is possible to transform φ into disjunctive normal form, i.e. into a formula of the form $\bigvee_{j=1}^{d} \alpha_j$ where α_j is a constraint (for $j = 1, ..., d$ and $d \geq 1$), we can consider the set $AU'_{QF} := AU_{QF} \setminus \{\varphi \rightsquigarrow r\} \cup \{\alpha_j \rightsquigarrow r \mid j = 1, ..., d\}$. Clearly, the user-role assignment relations induced by AU_{QF} and by AU'_{QF} are the same. By iterating the process, we derive a set of syntactic authorization rules whose conditions are constraints only.

4.2   The MS-AU Problem

We consider the MS-AU problem (recall its definition at the end of Section 2.2) for RBAC-DR policies with a background theory of the attributes, i.e. given a

```
Input: c: array[0..N-1] of constraints
       r: array[0..N-1] of (signed) roles
1: S := ∅;
2: FOR i=0 TO N-2 DO
3:   FOR j=i+1 TO N-1 DO
4:     f1 := opposite(r[i],r[j]); f2 := related(c[i],c[j])
5:     IF f2 THEN
6:       IF find(i)≠find(j) THEN
7:         i2j := check_T_A (c[i] ∧ ¬ c[j])
8:         IF i2j=unsat THEN S := S∪{(i,j,f1,f2)};
9:         j2i := check_T_A (c[j] ∧ ¬ c[i])
10:        IF j2i=unsat THEN S := S∪{(j,i,f1,f2)};
11:        IF (i2j=unsat AND j2i=unsat) THEN union(i,j)
12:      ELSE S := S∪{(i,j,f1,f2),(j,i,f1,f2)};
13:    ELSE S := S∪{(i,j,f1,f2)}
```

Fig. 1. Solving the MS-AU and CD-SAU problems

RBAC-DR family of policies $(U, R, P, \underline{a}, AU_{QF}, PA, RH)$ with background theory T_A of the attribute, the MS-AU problem amounts to checking whether $\rho_1 \sqsupseteq \rho_2$ for every pair ρ_1 and ρ_2 of syntactic authorization rules in AU_{QF}.

The key observation is that we can reduce the problem of establishing whether $\rho_1 \sqsupseteq \rho_2$ to an SMT(T_A) problem involving the conditions of the rules ρ_1 and ρ_2. Before being able to formally state this, we need to introduce the following notion. Two rules $(\alpha_1(\underline{b}_1) \rightsquigarrow r_1)$ and $(\alpha_2(\underline{b}_2) \rightsquigarrow r_2)$ are *syntactically related* if $\underline{b}_1 \cap \underline{b}_2 \neq \emptyset$ (by abuse of notation, we consider the sequence $\underline{b}_i$ as a set, $i = 1, 2$); otherwise, they are *syntactically unrelated*.

Proposition 2. *Let $(R, P, \underline{a}, AU_{QF}, PA, RH)$ be a RBAC-RD family of policies with background theory T_A of the attributes and $\rho_i = (\alpha_i(\underline{b}_i) \rightsquigarrow r_i)$ be an authorization rule in AU_{QF} for $i = 1, 2$. Then, the following facts hold:*

(a) *if ρ_1 and ρ_2 are syntactically unrelated, then $\rho_1 \not\sqsupseteq \rho_2$ and $\rho_2 \not\sqsupseteq \rho_1$ and*
(b) *if ρ_1 and ρ_2 are syntactically related, then $\rho_1 \sqsupseteq \rho_2$ iff $\alpha_1(\underline{b}_1) \rightarrow \alpha_2(\underline{b}_2)$ is T_A-valid (or, equivalently, $\alpha_1(\underline{b}_1) \wedge \neg\alpha_2(\underline{b}_2)$ is T_A-unsatisfiable).*

Assuming that the SMT(T_A) problem is decidable, it is possible to automatically check whether a rule is more senior than the other by Proposition 2. This is the idea underlying the algorithm in Figure 1 for automatically solving the MS-AU problem. Let $(R, P, \underline{a}, AU_{QF}, PA, RH)$ be a RBAC-RD family of policies with background theory T_A where $|AU_{QF}| = $ N. The input to the algorithm are two arrays c and r such that for each rule $\alpha \rightsquigarrow r$ in AU_{QF} there exists i in the range [0..N-1] such that c[i] $= \alpha$ and r[i] $= r$ (the latter will contain signed roles only when solving CD-SAU problems, see Section 4.3 below). The variable S $\subseteq [0, N-1] \times [0, N-1] \times \{true, false\} \times \{true, false\}$ stores increasingly precise approximations of the relation $\sqsupseteq$ in a sense to be made precise below. The Boolean function related(c[i], c[j]) returns true iff the rules c[i] $\rightsquigarrow$ r[i] and c[j] $\rightsquigarrow$ r[j] are syntactically related. The Boolean function opposite is the constant function false so that the third component of the tuples in S is always set to `false` (this will be important only when solving CD-SAU problems, see Section 4.3 below). The function `find` and the procedure `union` form the interface to a union-find data structure (see, e.g., [30]) to maintain a set of equivalence classes: `find(i)` returns the representant of the equivalence class to which the rule $c[i] \rightsquigarrow r[i]$ belongs to and `union(i, j)` merges the equivalence classes to which the rules $c[i] \rightsquigarrow r[i]$ and $c[j] \rightsquigarrow r[j]$ belong to. The function $\text{check}_{T_A}(\varphi)$ returns `sat` (`unsat`, resp.) iff the quantifier-free Σ_A-formula φ is T_A-satisfiable (T_A-unsatisfiable, resp.) and it is implemented by invoking an available solver supporting the solution of $\text{SMT}(T_A)$ problems.

The algorithm works by enumerating pairs of rules (the two nested loops at lines 2 and 3) to establish whether they are syntactically related (flag `f2` line 4). In case they are not, the tuple $(\text{i}, \text{j}, \textit{false}, \textit{false})$ is added to `S` (line 13) meaning that the pair (i, j) of rules cannot be compared with respect to $\sqsupseteq$. Otherwise, the union-find data structure is queried (line 6) in order to establish if the rules identified by `i` and `j` are already in the same equivalence class. If this is the case, the tuples $(\text{i}, \text{j}, \textit{false}, \textit{true})$ and $(\text{j}, \text{i}, \textit{false}, \textit{true})$ are added to `S` (line 12) meaning that both $\text{i} \sqsupseteq \text{j}$ and $\text{j} \sqsupseteq \text{i}$. Otherwise, it is checked if $\text{i} \sqsupseteq \text{j}$ (line 7, by testing if $c[\text{i}]$ implies $c[\text{j}]$ modulo T_A or, equivalently, the conjunction of $c[\text{i}]$ with the negation of $c[\text{j}]$ is T_A-unsatisfiable) and the tuple $(\text{i}, \text{j}, \textit{false}, \textit{true})$ is added to `S` (line 8). The same is done to establish if $\text{j} \sqsupseteq \text{i}$ (lines 9 and 10). If both previous tests (lines 8 and 10) have been successful, then the two equivalence classes to which `i` and `j` belong to are merged (line 11).

Lemma 1. *Let $(R, P, \underline{a}, AU_{QF}, PA, RH)$ be a family of RBAC-RD policies with background theory T_A of the attributes whose SMT(T_A) problem is decidable and $|AU_{QF}| = \text{N}$. Then, the following facts hold after the execution of the algorithm in Figure 1: (a) $(\text{i}, \text{j}, \textit{false}, \textit{true}) \in \text{S}$ iff $c[\text{i}] \rightsquigarrow r[\text{i}] \sqsupseteq c[\text{j}] \rightsquigarrow r[\text{j}]$, (b) $\text{find(i)} = \text{find(j)}$ iff $c[\text{i}] \rightsquigarrow r[\text{i}] \sqsupseteq c[\text{j}] \rightsquigarrow r[\text{j}]$ and $c[\text{j}] \rightsquigarrow r[\text{j}] \sqsupseteq c[\text{i}] \rightsquigarrow r[\text{i}]$, i.e. the union-find data structure stores the equivalence classes of the "more senior than" relation over AU_{QF}, and (c) the number of invocations to check_{T_A} is at most $\text{N}(\text{N} - 1)$.*

The cost of invoking the function `opposite` is constant (recall that this is a constant function returning `false` when solving the MS-AU problem) while that of `related` is linear in $\underline{a}$ by using bit-strings to represent sequences of variables occurring in the conditions of the rules. The union-find algorithm in [30] takes almost constant (amortized) time for invoking both the `find` and `union` operations (more precisely, it takes $O(A^{-1}(N))$ where A^{-1} is the inverse of the Ackermann function; since for any practical values of N, $A^{-1}(N)$ is bounded by 5, each invocation to `find` and `union` can be considered as constant). Thus, the complexity of the algorithm is clearly dominated by the number of invocations to check_{T_A}. Notice that the function is invoked on quantifier-free formulae obtained by conjoining a constraint α_i with the negation of a constraint $\neg\alpha_j$ (i.e. a clause). This allows us to characterize the complexity of the MS-AU problem. Formally, we introduce the following notion. A theory T is *convex* (see, e.g., [23]) iff whenever a constraint implies a clause, it also implies one of the literals in the clause. Examples of convex theories are $\mathcal{EUF}$, $\mathcal{ED}$, $\mathcal{LAR}$, and $\mathcal{RDL}$ whose constraint satisfiability problem is polynomial (convexity of $\mathcal{LAR}$ derives from the convexity of linear algebra). Instead, both $\mathcal{LAI}$ and its sub-theory $\mathcal{IDL}$ are not convex (e.g., $x \geq 0 \wedge x \leq 1 \wedge \neg(x \neq 0 \wedge x \neq 1)$ is $\mathcal{LAI}$-unsatisfiable but neither $x \geq 0 \wedge x \leq 1 \wedge x = 0$ nor $x \geq 0 \wedge x \leq 1 \wedge x = 1$ is $\mathcal{LAI}$-unsatisfiable).

Theorem 1. *If the SMT problem of the background theory T_A of the attributes is decidable, then the MS-AU problem is decidable. If furthermore T_A is convex and the constraint T_A-satisfiability problem is polynomial, then the MS-AU problem is also polynomial.*

When the background theory of the attributes is $\mathcal{EQ}$, $\mathcal{LAR}$, $\mathcal{RDL}$, or we consider composite constraints of these theories, the MS-AU problem is decidable and polynomial since all these theories are decidable and polynomial (see Section 3). When the background theory of the attributes is $\mathcal{ED}$, $\mathcal{LAI}$, or $\mathcal{IDL}$, the MS-AU problem is decidable and NP-complete. While this is obvious for $\mathcal{ED}$ and $\mathcal{LAI}$ (whose constraint satisfiability problem is already NP-complete), it is less so for $\mathcal{IDL}$ since its constraint satisfiability problem is polynomial. Since $\mathcal{IDL}$ is not convex, the problem of checking the satisfiability of a constraint with a clause becomes NP-complete [17]. Thus, besides polynomial constraint satisfiability, convexity is crucial to have a polynomial MS-AU problem.

4.3   The CD-SAU Problem

We consider the CD-SAU problem (recall its definition at the end of Section 2.3) for RBAC-NDR policies with a background theory of the attributes, i.e. given a RBAC-NDR family of policies $(R, P, \underline{a}, AU_{QF}, PA, RH)$ with background theory T_A of the attributes, the CD-SAU problem amounts to checking whether for each signed authorization rule $\alpha_1(\underline{b}_1) \rightsquigarrow +r$ in AU_{QF}, there is no rule $\alpha_2(\underline{b}_2) \rightsquigarrow -r$ in AU_{QF} such that a user u satisfies both $\alpha_1(\underline{b}_1)$ and $\alpha_2(\underline{b}_2)$ under some user-attribute mapping $\underline{s}$. We say that there is a *conflict* between rule $\alpha_1(\underline{b}_1) \rightsquigarrow +r$ and $\alpha_2(\underline{b}_2) \rightsquigarrow -r$ if there exists a user u satisfying both $\alpha_1(\underline{b}_1)$ and $\alpha_2(\underline{b}_2)$ under some user-attribute mapping $\underline{s}$. Interestingly, it is possible to distinguish two types of conflicts depending on the fact that the rules can or cannot be compared by the "more senior than" relation. Formally, the rules ρ_1 and ρ_2 are *relevant* iff $\rho_1 \sqsupseteq \rho_2$ or $\rho_1 \sqsubseteq \rho_2$, and are *irrelevant* otherwise. If two rules are syntactically unrelated, they are also irrelevant by Proposition 2 **(a)**.

   We solve the CD-SAU problem by using again the algorithm in Figure 1 where the array $\mathbf{r}$ stores signed roles, the auxiliary function $\mathtt{opposite}(\mathbf{r}[\mathbf{i}], \mathbf{r}[\mathbf{j}])$ returns $\mathbf{true}$ iff $\mathbf{r}[\mathbf{i}] = +r$ and $\mathbf{r}[\mathbf{j}] = -r$ or $\mathbf{r}[\mathbf{i}] = -r$ and $\mathbf{r}[\mathbf{j}] = +r$, and all the other data structures and functions are as described in Section 4.2. As anticipated in Section 4.2, the third component of the relation S is important and distinguishes between relevant and irrelevant pairs of rules.

   As before, the algorithm enumerates pairs of rules and establishes whether they have signed roles $+r$ and $-r$ and they are syntactically related (flags $\mathtt{f1}$ and $\mathtt{f2}$ at line 4, respectively). In case the two rules are not syntactically related, the tuple $(\mathtt{i}, \mathtt{j}, \mathtt{f_1}, \textit{false})$ is added to S (line 13) meaning that the rules identified by $\mathtt{i}$ and $\mathtt{j}$ are irrelevant and there is a conflict between them when $\mathtt{f_1}$ is *true*.

Otherwise, the union-find data structure is queried (line 6) in order to establish if the rules identified by i and j are already in the same equivalence class. If this is the case, the tuples $(i, j, f_1, true)$ and $(j, i, f_1, true)$ are added to S (line 12) meaning than the rules identified by i and j are relevant and there is a conflict between them when f_1 is *true*. Otherwise, it is checked if $c[i] \rightsquigarrow r[i] \sqsupseteq c[j] \rightsquigarrow r[j]$ (line 7) and the tuple $(i, j, f_1, true)$ is added to S (line 8), the same is done to establish if $c[j] \rightsquigarrow r[j] \sqsupseteq c[i] \rightsquigarrow r[i]$ (lines 9 and 10): the rules are relevant and there is a conflict between them when f_1 is *true*. If both previous tests (lines 8 and 10) have been successful, then the two equivalence classes to which i and j belong to are merged (line 11).

Lemma 2. *Let $(R, P, \underline{a}, AU_{QF}, PA, RH)$ be a family of RBAC-NDR policies with background theory T_A of the attributes whose SMT(T_A) problem is decidable and $|AU_{QF}| = $ N. Then, after the execution of the algorithm in Figure 1 the following holds: if $(i, j, \tau, true) \in$ S, then there is a conflict between rules i and j and the two rules are relevant (irrelevant, resp.) when τ is true (false, resp.) and the number of invocations to $check_{T_A}$ is at most $N(N-1)$.*

We state the second main result of the paper.

Theorem 2. *If the SMT problem of the background theory T_A of the attributes is decidable, then the CD-SAU problem is decidable. If furthermore T_A is convex and the constraint T_A-satisfiability problem is polynomial, then the CD-SAU problem is also polynomial.*

4.4 Experiments

To test the practical viability of our techniques, we have implemented the algorithm of Figure 1 in C. To implement $check_{T_A}$, we have chosen Yices [31] for its easy-to-use API, although many other state-of-the-art SMT solvers (e.g., Z3 [34]) can be used. We have also implemented a generator of synthetic MS-AU problems; we expect similar performances on CD-SAU problems since the algorithm of Figure 1 can solve also these problems with minimal variations. The starting point is the policy of Example 1. The idea is to randomly generate composite constraints for the conditions of the rules taken from an enumerated data-type theory $\mathcal{ED}(\{v_1, ..., v_V\}, E)$ and $\mathcal{IDL}$: the former corresponds to the set of countries in the world and the latter to the age of users. We randomly generate composite constraints whose literals are $e = v_i$, $e \neq v_i$ (for $i \in \{1, ..., V\}$), $a = k$, $a > k$, and $a < k$ (for $k \in \mathbb{Z}$) where e and a are variables of sort E and $\mathbb{Z}$, respectively. According to Theorem 1, the resulting MS-AU problem is NP-complete as the constraint $\mathcal{ED}$-satisfiability problem is NP-complete and that of $\mathcal{IDL}$ is polynomial but $\mathcal{IDL}$ is not convex. Thus the invocations to $check_{T_A}$ can be computationally expensive. The random generation of literals in the composite constraints is inspired to Gorrilla [15] that is known to generate difficult problems for many state-of-the-art SMT solvers. The method of [15] has been adapted to satisfy assumption **(A1)** of Section 4.1 and to generate constraints that can be considered as "realistic" conditions of authorization rules, i.e. roughly similar to those in Example 2.

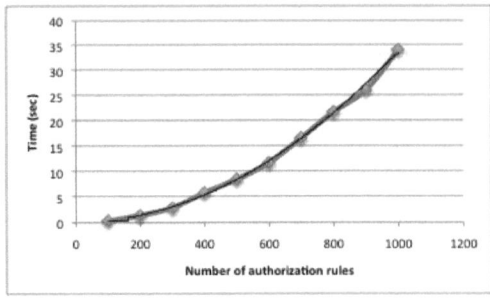

Fig. 2. Experimental results obtained on a MacBook with Intel Core i5 2.53 GHz and 4
GB of RAM running Mac OS X 10.6.6. Timings are averages over 10 random instances

There are three inputs to the benchmark generator: the number V of values
of $\mathcal{ED}$, a positive integer M bounding the constant k occurring in the arithmetic literals, and the number N of conditions to be considered. The first two
parameters V and M have negligible influence on the performance of our implementation of the algorithm in Figure 1. As a consequence, Figure 2 shows the
plot of the time taken to solve an MS-AU problem for the last parameter, i.e.
an increasing number N of rules (ranging from 100 to 1,000).[2] The behaviour of
the algorithm is clearly quadratic. This is possible because we invoke the SMT
solver only (around) $1/4$ of the potential N^2 calls, thanks to the notion of two
rules being syntactically unrelated (see before Proposition 2 for the definition)
and the use of the union-find data structure that provides us with a computationally cheap method to deduce the transitive chains of the "more senior than
relation." We believe that this gives a first important evidence of the practical
viability of our technique.

5   Related Work and Discussion

The problem of combining RBAC with more flexible methods of assigning users
to roles has been considered several times (see, e.g., [21,6]). In this line of work,
to the best of our knowledge, no automated analysis technique has been proposed to assist designers to detect redundancies and inconsistencies of policies
as we do here. We have presented an abstract rule-based framework for the integration of attributes in RBAC inspired to [2]. The MS-AU and the CD-SAU
problems are inspired to problems informally characterized in [3] and [4]. The
idea of using SMT solvers for their efficient solution together with their decidability and complexity results are new; except for [7,5] that exploit SMT solvers
to solve administrative RBAC problems. In [32,14], techniques to solve (what
we call) CD-SAU problems by using description logic and tableaux reasoners,
respectively, are described. Neither a complexity characterization nor an extensive experimental evaluation are proposed and the techniques do not support

---

[2] The sources of the program used to perform the experiments are available at
http://st.fbk.eu/SilvioRanise#Papers.

rich background theories of the attributes as we do here. The policy specification languages in [33,19] can express authorization policies with attributes. They support policies that go beyond those considered in this paper but do not propose automated techniques for detecting inconsistencies and redundancies as we do here. Furthermore, they adopt a closed (deny-by-default) policy model (where access is allowed if there exists a corresponding (positive) authorization and is denied otherwise [13,26]) rather than an open (allow-by-default) policy as we do in this paper. We leave it to future work the study of the impact of using a closed policy model on the decidability and complexity of the MS-AU and CD-SAU problems. We notice that our framework (including Theorems 1 and 2) can be easily extended to cope with conditions about the environment (e.g., time of the day) as done in [33].

Several works [29,1,22,12,11,8,20] have proposed techniques to detect inconsistencies and redundancies in XACML or extensions of RBAC policies by leveraging a variety of verification engines. None of these works provides decidability and complexity results of the analysis techniques as we do in this paper.

Acknowledgements. This work was partially supported by the "Automated Security Analysis of Identity and Access Management Systems (SIAM)" project funded by Provincia Autonoma di Trento in the context of the "team 2009 - Incoming" COFUND action of the European Commission (FP7).

References

1. Adi, K., Bouzida, Y., Hattak, I., Logrippo, L., Mankovskii, S.: Typing for Conflict Detection in Access Control Policies. In: Babin, G., Kropf, P., Weiss, M. (eds.) MCETECH 2009. LNBIP, vol. 26, pp. 212–226. Springer, Heidelberg (2009)
2. Al-Kahtani, M., Sandhu, R.: A Model for Attribute-Based User-Role Assignment. In: Proc. of 18th Annual Comp. Sec. App. Conf., Las Vegas, Nevada (2002)
3. Al-Kahtani, M., Sandhu, R.: Induced Role Hierarchies with Attribute-Based RBAC. In: Proc. of 8th ACM SACMAT (2003)
4. Al-Kahtani, M., Sandhu, R.: Rule-based RBAC with negative authorization. In: Proc. of 20th Annual Comp. Sec. App. Conf., pp. 405–415 (2004)
5. Alberti, F., Armando, A., Ranise, S.: Efficient Symbolic Automated Analysis of Administrative Role Based Access Control Policies. In: Proc. of 6th ACM Symp. on Info., Computer and Comm. Security, ASIACCS 2011 (2011)
6. Ardagna, C., De Capitani di Vimercati, S., Paraboschi, S., Pedrini, E., Samarati, P., Verdicchio, M.: Expressive and Deployable Access Control in Open Web Service Applications. IEEE Trans. on Serv. Comp. (TSC) 4(2), 96–109 (2011)
7. Armando, A., Ranise, S.: Automated Symbolic Analysis of ARBAC-Policies. In: Cuellar, J., Lopez, J., Barthe, G., Pretschner, A. (eds.) STM 2010. LNCS, vol. 6710, pp. 17–34. Springer, Heidelberg (2011)
8. Autrel, F., Cuppens, F., Cuppens, N., Coma, C.: MotOrBAC 2: a security policy tool. In: 3rd Conf. SARSSI, pp. 13–17 (2008)
9. De Moura, L., Bjørner, N.: Satisfiability modulo theories: introduction and applications. Commun. ACM 54, 69–77 (2011)
10. Enderton, H.B.: A Mathematical Introduction to Logic. Academic Press, New York (1972)

11. Fisler, K., Krishnamurthi, S., Meyerovich, L.A., Tschantz, M.C.: Verification and change-impact analysis of access control policies. In: Int. Conf. on Sw Eng. (ICSE), pp. 196–206 (2005)
12. Hughes, G., Bultan, T.: Automated Verification of Access Control Policies Using a SAT Solver. Int. J. on Sw Tools for Tech. Trandf. (STTT) 10(6), 473–534 (2008)
13. Jajodia, S., Samarati, P., Sapino, M.L., Subrahmanian, V.S.: Flexible support for multiple access control policies. ACM Trans. DB Syst. 26, 214–260 (2001)
14. Kamoda, H., Yamaoka, M., Matsuda, S., Broda, K., Sloman, M.: Access Control Policy Analysis Using Free Variable Tableaux. Trans. of Inform. Proc. Soc. of Japan, 207–221 (2006)
15. Korovin, K., Voronkov, A.: GoRRiLA and Hard Reality. In: Clarke, E., Virbitskaite, I., Voronkov, A. (eds.) PSI 2011. LNCS, vol. 7162, pp. 243–250. Springer, Heidelberg (2012)
16. Kuhn, D.R., Coyne, E.J., Weil, T.R.: Adding Attributes to Role Based Access Control. IEEE Computer 43(6), 79–81 (2010)
17. Lahiri, S.K., Musuvathi, M.: An Efficient Decision Procedure for UTVPI Constraints. In: Gramlich, B. (ed.) FroCos 2005. LNCS (LNAI), vol. 3717, pp. 168–183. Springer, Heidelberg (2005)
18. Li, N., Mitchell, J.C.: DATALOG with Constraints: A Foundation for Trust Management Languages. In: Dahl, V. (ed.) PADL 2003. LNCS, vol. 2562, pp. 58–73. Springer, Heidelberg (2003)
19. Li, N., Mitchell, J.C.: RT: A Role-based Trust-management Framework. In: 3rd DARPA Infor. Surv. Conf. and Exp. (DISCEX III), pp. 201–212 (2003)
20. Lin, D., Rao, P., Bertino, E., Li, N., Lobo, K.: EXAM: a comprehensive environment for the analysis of access control policies. IJIS 9, 253–273 (2010)
21. Lupu, E., Sloman, M.: Reconciling Role Based Management and Role Based Access Control. In: 2nd ACM Ws. on Role Based Acc. Contr., pp. 135–142 (1997)
22. Mankai, M., Logrippo, L.: Access Control Policies: Modeling and Validation. In: Proc. of NOTERE, pp. 85–91 (2005)
23. Nelson, C.G., Oppen, D.: Simplification by Cooperating Decision Procedures. ACM Trans. on Programming Languages and Systems 1(2), 245–257 (1979)
24. Ranise, S., Tinelli, C.: The SMT-LIB Standard: Version 1.2, http://goedel.cs.uiowa.edu/smtlib/papers/format-v1.2-r06.08.30.pdf
25. Ribeiro, C., Zúquete, A., Ferreira, P., Guedes, P.: Security Policy Consistency. In: 1st Ws. on Rule-Based Constr. Reas. and Progr. CoRR cs.LO/0006045 (2000)
26. Samarati, P., De Capitani di Vimercati, S.: Access Control: Policies, Models, and Mechanisms. In: Focardi, R., Gorrieri, R. (eds.) FOSAD 2000. LNCS, vol. 2171, pp. 137–196. Springer, Heidelberg (2001)
27. Sandhu, R., Coyne, E., Feinstein, H., Youmann, C.: Role-Based Access Control Models. IEEE Computer 2(29), 38–47 (1996)
28. Sebastiani, R.: Lazy Satisfiability Modulo Theories. Journal on Satisfiability, Boolean Modeling and Computation, JSAT 3, 141–224 (2007)
29. Shaikh, R., Adi, K., Logrippo, L., Mankovski, S.: Inconsistency Detection Method for Access Control Policies. In: IEEE 6th IAS, pp. 204–209 (2010)
30. Tarjan, R.E.: Efficiency of a Good But Not Linear Set Union Algorithm. Journal of the ACM 22(2), 215–225 (1975)
31. Yices, http://yices.csl.sri.com/
32. Yu, H., Xie, Q., Che, H.: Research on Description Logic Based Conflict Detection Methods for RB-RBAC Model. In: 4th Int. Conf. on AMT, pp. 335–339 (2006)
33. Yuan, E., Tong, J.: Attributed Based Access Control (ABAC) for Web Services. In: Proc. of IEEE ICWS, pp. 561–569 (2005)
34. Z3, http://research.microsoft.com/en-us/um/redmond/projects/z3

A Unified Attribute-Based Access Control Model Covering DAC, MAC and RBAC

Xin Jin[1], Ram Krishnan[2], and Ravi Sandhu[1]

[1] Institute for Cyber Security & Department of Computer Science
[2] Institute for Cyber Security & Dept. of Elect. and Computer Engg.
xjin@cs.utsa.edu, {ram.krishnan,ravi.sandhu}@utsa.edu

Abstract. Recently, there has been considerable interest in attribute based access control (ABAC) to overcome the limitations of the dominant access control models (i.e, discretionary-DAC, mandatory-MAC and role based-RBAC) while unifying their advantages. Although some proposals for ABAC have been published, and even implemented and standardized, there is no consensus on precisely what is meant by ABAC or the required features of ABAC. There is no widely accepted ABAC model as there are for DAC, MAC and RBAC. This paper takes a step towards this end by constructing an ABAC model that has "just sufficient" features to be "easily and naturally" configured to do DAC, MAC and RBAC. For this purpose we understand DAC to mean owner-controlled access control lists, MAC to mean lattice-based access control with tranquility and RBAC to mean flat and hierarchical RBAC. Our central contribution is to take a first cut at establishing formal connections between the three successful classical models and desired ABAC models.

Keywords: Attribute, XACML, DAC, MAC, RBAC, ABAC.

1 Introduction

Starting with Lampson's access matrix in the late 1960's, dozens of access control models have been proposed. Only three have achieved success in practice: discretionary access control (DAC) [24], mandatory access control (MAC, also known as lattice based access control or multilevel security) [22] and role-based access control (RBAC) [11,23]. While DAC and MAC emerged in the early 1970's it took another quarter century for RBAC to develop robust foundations and flourish. RBAC emerged due to increasing practitioner dissatisfaction with the then dominant DAC and MAC paradigms, inspiring academic research on RBAC. Since then RBAC has become the dominant form of access control in practice.

Recently there has been growing practitioner concern with the limitations of RBAC, which has been met by researchers in two different ways. On one hand researchers have diligently and creatively extended RBAC in numerous directions. Conversely there is growing appreciation that a more general model, specifically attribute-based access control (ABAC), could encompass the demonstrated benefits of DAC, MAC and RBAC while transcending their limitations. Identities,

N. Cuppens-Boulahia et al. (Eds.): DBSec 2012, LNCS 7371, pp. 41–55, 2012.
© IFIP International Federation for Information Processing 2012

clearances, sensitivity, roles and other properties of users, subjects and objects can all be expressed as attributes. Languages for specifying permitted accesses based on the values and relationships among these attributes provide policy flexibility and customization. However, the proliferation and flexibility of policy configuration points in ABAC leads to greater difficulty in policy expression and comprehension relative to the simplicity of DAC, MAC and RBAC. It will require strong and comprehensive foundations for ABAC to flourish.

Intuitively, an attribute is a property expressed as a name:value pair associated with any entity in the system, including users, subjects and objects. Appropriate attributes can capture identities and access control lists (DAC), security labels, clearances and classifications (MAC) and roles (RBAC). As such ABAC supplements and subsumes rather than supplants these currently dominant models. Moreover any number of additional attributes such as location, time of day, strength of authentication, departmental affiliation, qualification, and frequent flyer status, can be brought into consideration within the same extensible framework of attributes. Thus the proliferation of RBAC extensions might be unified by adding appropriate attributes within a uniform framework, solving many of these shortcomings of core RBAC. At the same time we should recognize that ABAC with its flexibility may further confound the problem of role design and engineering. Attribute engineering is likely to be a more complex activity, and a price we may need to pay for added flexibility.

Much as RBAC concepts were around for decades before their formalization [13], nascent ABAC notions have been around for a while (see related work). The ABAC situation today is analogous to RBAC in its pre-1992 pre-RBAC and 1992-1996 early-RBAC periods [13]. Although considerable literature has been published, there is no agreement on what ABAC means. Fundamental questions such as components of core models lack authoritative answers, let alone a widely accepted ABAC model.

In this paper, we take a first step towards our eventual goal of developing an authoritative family of foundational models for attribute based access control. We believe this goal can be achieved only by means of incremental steps that advance our understanding. ABAC is a rich platform. Addressing it in its full scope from the beginning is infeasible. There are simply too mnay moving parts. A reasonable first step is to develop a formal ABAC model that is just sufficiently expressive to capture DAC, MAC and RBAC. This provides us a well-defined scope while ensuring that the resulting model has practical relevance. There have been informal demonstrations, such as [8,21], of the classical models using attributes. Our goal is to develop more complete and formal constructions.

The paper is organized as follows. We review previous work in section 2. In section 3, we characterize the three classical models from an ABAC perspective and informally identify the minimal features of an unifying ABAC model. In section 4, we give an overview of the ABAC$_\alpha$ model. In section 5, we present the formal definition of the model as well as functional specifications. In section 6, we show the configurations for DAC, MAC and RBAC in ABAC$_\alpha$. Section 7 concludes the paper.

2   Related Work

Extensions to RBAC by combining attributes and roles have been widely studied. [15] defines parameterized privileges to restrict access to a subset of objects. Similar literature such as parameterized role [3,10,14], object sensitive role [12] and attributed role [27] are also proposed. RB-RBAC model [4] use attributes to assist automatic user-role assignment.

Several attribute based access control systems and models have been proposed. The UCON model [21] focuses on usage control where authorizations are based on the attributes of the involved components. It is attribute-based but, rather than dealing with core ABAC concepts, it focuses on advanced access control features such as mutable attributes, continuous enforcement, obligations and conditions. UCON more or less assumes that an ABAC model is in place on top of which the UCON model is constructed. [21] sketches out instantiation of DAC, MAC and RBAC in UCON but the constructions are informal and not complete. Informal mappings of an ABAC system into DAC, MAC and RBAC are also described in [8]. Damiani et al [9] describe an informal framework for attribute based access control in open environments. Bonatti et al [6,7] present a uniform structure to logically formulate and reason about both service access and information disclosure constraints according to related entity attributes. Similarly, [28,29,30] develop a service negotiation framework for requesters and providers to gradually expose their attributes. However, none of these investigates their connections with DAC, MAC and RBAC. Wang et al [26] proposes a framework that models an attribute-based access control system using logic programming with set constraints of a computable set theory. This work mainly focus on how set theory helps define the policy, rather than the model itself. Flexible access control system [16,5] can specify some features of attribute based access policies. Yuan and Tong [31] describe ABAC in the aspects of authorization architecture and policy formulation. This work focus on enforcement level rather than policy level of the model. Bo et al [19] mention that DAC, MAC and RBAC is configurable through ABAC. However, neither formal model nor details of the configurations are provided. Role-based trust management [20] is a flexible approach for access control in distributed systems where access control decisions are based on tracking chaining credentials. However, its core idea is extensions to role based access control. XACML [1] and SAML [2] are access control-related web services standards that both support attribute-based access control. These standard languages are designed without a formal ABAC model.

3   ABAC$_\alpha$: Covering DAC, MAC and RBAC

Our goal is to develop an ABAC model that has "just sufficient" features to be "easily and naturally" configured to do DAC, MAC and RBAC. We recognize these terms are qualitative, hence the quotation marks. For clarity of reference we designate this model as ABAC$_\alpha$ and understand ABAC to denote the larger concept. Our goal is to eventually develop a family of ABAC models, analogous to RBAC96 [23], which will become the de facto standard for defining, refining and evolving ABAC. The contributions of this paper are one step towards this goal.

We very much expect ABAC to include advanced features that go significantly beyond $ABAC_\alpha$, e.g., mutable attributes [21], environment attributes [31] and connection attributes [18]. At this point it is premature to consider whether $ABAC_\alpha$ might be the core ABAC model, an advanced model or a special case of some model in a prospective ABAC family. Which features belong in a core ABAC model, which belong in advanced models and which are outside the scope of ABAC are crucial questions that researchers must eventually resolve. However, for the moment, we deliberately limit our scope to developing $ABAC_\alpha$.

$ABAC_\alpha$ is motivated by the fact that the three classical models have been widely deployed and remain in active widespread use. The value of ABAC has been perceived in benefits it provides beyond DAC, MAC and RBAC, such as dynamic access control [25]. Nonetheless, it is of interest to develop $ABAC_\alpha$ that captures these three without incorporating "extraneous" features. We anticipate that $ABAC_\alpha$ will eventually fit somewhere within the yet-to-be-developed authoritative family of ABAC models.

For purpose of $ABAC_\alpha$ we understand DAC to mean owner-controlled access control lists [24], MAC to mean lattice-based access control with tranquility [22] (i.e.,subject and object label do not change)and RBAC to mean core or flat RBAC ($RBAC_0$), and hierarchical RBAC ($RBAC_1$) [11,23]. Extensions beyond these interpretations of DAC, MAC and RBAC may or may not require extensions to $ABAC_\alpha$, comprehensive study of which is outside the scope of this paper.

Table 1. $ABAC_\alpha$ intrinsic requirements

	Subject attribute values constrained by creating user?	Object attribute values constrained by creating subject?	Attribute range ordered?	Attribute functions return set value?	Object attributes modification?	Subject attribute modification by creating user?
DAC	YES	YES	NO	YES	YES	NO
MAC	YES	YES	YES	NO	NO	NO
$RBAC_0$	YES	NA	NO	YES	NA	YES
$RBAC_1$	YES	NA	YES	YES	NA	YES
$ABAC_\alpha$	YES	YES	YES	YES	YES	YES

The intrinsic features of $ABAC_\alpha$ that follow from the above interpretation of DAC, MAC and RBAC are highlighted in Table 1. This table recognizes three kinds of familiar entities: users, subjects (or sessions in RBAC) and objects. Each user, subject and object has attributes associated with it. The range of each attribute is either atomic valued or set valued, with atomic values partially ordered or unordered and set values ordered by subset. Let us consider each column in turn.

Column 1. In all cases subject attribute values are constrained by attributes of the creating user. In MAC, users can only create subjects whose clearance is dominated by that of the user. In RBAC, subjects can only be assigned roles assigned to or inherited by the creating user. In DAC, MAC and RBAC, the subject's creator is set to be the creating user. Interestingly this is the only column with YES values for all rows.

Column 2. For object attributes in MAC a subject can only create objects with the same or higher classification as the subject's clearance. In DAC there is no constraint on the access control list associated with a newly created object. It is up to the creator's discretion. However, we recognize that DAC has a constraint on newly created objects in that root user usually has all access rights to every object and the owner can not forbid this. RBAC does not speak to object creation.

Column 3. In MAC clearances are values from a lattice of security labels. In $RBAC_1$ roles are partially ordered by permission inheritance. DAC and $RBAC_0$ do not require ordered attribute values.

Column 4. In MAC the clearance attribute is atomic valued as a single label from a lattice. In $RBAC_0$ and $RBAC_1$ attributes are sets of roles, and in DAC each access control list is a set of user identities.

Column 5. In DAC the user who created an object can modify its access control lists. MAC (with tranquility) does not permit modification of an object's classification. $RBAC_0$ and $RBAC_1$ do not speak to this issue.

Column 6. Modification of subject attributes by the creating user is explicitly permitted in $RBAC_0$ and $RBAC_1$ to allow dynamic activation and deactivation of roles. DAC and MAC do not require this feature.

Each column imposes requirements on $ABAC_\alpha$ so we have YES across the entire row. Table 1 is, of course, not a complete list of all required features to configure the classical models, but rather highlights the salient requirements that stem from each classical model.

4 $ABAC_\alpha$ Components

Based on the above analysis, we present a unified $ABAC_\alpha$ model informally in this section followed by its formalization in the next section. The structure of $ABAC_\alpha$ model is shown in Figure 1. The core components of this model are: users (U), subjects (S), objects (O), user attributes (UA), subject attributes (SA), object attributes (OA), permissions (P), authorization policies, and constraint checking policies for creating and modifying subject and object attributes.

An **attribute** is a function which takes an entity such as a user and returns a specific value from its range. An attribute range is given by a finite set of atomic

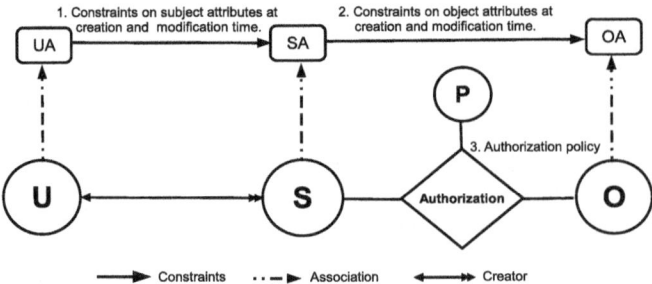

Fig. 1. Unified ABAC model structure

values. An atomic valued attribute will return one value from the range, while a set valued attribute will return a subset of the range. Each **user** is associated with a finite set of **user attribute** functions whose values are assigned by security administrators (outside the scope of the model). These attributes represent the user properties, such as name, clearance, roles and gender. **Subjects** are created by users to perform some actions in the system. For the purpose of this paper, subjects can only be created by a user and are not allowed to create other subjects. The creating user is the only one who can terminate a subject. Each subject is associated with a finite set of **subject attribute** functions which require an initial value at creation time. Subject attributes are set by the creating user and are constrained by policies established by security architects (discussed later). For example, a subject attribute value may be inherited from a corresponding user attribute. This is shown in Figure 1 as an arrow from user attributes to subject attributes. **Objects** are resources that need to be protected. Objects are associated with a finite set of **object attribute** functions. Objects may be created by a subject on behalf of its user. At creation, the object's attribute values may be set by the user via the subject. The values may be constrained by the corresponding subject's attributes. For example, the new object may inherit values from corresponding subject attributes. In Figure 1, the arrow from subject attributes to object attributes indicates this relationship.

Constraints are functions which return true when conditions are satisfied and false otherwise. Security architects configure constraints via policy languages. Constraints can apply at subject and object creation time, and subsequently at subject and object attribute modification time.

Permissions are privileges that a user can hold on objects and exercise via a subject. Permissions enable access of a subject to an object in a particular mode, such as read or write. Permissions definition is dependent on specific systems built using this model.

Authorization policy. Authorization policies are two-valued boolean functions which are evaluated for each access decision. An authorization policy for a specific permission takes a subject, an object and returns true or false based on attribute values. More generally, access decision may be three-valued, possibly returning "don't know" in addition to true and false. This is appropriate

Table 2. Basic sets and functions of ABAC$_\alpha$

U, S and O represent finite sets of *existing* users, subjects and objects respectively.

UA, SA and OA represent finite sets of user, subject and object attribute functions respectively. (Henceforth referred to as simply attributes.)

P represents a finite set of permissions.

For each *att* in UA ∪ SA ∪ OA, Range(*att*) represents the attribute's range, a finite set of *atomic* values.

SubCreator: $S \to U$. For each subject SubCreator gives its creator.

attType: UA ∪ SA ∪ OA → {set, atomic}. Specifies attributes as set or atomic valued.

Each attribute function maps elements in U, S and O to atomic or set values.

$$\forall ua \in \text{UA}.\, ua : \text{U} \to \begin{cases} \text{Range(ua) if attType}(ua) = \text{atomic} \\ 2^{\text{Range(ua)}} \;\; \text{if attType}(ua) = \text{set} \end{cases}$$

$$\forall sa \in \text{SA}.\, sa : \text{S} \to \begin{cases} \text{Range(sa) if attType}(sa) = \text{atomic} \\ 2^{\text{Range(sa)}} \;\; \text{if attType}(sa) = \text{set} \end{cases}$$

$$\forall oa \in \text{OA}.\, oa : \text{O} \to \begin{cases} \text{Range(oa) if attType}(oa) = \text{atomic} \\ 2^{\text{Range(oa)}} \;\; \text{if attType}(oa) = \text{set} \end{cases}$$

---

in multi-policy systems. It suffices for our purpose to consider just two values. Security architects are able to specify different authorization policies using the language offered in this model.

5   Formal ABAC$_\alpha$ Model

The basic sets and functions in ABAC$_\alpha$ are given in Table 2. U is the set of existing users and UA is a set of attribute function names for the users in U. Each attribute function in UA maps a user in U to a specific value. This could be atomic or set valued as determined by the type of the attribute function (attType). We specify similar sets and functions for subjects and objects. Sub-Creator is a distinguished attribute that maps each subject to the user who creates it (an alternate would be to treat this attribute as a function in SA). Finally, P is a set of permissions.

Policy Configuration Points. We define four policy configuration points as shown in Table 3. The first is for authorization policies (item 1 in table 3). The security architect specifies one authorization policy for each permission. The authorization function returns true or false based on attributes of the involved subject and object. The second configuration point is constraints for subject attribute assignment (item 2 in table 3). The third is constraints for object attributes assignment at the time of object creation (item 3 in table 3). The fourth is constraints for object attribute modification after the object has been

Table 3. Policy configuration points and languages of $ABAC_\alpha$

---

1. Authorization policies.
For each $p \in P$, $Authorization_p(s{:}S, o{:}O)$ returns true or false.
Language LAuthorization is used to define the above functions (one per permission),
where s and o are formal parameters.

---

2. Subject attribute assignment constraints.
Language LConstrSub is used to specify ConstrSub(u:U,s:S,saset:SASET), where u, s
and saset are formal parameters. The variable saset represents proposed attribute name
and value pairs for each subject attribute. Thus SASET is a set defined as follows:

$$SASET = \bigcup\nolimits_{\forall sa \in SA} \text{OneElement}(SASET_{sa})$$

$$\text{For each } sa \text{ in SA, } SASET_{sa} = \begin{cases} \{sa\} \times \text{Range}(sa) & \text{if attType}(sa) = \text{atomic} \\ \{sa\} \times 2^{\text{Range}(sa)} & \text{if attType}(sa) = \text{set} \end{cases}$$

We define OneElement to return a singleton subset from its input set.

---

3. Object attribute assignment constraints at object creation time.
Language LConstrObj is used to specify ConstrObj(s:S,o:O,oaset:OASET), where s,
o and oaset are formal parameters. The variable oaset represents proposed attribute
name and value pairs for each object attribute. Thus OASET is a set defined as follows:

$$OASET = \bigcup\nolimits_{\forall oa \in OA} \text{OneElement}(OASET_{oa})$$

$$\text{For each } oa \text{ in OA}, OASET_{oa} = \begin{cases} \{oa\} \times \text{Range}(oa) & \text{if attType}(oa) = \text{atomic} \\ \{oa\} \times 2^{\text{Range}(oa)} & \text{if attType}(oa) = \text{set} \end{cases}$$

---

4. Object attribute modification constraints.
Language LConstrObjMod is used to specify ConstrObjMod(s:S,o:O,oaset:OASET),
where s, o and oaset are formal parameters.

---

created (item 4 in table 3). Note that we have not provided a configuration
point for subject attribute modification after it has been created. For the stated
purposes in this paper, the function SubCreator captures necessary information.

Policy Configuration Languages. Each policy configuration point is expressed using a specific language. The languages specify what information is
available for the functions that configure the four points discussed above. For
example, in LConstrSub function, only attributes from the user who wants to
create the subject as well as the proposed subject attribute values are allowed.
Since all specification languages share the same format of logical structure while
differing only in the values they can use for comparison, we define a template
called Common Policy Language (CPL). CPL is not a complete language unless
terminals *set* and *atomic* are specified. It can be instantiated for specifying each
configuration point. CPL is defined in table 4.

Table 4. Definition of CPL

$\varphi ::= \varphi \wedge \varphi \mid \varphi \vee \varphi \mid (\varphi) \mid \neg \varphi \mid \exists\, x \in \text{set}.\varphi \mid \forall\, x \in \text{set}.\, \varphi \mid \text{set setcompare set} \mid$
  atomic $\in$ set $\mid$ atomic atomiccompare atomic

setcompare $::= \subset \mid \subseteq \mid \not\subseteq$

atomiccompare $::= < \mid = \mid \leq$

LAuthorization is a CPL instantiation for specifying authorization policies in which *set* and *atomic* are specified as follows:

> set::= setsa(s) | setoa(o)
> atomic::= atomicsa(s) | atomicoa(o)
> setsa $\in$ {sa | sa $\in$ SA $\wedge$ attType(sa) = set }
> setoa $\in$ {oa | oa $\in$ OA $\wedge$ attType(oa) = set }
> atomicoa $\in$ {oa | oa $\in$ OA $\wedge$ attType(oa) = atomic }
> atomicsa $\in$ {sa | sa $\in$ SA $\wedge$ attType(sa) = atomic }

LAuthorization allows one to specify policies based only on the value of involved subject and object. Parameters such as s and o in this and following languages are formal parameters as introduced in table 3.

LConstrSub is a CPL instantiation for specifying ConstrSub where:

> set::= setua(u) | value

> atomic::= atomicua(u) | value
> setua $\in$ {ua | ua $\in$ UA $\wedge$ attType(ua) = set }
> atomicua $\in$ {ua | ua $\in$ UA $\wedge$ attType(ua)= atomic }
> value $\in$ {val | (sa, val) $\in$ saset $\wedge$ sa $\in$ SA}

This instance is different from above because in the constraint function for subject attributes, only the attribute of user who wants to create the subject and the proposed values for subject attributes are allowed.

LConstrObj is a CPL instantiation for specifying ConstrObj where:

> set::= setsa(s) | value
> atomic::= atomicsa(s) | value
> setsa $\in$ {sa | sa $\in$ SA $\wedge$ attType(sa) = set }
> atomicsa $\in$ {sa | sa $\in$ SA $\wedge$ attType(sa)= atomic }
> value $\in$ {val | (oa, val) $\in$ oaset $\wedge$ oa $\in$ OA}

Here we use subject attributes instead of user attributes.

LConstrObjMod, used to specify ConstrObjMod, is the same as above except: set::= setsa(s) | setoa(o) | value and atomic::= atomicsa(s) | atomicoa(o) | value. Note that this language allows one to compare proposed new attribute values with current attribute values of an object unlike LConstrObj.

Functional Specifications. The $ABAC_\alpha$ functional specification, as shown in Table 5, outlines the semantics of various functions that are required for creation and maintenance of the $ABAC_\alpha$ model components. Our intention here is to only provide a sample set of key functions due to space limitations. The first column

Table 5. Functional specification

Functions	Conditions	Updates
Administrative functions: Creation and maintenance of user and their attributes. UASET is a set containing name and value pairs for each user attribute. $$\text{UASET} = \bigcup_{\forall ua \in UA} \text{OneElement}(\text{UASET}_{ua})$$ $$\forall ua \in \text{UA}.\ UASET_{ua} = \begin{cases} \{ua\} \times \text{Range(ua)} & \text{if attType}(ua) = \text{atomic} \\ \{ua\} \times 2^{\text{Range(ua)}} & \text{if attType}(ua) = \text{set} \end{cases}$$		
AddUser (u:NAME,$uaset$:UASET)	$u \notin U$	$U'=U \cup \{u\}$ **forall** (ua,va)$\in uaset$ **do** ua(u)=va
DeleteUser(u:NAME) /*delete all u's subjects*/	$u \in U$	$S'=S \setminus \{s \mid \text{SubCreator}(s)=u\}$ $U'=U \setminus \{u\}$
ModifyUserAtt (u:NAME,$uaset$:UASET) /*delete all u's subjects*/	$u \in U$	**forall** (ua,va)$\in uaset$ **do** ua(u)=va $S'=S \setminus \{s \mid \text{SubCreator}(s)=u\}$
System functions: User level operations.		
CreateSubject (u, s:NAME,$saset$:SASET)	$u \in U \wedge s \notin S \wedge$ ConstrSub($u, s, saset$)	$S'=S \cup \{s\}$; SubCreator(s)=u **forall** (sa,va)$\in saset$ **do** sa(s)=va
DeleteSubject (u, s:NAME)	$s \in S \wedge u \in U \wedge$ SubCreator(s)=u	$S'=S \setminus \{s\}$
ModifySubjectAtt (u, s:NAME,$saset$:SASET)	$s \in S \wedge u \in U \wedge$ SubCreator(s)=$u \wedge$ ConstrSub($u, s, saset$)	**forall** (sa,va)$\in saset$ **do** sa(s)=va
CreateObject (s, o:NAME,$oaset$:OASET)	$s \in S \wedge o \notin O \wedge$ ConstrObj($s, o, oaset$)	$O'=O \cup \{o\}$ **forall** (oa,va)$\in oaset$ **do** oa(o)=va
ModifyObjectAtt (s, o:NAME,$oaset$:OASET)	$s \in S \wedge o \in O \wedge$ ConstrObjMod($s, o, oaset$)	**forall** (oa,va)$\in oaset$ **do** oa(o)=va
$\forall\ p \in$ **P. Authorization$_p$;** **ConstrSub; ConstrObj; ConstrObjMod**	/*Left to be specified by security architects*/	

lists all the function names as well as required parameters. The second column represents the conditions which need to be satisfied before the updates, which are listed in the third column, can be executed. $NAME$ refers to set of all names for various entities in the system.

The first kind of functions are administrative in nature which are designed to be invoked only by security administrators. We do not specify the authorization conditions for administrative functions which are outside the scope of ABAC$_\alpha$. They mainly deal with user and user attribute management. One important issue with the user management is that the subjects created by a user are forced to be terminated whenever user attributes are modified or the user is deleted. We understand there are various options here (discussion on this question is out

Table 6. DAC (Owner-controlled access control lists) configuration

---

Basic sets and functions

UA={}, SA={}, OA={*reader, writer, createdby*}

P={*read, write*}

Range(*reader*)=Range(*writer*)=Range(*createdby*)=U

attType(*reader*)=attType(*writer*)=set

attType(*createdby*)=atomic

Thus, reader: O $\rightarrow 2^{U}$, writer: O $\rightarrow 2^{U}$, createdby: O $\rightarrow$ U

The function SubCreator is defined in Table 2.

Configuration points

1. Authorization policy

Authorization$_{read}$(s:S, o:O)$\equiv$SubCreator(s)$\in$reader(o)

Authorization$_{write}$(s:S, o:O)$\equiv$SubCreator(s)$\in$writer(o)

2. Constraint for subject attribute is not required

Note that SubCreator is implicitly captured in function CreateSubject in Table 5.
Function ConstrSub(u:U, s:S, {}:SASET) is defined to return true.

3. Constraint for object attribute at creation time

ConstrObj(s:S, o:O, {(*reader*,val1), (*writer*,val2), (*createdby*,val3)}:OASET)$\equiv$
val3=SubCreator(s)

4. Constraint for object attribute at modification time

ConstrObjMod(s:S, o:O, {(*reader*,val1), (*writer*,val2), (*createdby*,val3)}:OASET)$\equiv$
createdby(o)=SubCreator(s)

---

of scope due to lack of space). The second kind of functions are system functions which can be invoked by subjects and users. By default, the first function parameter is the invoker of each function. For example, CreateSubject is invoked by user u and ModifyObjectAtt is invoked by subject s. The third kind of functions are authorization policies and subject and object attribute constraint functions which are left to be configured by security architects.

6   ABAC$_\alpha$: Configuring DAC, MAC and RBAC

In this section, we show the capability of ABAC$_\alpha$ in configuring DAC, MAC and RBAC. For this illustration, we set P={*read, write*}.

DAC (Table 6). Each object is associated with the same number of set-valued attributes as that of permissions and there is a one to one semantic mapping between them. An object attribute returns the list of users that hold the permission indicated by the object attribute name. Object attribute *createdby* is set to be the owner of this object.

MAC (Table 7). Each user is associated with an atomic-valued attribute *uclearance*. Each subject is also associated with an atomic-valued attribute *sclearance*. Each object is associated with an atomic-valued attribute *sensitivity*. Similar to MAC, the user and subject attributes represent their clearance in the system. The *sensitivity* attribute of the object represents the object's

Table 7. MAC configuration

Basic sets and functions
UA={$uclearance$}, SA={$sclearance$}, OA={$sensitivity$}
P={$read, write$}
Range($uclearance$)=Range($sclearance$)=Range($sensitivity$)=L
L is a lattice defined by system.
attType($uclearance$)=attType($sclearance$)=attType($sensitivity$)= atomic
Thus, uclearance: U $\rightarrow$ L, sclearance: S $\rightarrow$ L, sensitivity: O $\rightarrow$ L.
Configuration points
1. Authorization policies
Authorization$_{read}$(s:S, o:O)$\equiv sensitivity(o) \leq sclearance(s)$
Liberal Star: Authorization$_{write}$(s:S, o:O)$\equiv sclearance(s) \leq sensitivity(o)$
Strict Star: Authorization$_{write}$(s:S, o:O)$\equiv sclearance(s)=sensitivity(o)$
2. ConstrSub(u:U, s:S, {($sclearance$,value)}:SASET)$\equiv$value$\leq uclearance(u)$
3. ConstrObj(s:S, o:O, {($sensitivity$, value)}:OASET)$\equiv sclearance(s) \leq$value
4. ConstrObjMod(s:S, o:O, {($sensitivity$, value)}:OASET) returns false.

Table 8. RBAC configurations

RBAC$_0$ configuration
Basic sets and functions
UA={$urole$}, SA={$srole$}, OA={$rrole, wrole$}
P={$read, write$}
Range($urole$)=Range($srole$)=Range($rrole$)=Range($wrole$)=R
R is a set of atomic roles define by the system.
attType($urole$)=attType($srole$)=attType($rrole$)=attType($wrole$)=set
Thus, urole: U $\rightarrow 2^R$, srole: S $\rightarrow 2^R$, rrole: O $\rightarrow 2^R$, wrole: O $\rightarrow 2^R$
Configuration points
1. Authorization policy
Authorization$_{read}$(s:S, o:O)$\equiv \exists r \in srole(s).r \in rrole(o)$
Authorization$_{write}$(s:S, o:O)$\equiv \exists r \in srole(s).r \in wrole(o)$ (same as above)
2. ConstrSub(u:U, s:S, {($srole$,val1)}:SASET)$\equiv$val1$\subseteq urole(u)$
3. ConstrObj(s:S, o:O, {(rrole,val1),(wrole,val2)}:OASET) returns false.
4. ConstrObjMod(s:S, o:O, {(rrole,val1),(wrole,val2)}:OASET) returns false.

RBAC$_1$ configuration
Basic sets and functions
The basic sets and functions are the same as RBAC$_0$ except:
R is a partially ordered set defined by the system.
Configuration points
1. Authorization policy
Authorization$_{read}$(s:S, o:O)$\equiv \exists r1 \in srole(s). \exists r2 \in rrole(o).r2 \leq r1$
Authorization$_{write}$(s:S, o:O)$\equiv \exists r1 \in srole(s). \exists r2 \in wrole(o).r2 \leq r1$ (same as above)
2. ConstrSub(u:U, s:S, {($srole$,val1)}:SASET)$\equiv \forall r1 \in$val1.$\exists r2 \in urole(u).r1 \leq r2$
3. ConstrObj(s:S, o:O, {(rrole,val1),(wrole,val2)}:OASET) returns false.
4. ConstrObjMod(s:S, o:O, {(rrole,val1),(wrole,val2)}:OASET) returns false.

---

classification in MAC. The 3 attributes share the same range which is represented by a system maintained lattice L.

RBAC (Table 8). Each user and subject is associated with set-valued attributes *urole* and *srole* respectively. Each object is associated with the same number of set-valued attributes as that of permissions and there is a one to one semantic mapping between them. Each attribute returns the role that is assigned the permission on this specific object. For example, *rrole* of object *obj* returns the role which is assigned the permission of reading *obj*. The ranges of all attributes are the same as that of a system defined set of role names R which are unordered for $RBAC_0$ and partially ordered for $RBAC_1$. Note that subjects model sessions in RBAC.

7 Conclusion and Future Work

In this paper, we proposed a unified $ABAC_\alpha$ model and showed that it can be used to naturally configure the three classical models. We believe the insights gained in this paper will assist understanding the connections between desired ABAC model and widely-deployed classical models. In addition, we hope this work will inspire further research in formally designing foundational ABAC models.

Some extensions of classical models can also be accommodated. In MAC, it is useful to categorize subjects into different types as read_only and read_write for both security and availability. The rule governing their actions can be different in that read_only subjects are allowed to read all levels of objects. While read_write subjects' action is strictly regulated. Another example is in RBAC, certain level of automatic permission-role assignment can be achieved by interpreting permissions as accessing a group of objects with the same attribute expression. Organization based access control model (OrBAC)[17] is another example of abstracting activities, objects and so on.

The first aspect of future work is to extend and consolidate the proposed model. Examples are to accommodate static/dynamic separation of duty in RBAC and subjects carrying additional attributes other than the corresponding users to reflect contextual information. Security properties and expressive power of this model are important questions for further theoretical analysis. On the other hand, useful instances of this model with various relationships between user, subject and object attributes can be developed for specific groups of application. For example, usable $ABAC_\alpha$ instance in organizations offer better guidance than general $ABAC_\alpha$. In future work, we plan to develop XACML profiles for ABAC models as we develop them. By design XACML does not recognize user-subject mapping but assumes that subject attributes are correctly produced from user attributes prior to making access decisions. Modeling this process will therefore require extensions to XACML.

Acknowledgment. The authors are partially supported by grants from AFOSR MURI and the State of Texas Emerging Technology Fund.

References

1. OASIS, Extensible access control markup language (XACML), v2.0 (2005)
2. OASIS, Security assertion markup language (SAML), v2.0 (2005)
3. Abdallah, A.E., Khayat, E.J.: A formal model for parameterized role-based access control. In: Formal Aspects in Security and Trust (2004)
4. Al-Kahtani, M.A., Sandhu, R.S.: A model for attribute-based user-role assignment. In: ACSAC (2002)
5. Bertino, E., Catania, B., Ferrari, E., Perlasca, P.: A logical framework for reasoning about access control models. In: SACMAT (2001)
6. Bonatti, P.A., Samarati, P.: Regulating service access and information release on the web. In: ACM CCS (2000)
7. Bonatti, P.A., Samarati, P.: A uniform framework for regulating service access and information release on the web. J. Comp. Secur. (2002)
8. Chadwick, D.W., Otenko, A., Ball, E.: Role-based access control with X.509 attribute certificates. IEEE Internet Computing (2003)
9. Damiani, E., di Vimercati, S.D.C., Samarati, P.: New paradigms for access control in open environments. In: Int. Sym. on Sig. Proc. and Info. Tech. (2005)
10. Evered, M.: Supporting parameterised roles with object-based access control. In: HICSS (2003)
11. Ferraiolo, D.F., Sandhu, R., Gavrila, S., Richard Kuhn, D., Chandramouli, R.: Proposed nist standard for role-based access control. ACM Trans. Inf. Syst. Secur. (2001)
12. Fischer, J., Marino, D., Majumdar, R., Millstein, T.: Fine-Grained Access Control with Object-Sensitive Roles. In: Drossopoulou, S. (ed.) ECOOP 2009. LNCS, vol. 5653, pp. 173–194. Springer, Heidelberg (2009)
13. Fuchs, L., Pernul, G., Sandhu, R.: Roles in information security: A survey and classification of the research area. Comp. and Secur. (2011)
14. Ge, M., Osborn, S.L.: A design for parameterized roles. In: DBSec (2004)
15. Giuri, L., Iglio, P.: Role templates for content-based access control. In: ACM Workshop on RBAC (1997)
16. Jajodia, S., Samarati, P., Sapino, M.L., Subrahmanian, V.S.: Flexible support for multiple access control policies. ACM Trans. Database Syst. (2001)
17. El Kalam, A.A., Benferhat, S., Miège, A., El Baida, R., Cuppens, F., Saurel, C., Balbiani, P., Deswarte, Y., Trouessin, G.: Organization based access control. In: POLICY (2003)
18. Kandala, S., Sandhu, R., Bhamidipati, V.: An attribute based framework for risk-adaptive access control models. In: ARES (2011)
19. Lang, B., Foster, I.T., Siebenlist, F., Ananthakrishnan, R., Freeman, T.: A flexible attribute based access control method for grid computing. J. Grid Comput. (2009)
20. Li, N., Mitchell, J.C., Winsborough, W.H.: Design of a role-based trust management framework. In: 2002 IEEE S&P (2002)
21. Park, J., Sandhu, R.: The UCONabc usage control model. ACM Trans. Inf. Syst. Secur. (2004)
22. Sandhu, R.S.: Lattice-based access control models. IEEE Computer (1993)
23. Sandhu, R.S., Coyne, E.J., Feinstein, H.L., Youman, C.E.: Role-based access control models. IEEE Computer (1996)
24. Sandhu, R.S., Samarati, P.: Access control: Principles and practice. IEEE Com. Mag. (1994)

25. Schläger, C., Sojer, M., Muschall, B., Pernul, G.: Attribute-Based Authentication and Authorisation Infrastructures for E-Commerce Providers. In: Bauknecht, K., Pröll, B., Werthner, H. (eds.) EC-Web 2006. LNCS, vol. 4082, pp. 132–141. Springer, Heidelberg (2006)
26. Wang, L., Wijesekera, D., Jajodia, S.: A logic-based framework for attribute based access control. In: 2nd ACM Workshop on FMSE (2004)
27. Yong, J., Bertino, E., Toleman, M., Roberts, D.: Extended RBAC with role attributes. In: 10th Pacific Asia Conf. on Info. Sys. (2006)
28. Yu, T., Ma, X., Winslett, M.: Prunes: an efficient and complete strategy for automated trust negotiation over the internet. In: ACM CCS (2000)
29. Yu, T., Winslett, M., Seamons, K.E.: Interoperable strategies in automated trust negotiation. In: ACM CCS (2001)
30. Yu, T., Winslett, M., Seamons, K.E.: Supporting structured credentials and sensitive policies through interoperable strategies for automated trust negotiation. ACM Trans. Inf. Syst. Secur. (2003)
31. Yuan, E., Tong, J.: Attributed based access control (ABAC) for web services. In: Intl. ICWS (2005)

Signature-Based Inference-Usability Confinement for Relational Databases under Functional and Join Dependencies*

Joachim Biskup[1], Sven Hartmann[2], Sebastian Link[3], Jan-Hendrik Lochner[1], and Torsten Schlotmann[1]

[1] Fakultät für Informatik, Technische Universität Dortmund, Germany
{joachim.biskup,jan-hendrik.lochner,torsten.schlotmann}@cs.tu-dortmund.de
[2] Institut für Informatik, Technische Universität Clausthal, Germany
sven.hartmann@tu-clausthal.de
[3] Department of Computer Science, The University of Auckland, New Zealand
s.link@auckland.ac.nz

Abstract. Inference control of queries for relational databases confines the information content and thus the usability of data returned to a client, aiming to keep some pieces of information confidential as specified in a policy, in particular for the sake of privacy. In general, there is a tradeoff between the following factors: on the one hand, the expressiveness offered to administrators to declare a schema, a confidentiality policy and assumptions about a client's a priori knowledge; on the other hand, the computational complexity of a provably confidentiality preserving enforcement mechanism. We propose and investigate a new balanced solution for a widely applicable situation: we admit relational schemas with functional and join dependencies, which are also treated as a priori knowledge, and select-project sentences for policies and queries; we design an efficient signature-based enforcement mechanism that we implement for an Oracle/SQL-system. At declaration time, the inference signatures are compiled from an analysis of all possible crucial inferences, and at run time they are employed like in the field of intrusion detection.

Keywords: a priori knowledge, confidentiality policy, functional dependency, inference control, inference-usability confinement, interaction history, join dependency, refusal, relational database, select-project query, inference signature, SQL, template dependency.

1 Introduction

Inference control for information systems in general and relational databases in particular is a mechanism to confine the information content and thus the usability of data made accessible to a client to whom some piece(s) of information

---

* This work has been partially supported by the Deutsche Forschungsgemeinschaft under grant BI 311/12-2 and under grant SFB 876/A5 for the Collaborative Research Center "Providing Information by Resource-Constrained Data Analysis".

N. Cuppens-Boulahia et al. (Eds.): DBSec 2012, LNCS 7371, pp. 56–73, 2012.

should be kept confidential. Thus inference control aims at protecting *information* rather than just the underlying *data*, as achieved by traditional access control or simple encryption. Though protection of information is a crucial requirement for many public and commercial applications, the actual enforcement is facing great challenges arising from conceptual and computational problems.

On the conceptual side, among others the following main factors have to be considered: a client-specific and declaratively expressed *confidentiality policy* which might be balanced with availability demands; an assumption about the client's *a priori knowledge* regarding the information managed by the information system, which will include schema information in many cases; the client's postulated *system awareness* regarding the semantics of both the underlying information system and the monitoring control mechanism.

On the computational side, the high runtime complexity is a major concern. In fact, the fundamental semantics of a well-designed information system can be defined in terms of an appropriate logic. In particular, a *relational database* comes along with the *relational calculus* for querying and some class of *dependencies* (semantic constraints) for declaring schemas [1]. Thus, data managed by such a system can be interpreted as sentences in the underlying first-order logic. Accordingly, confining the usability of data comprises the task of monitoring all options for inferring implied (entailed) sentences from the sentences available to a client, at any point in time while the client is interacting with the system. Unfortunately, as well-known from the discipline of *theorem proving*, the computational treatment of entailment problems might be inherently complex.

Consequently, a major research task regarding information protection is to identify practically relevant situations that still enable a reasonably efficient enforcement of suitably restricted conceptual requirements. In our previous work [8] we already introduced and theoretically analyzed the following situation as highly promising: Using the *refusal* approach, where harmful correct answers are replaced by a refusal notification denoted by mum, we protect *select-project sentences* under closed (yes/no-)*select-project queries* evaluated for relational database instances of a schema with *functional and join dependencies*. In the present article, we present a successful elaboration of the proposed approach:

- Based on the theoretical analysis, we have designed, implemented and investigated a practical, SQL-conforming *signature-based* enforcement method.
- The inference signatures are *compiled* from an analysis of all crucial inferences that are possible for the given situation, and later on *monitored* like in the field of intrusion detection.

To set up a larger perspective, we observe that the conceptual requirements of inference-usability confinement can be captured by an invariant that a control mechanism has to guarantee for all sequences of query-response interactions. Such an *invariant* might have several forms, which are equivalent under careful formalizations [13,10]. E.g., the invariant might require that for any sentence in

the confidentiality policy, based on his current knowledge, which results from
the a priori knowledge and previous interactions, and his system awareness,

- the client cannot exclude that this sentence is *not* valid in the instance;
- the client does not know that this sentence is valid in the instance.

To enforce such an invariant, a control mechanism has to inspect each query
considered as an interaction request and the answer to be returned whether or
not they satisfy an adequate *control condition*. Clearly, a *necessary* control condition is that the current knowledge updated with the answer will not entail any
sentence in the confidentiality policy. Unfortunately, however, this condition is
not sufficient in general, since it neglects the impact of a client's system awareness, which might enable so-called meta-inferences. Thus, in general, we have to
strengthen this condition to become *sufficient*, while preferably remaining to be
necessary for the sake of availability. Moreover, as far as achievable, checking a
sufficient (and necessary) control condition should be *computationally feasible*.

Several sufficient and "reasonably necessary" control conditions for comprehensive and general situations have been proposed in the past [3], which, however,
inevitably tend to be infeasible in the worst case. Moreover, dedicated narrower
situations have been investigated to find effective control conditions that are
also efficiently testable. The contribution of the present article is particularly
related to the following situations, all of which consider the refusal approach for
relational databases, assuming that the client knows the confidentiality policy.

- *Situation 1.* As long as decidability is achieved, any a priori knowledge,
  confidentiality policy and closed (yes/no-)queries (of the relational calculus)
  are admitted. For the confinement, while maintaining and employing a log file
  that represents the a priori knowledge and the answers to previous queries,
  we have to ensure that adding neither the correct answer to the current
  query nor the negation of the correct answer will be harmful; additional
  refusals for harmful negations of correct answers guarantee that an observed
  "refused" answer mum cannot be traced back to its actual cause by exploiting
  the system awareness [4].
- *Situation 2.* The a priori knowledge may only comprise a schema declaration
  with functional dependencies that lead to Boyce-Codd normal form with a
  unique minimal key. Confidentiality policies are restricted to select-project
  sentences of a special kind referring to "facts", and queries are restricted to
  arbitrary select-project sentences. For the confinement, it suffices to ensure
  that the query sentence does not "cover" any policy element [6].
- *Situation 3.* The a priori knowledge may only comprise a schema declaration
  with functional dependencies. Confidentiality policies are restricted to select-
  project sentences, whereas queries must be closed select-queries. For the
  confinement, it suffices to ensure that the query sentence does not "cover"
  any policy element [9].
- *Situation 4.* The a priori knowledge may only comprise a schema declaration with functional dependencies and full join dependencies (without any
  further restrictions). Confidentiality policies and queries are restricted to

select-project sentences. For the confinement, we have to ensure two conditions: (1) The query sentence does not "cover" any policy element. (2) Previous positive answers together with a positive answer to the current query do not "instantiate" any template dependency implied by the schema dependencies and "covering" an element of the confidentiality policy [8]. In the present article, we will show how this requirement can be converted into an efficient enforcement mechanism.

Notably, Situations 2 to 4 postulate that the client's a priori knowledge only comprises *schema declarations*. Accordingly, if additional a priori knowledge was assumed, further potential sources of inferences should be considered, and thus the respective confinement method would have to be appropriately enhanced.

The control conditions sketched so far are devised to be used *dynamically* at run time to detect current options for crucial entailments. To avoid the runtime overhead, one might prefer a *static* approach [11]: We then interpret the confidentiality policy and the a priori knowledge as constraints to be satisfied by an alternative instance that minimally distorts the actual instance, precompute a solution to such a constrained optimization problem, and let the solution instance be queried by the client without any further control.

We might also follow a *mixed* approach that suitably splits the workload among (1) some precomputations before the client is involved at all, (2) appropriate dynamic control operations after receiving a specific client request and before returning an answer, and (3) some follow-up adaptation actions between two requests [2]. The present article follows a mixed approach for the specific relational framework of [8] described as Situation 4 above; roughly outlined, our new signature-based enforcement mechanism consists of a two-phase protocol:

- At declaration time, we compile inference signatures as representatives of "forbidden structures" in the sense of [8]: "instantiations" of template dependencies implied by the schema dependencies and "covering" a policy element.
- At run time, we monitor these inference signatures for the actual queries.

2 Formal Framework

In this section we summarize our formal framework and restate the theorem that justifies our signature-based enforcement mechanism, referring the reader to [1,8] for more details. Examples can be found in the next section.

A *relation schema* $RS = \langle R, \mathcal{U}, \Sigma \rangle$ consists of a *relation symbol* R, a finite set $\mathcal{U}$ of *attributes*, and a finite set Σ of dependencies (semantic constraints). Σ comprises either functional dependencies, assumed to be a minimal cover, or full join dependencies, or both kinds of dependencies. An *instance* r is a finite dependency-satisfying Herbrand interpretation of the schema, considering the relation symbol as a predicate. A *tuple* is denoted by $\mu = R(a_1, \ldots, a_n)$ where $n = |\mathcal{U}|$ and $a_i \in Const$, an infinite set of constants. If μ is an element of r, we write $r \models_M \mu$. More generally, $\models_M$ denotes the *satisfaction* relation between an interpretation and a sentence. The corresponding notion of logical *implication* (entailment) between sentences is denoted by $\models$.

Let $\mathcal{A}, \mathcal{B} \subseteq \mathcal{U}$ be attribute sets. A relation r over $\mathcal{U}$ satisfies the *functional dependency* $\mathcal{A} \to \mathcal{B}$ if any two tuples that agree on the values of attributes in $\mathcal{A}$ also agree on the values of the attributes in $\mathcal{B}$.

Let $\mathcal{C}_1, \ldots, \mathcal{C}_l \subseteq \mathcal{U}$ be attribute sets such that $\mathcal{C}_1 \cup \ldots \cup \mathcal{C}_l = \mathcal{U}$. A relation r over $\mathcal{U}$ satisfies the (full) *join dependency* $\bowtie [\mathcal{C}_1, \ldots, \mathcal{C}_l]$ if whenever there are tuples $\mu_1, \ldots, \mu_l$ in r with $\mu_i[\mathcal{C}_i \cap \mathcal{C}_j] = \mu_j[\mathcal{C}_i \cap \mathcal{C}_j]$ for $1 \le i, j \le l$, there is also a tuple μ_{l+1} in r with $\mu_{l+1}[\mathcal{C}_i] = \mu_i[\mathcal{C}_i]$ for $1 \le i \le l$.

Join dependencies are a special case of *template dependencies*. A *template dependency* $TD[h_1, \ldots, h_l|c]$ over $\mathcal{U}$ has one or more *hypothesis rows* $h_1, \ldots, h_l$ and a *conclusion row* c. Each row consists of abstract symbols (best seen as variables), one symbol per attribute in $\mathcal{U}$. A symbol may appear more than once but only for one attribute, i.e., in a *typed* way. For t and t' denoting tuples or rows, respectively, over $\mathcal{U}$, $ag(t, t') := \{A \mid A \in \mathcal{U} \text{ and } t(A) = t'(A)\}$ is the *agree set* of these tuples or rows, respectively. The aggregated agree sets of the conclusion $\cup_{j=1}^{l} ag(c, h_j)$ form the *scheme* of the template dependency. A symbol occurring in the conclusion for an attribute A in the scheme is often called "distinguished" (free variable) and denoted by a_A. Any other symbol in the template dependency is often called "nondistinguished" (existentially quantified variable) and denoted by b_i, where each such symbol gets a different index i.

A relation r over $\mathcal{U}$ *satisfies* the template dependency $TD[h_1, \ldots, h_l|c]$ if whenever there are tuples $t_1, \ldots, t_l$ in r with $ag(h_i, h_j) \subseteq ag(t_i, t_j)$ for all $i, j \in \{1, \ldots, l\}$ there is also a tuple t in r with $ag(c, h_i) \subseteq ag(t, t_i)$ for $i = 1, \ldots, l$.

A template dependency $TD[h_1, \ldots, h_l|c]$ is called (hypothesis-)*minimal* with respect to Σ if dropping a full hypothesis row h_i or replacing any symbol in a hypothesis by a new symbol, different from all others – thus (potentially) deleting an equality condition – would result in a template dependency that is not implied by Σ. Moreover, a minimal template dependency is called (conclusion-)*maximal* with respect to Σ if the following additionally holds: if we replace a symbol in the conclusion row c that so far is not involved in any agree set with a hypothesis by another symbol that already occurs in some hypothesis, then we would obtain a template dependency that is not implied by Σ. Finally, a template dependency enjoying both optimization properties is called a *basic implication* of Σ.

Queries and elements of a confidentiality policy *psec*, called *potential secrets*, are expressed in a fragment of the relational calculus, using a set of variables *Var*. This fragment is given by the *language* $\mathcal{L}$ of *existential-R-sentences*, or *select-project sentences* which are *sentences* (closed formulas) of the form $(\exists X_1) \ldots (\exists X_l) R(v_1, \ldots, v_n)$ with $0 \le l \le n$, $X_i \in Var$, $v_i \in Const \cup Var$, $\{X_1, \ldots, X_l\} \subseteq \{v_1, \ldots, v_n\}$, and $v_i \ne v_j$ if $v_i, v_j \in Var$ and $i \ne j$; these properties and the closedness imply that $\{X_1, \ldots, X_l\} = \{v_1, \ldots, v_n\} \cap Var$. For $\Phi \in \mathcal{L}$, we define the *scheme* $\mathcal{P}$ of Φ as the set of attributes for which a constant appears. For a sentence in $\mathcal{L}$ let its corresponding "generalized tuple" denote the sentence without its prefix of existential quantifiers. In this case we think of the variables in the generalized tuple as the null value "exists but unknown".

A sentence (generalized tuple) Φ is defined to *cover* a sentence (generalized tuple) Ψ if every constant c that appears in Ψ appears in Φ at the same position. The following equivalence can be easily verified: Φ covers Ψ if and only if $\Phi \models \Psi$.

Restating Theorem 2 of [8] below, we specify a "forbidden structure" an instantiation of which is necessary for any violation of a policy element by exploiting the a priori knowledge about the dependencies. Additionally, Theorem 1 of [8] indicates that an occurrence of such a structure is also sufficient for exploiting the dependencies. Accordingly, we obtain a necessary and "reasonably sufficient" *control condition* by avoiding both an immediate violation by "covering" a potential secret and an instantiated "forbidden structure"; the latter consists of an implied template dependency whose scheme comprises the scheme of a potential secret, while the schemes of the query answers "uniformly cover" the hypotheses. Our mechanism will be based on that control condition.

Theorem 1 (forbidden structures, necessary for a violation by exploiting the dependencies). *Let $RS = \langle R, \mathcal{U}, \Sigma \rangle$ be a relation schema where the dependency set Σ consists of functional and full join dependencies, and r an instance of RS. Let $\Psi \in \mathscr{L}$ be a potential secret with scheme $\mathcal{P} \subseteq \mathcal{U}$, and $\Phi_1, \ldots, \Phi_l \in \mathscr{L}$ queries with schemes $\mathcal{F}_1, \ldots, \mathcal{F}_l$ such that:*

1. *$\Phi_i \not\models \Psi$, for $i = 1, \ldots, l$, i.e., all queries do not cover the potential secret;*
2. *$r \models_M \Phi_i$, for $i = 1, \ldots, l$, i.e., all queries are true in the instance r;*
3. *$\Sigma \cup \{\Phi_1, \ldots, \Phi_l\} \models \Psi$, i.e., the answers violate the confidentiality policy.*

Then there exists a nontrivial template dependency $TD[h_1, \ldots, h_l | c]$ implied by Σ such that $\mathcal{P} = \cup_{j=1}^{l} ag(c, h_j)$ and $\cup_{j \in \{1, \ldots, l\} \smallsetminus \{i\}} ag(h_i, h_j) \subseteq \mathcal{F}_i$, for $i = 1, \ldots, l$.

3 Examples

We will outline the fundamental features of the signature-based enforcement mechanism by means of two examples. Though only dealing with functional dependencies specifying a key, the first example is beyond the scope of the Situations 2 and 3 sketched before and thus cannot be treated by the mechanisms of [6,9]. The second example introduces join dependencies as a priori knowledge.

Example 1. At *declaration time*, we consider the following items: a relation schema $RS = \langle R, \mathcal{U}, \Sigma \rangle$ with attribute set $\mathcal{U} = \{K, A, B\}$ and dependencies $\Sigma = \{K \to A, K \to B\}$, i.e., attribute K is the unique minimal key; an instance $r = \{R(c_K, c_A, c_B), R(\tilde{c}_K, c_A, c_B)\}$, where c_K, $\tilde{c}_K$, c_A and c_B are constants in *Const*; and a single potential secret $\Psi = (\exists X_K) R(X_K, c_A, c_B)$. We will compile inference signatures in four steps.

In step 1, we see that Σ entails the template dependency
$$td := TD[\ a_K\ a_A\ b_1\ ,\ a_K\ b_2\ a_B\ |\ a_K\ a_A\ a_B\ ]$$
as a "forbidden structure" that must not be instantiated by the potential secret and query answers according to the instance.

In step 2, first treating the potential secret, we find that the scheme KAB of the template dependency td covers the scheme AB of the potential secret Ψ.

In step 3, we specialize the conclusion (a_K a_A a_B) of td with the constants appearing in the potential secret Ψ, yielding (a_K c_A c_B). Then we propagate this specialization to the hypotheses on common attributes, i.e., according to agree sets, getting (a_K c_A b_1) for the first hypothesis, and (a_K b_2 c_B) for the second hypothesis. In this way we get the instantiated template dependency

$td[\Psi] := TD[\, a_K\ c_A\ b_1\, ,\, a_K\ b_2\ c_B \,|\, a_K\ c_A\ c_B \,].$

In step 4, finally considering the instance r, we further uniformly instantiate the hypotheses on the distinguished symbol a_K for the further agree set $ag(h_1, h_2) = \{K\}$ with $h_1 = (\, a_K\ c_A\ b_1\,)$ and $h_2 = (\, a_K\ b_2\ c_B\,)$ according to tuples in the instance r. For the single tuple $R(c_K, c_A, c_B) \in r$ used twice, we get

$Sig_1 := TD[\, c_K\ c_A\ b_1\, ,\, c_K\ b_2\ c_B \,|\, c_K\ c_A\ c_B \,]$

as an inference signature; similarly, for the tuple $R(\tilde{c}_K, c_A, c_B) \in r$ we get

$Sig_2 := TD[\, \tilde{c}_K\ c_A\ b_1\, ,\, \tilde{c}_K\ b_2\ c_B \,|\, \tilde{c}_K\ c_A\ c_B \,]$

as another inference signature. Each of them indicates that the user must not learn all of its hypotheses, and thus later on we can ignore its conclusion.

Once the inference signatures have been compiled at declaration time, they have to be monitored at *run time* according to the queries requested by the pertinent client. Suppose the client issues

$\Phi_1 := R(c_K, c_A, c_B),$
$\Phi_2 := (\exists X_B)R(c_K, c_A, X_B),$ and
$\Phi_3 := (\exists X_A)R(c_K, X_A, c_B).$

Covering the potential secret Ψ, the first query Φ_1 is immediately refused. Though the second query Φ_2 does not cover Ψ, it nevertheless might contribute to a forbidden structure together with other queries. So we consider the inference signatures: Φ_2 only covers the first hypothesis (c_K c_A b_1) of Sig_1. Observing that Sig_1 has another hypothesis still uncovered, we can determine the correct query evaluation, yielding a positive answer $(\exists X_B)R(c_K, c_A, X_B)$ to be returned to the client. Moreover, we have to mark the covered hypothesis as already hit.

The third query Φ_3 again does not cover Ψ, but the second hypothesis (c_K b_2 c_B) of Sig_1: independently of the correct query evaluation we have to refuse the answer for the following reasons. If the correct answer is positive, the knowledge about all hypotheses of the inference signature would enable the client to *directly infer* the validity of the conclusion and thus of the potential secret Ψ. If the correct answer is negative, this additional knowledge does not directly lead to the crucial inference; however, *only* refusing a positive answer would enable a *meta-inference* of the following kind: "the only reason for the refusal is a positive answer, which thus is valid".

Example 2. To further exemplify the compiling phase in some more detail, we now consider the relation schema $RS = \langle R, \mathcal{U}, \Sigma \rangle$ with attribute set $\mathcal{U} = \{S(ymptom), D(iagnosis), P(atient)\}$ and two join dependencies in $\Sigma = \{\bowtie [SD, SP], \bowtie [DS, DP]\}$, the instance r comprising the four tuples $R(Fever, Cancer, Smith)$, $R(Fever, Fraction, Smith)$, $R(Fever, Cancer, Miller)$, and $R(Fever, Fraction, Miller)$, and the confidentiality policy $psec = \{\Psi\}$ containing the single potential secret $\Psi = (\exists X_S)R(X_S, Cancer, Smith)$.

S	D	P
a_S	a_D	b_1
a_S	b_2	a_P
a_S	a_D	a_P

S	D	P
a_S	a_D	b_1'
b_2'	a_D	a_P
a_S	a_D	a_P

S	D	P
(a_S,a_S)	(a_D,a_D)	(b_1,b_1')
$(\mathbf{a_S},\mathbf{b_2'})$	(a_D,a_D)	(b_1,a_P)
(a_S,a_S)	$(\mathbf{b_2},\mathbf{a_D})$	(a_P,b_1')
$(\mathbf{a_S},\mathbf{b_2'})$	$(\mathbf{b_2},\mathbf{a_D})$	(a_P,a_P)
(a_S,a_S)	(a_D,a_D)	(a_P,a_P)

S	D	P
a_S	a_D	b_1
$\mathbf{b_2}$	a_D	b_3
a_S	$\mathbf{b_4}$	b_5
$\mathbf{b_2}$	$\mathbf{b_4}$	a_P
a_S	a_D	a_P

Fig. 1. $\bowtie[SD, SP]$, $\bowtie[DS, DP]$, and their direct product as tableaus

The upper part of Figure 1 shows the dependencies as template dependencies in graphical notation known as tableau. Intuitively, the first dependency expresses the following: a symptom a_S that both contributes to a diagnosis a_D for some patient b_1, whose identity does not matter, and applies for the patient a_P contributes to the diagnosis a_D for the patient a_P as well. The meaning of the second dependency has a similar flavor. As proved in [12], the two join dependencies together are equivalent to their *direct product* exhibited in the lower part of Figure 1, both as constructed by definition and rewritten by substituting each pair of variables by a single variable.

By Theorem 1, we have to consider all template dependencies that are implied by Σ. However, it suffices to finally employ only the basic implications. Unfortunately, so far we do not know an efficient algorithm to compute the set Σ^{+basic} of all basic implications, which even might be infinite. But, we can somehow successively generate all candidates, in turn check each candidate whether it is an implication by applying the chase procedure, see [16,12,1], and finally minimize the hypotheses and maximize the conclusion of the implied candidates.

Figure 2 shows those elements of Σ^{+basic} that have at most three hypotheses: we get the two declared dependencies and two basic versions of their direct product, obtained by deleting the third or the second hypothesis, respectively. In step 2 of the compiling phase, we identify those dependencies in Σ^{+basic} such that the scheme $\{D(iagnosis), P(atient)\}$ of the potential secret Ψ is contained in the scheme of the dependency; in this example, so far the condition is always

S	D	P
a_S	a_D	b_1
a_S	b_2	a_P
a_S	a_D	a_P

S	D	P
a_S	a_D	b_1
b_2	a_D	a_P
a_S	a_D	a_P

S	D	P
a_S	a_D	b_1
$\mathbf{b_2}$	a_D	b_3
$\mathbf{b_2}$	b_4	a_P
a_S	a_D	a_P

S	D	P
a_S	a_D	b_1
a_S	$\mathbf{b_4}$	b_5
b_2	$\mathbf{b_4}$	a_P
a_S	a_D	a_P

Fig. 2. All basic implications of Σ having two or three hypotheses

S	D	P
Fever	*Cancer*	b_1
Fever	b_2	*Smith*
Fever	*Cancer*	*Smith*

S	D	P
Fever	*Cancer*	b_1
Fever	*Fraction*	b_5
b_2	*Fraction*	*Smith*
Fever	*Cancer*	*Smith*

Fig. 3. Instantiated signatures

satisfied. Accordingly, for each element of $\Sigma^{+\,basic}$ determined so far, attributes D and P in the conclusion are instantiated with the constants *Cancer* and *Smith* occurring in Ψ. These instantiations are then propagated to the hypotheses. In the next step 3, the hypotheses must be further instantiated according to the instance r. We have to determine minimal sets of tuples such that all equalities expressed in the respective template dependency are satisfied and their values equal the values of already instantiated entries. Finally, all remaining agree sets not considered so far are instantiated with the values of those tuples.

The instantiated template dependencies obtained so far are candidates to become inference signatures. However, we do not have to retain all them. Firstly, if an instantiated hypothesis of a candidate covers a potential secret, we can discard the candidate, since in the monitoring phase a query whose answer would reveal such a hypothesis would be refused anyway. So, in the example the instantiation of the second basic implication is discarded. Secondly, if the hypotheses of a candidate constitute a superset of the hypotheses of another candidate, then the former candidate is redundant and can be discarded as well. So, in the example the instantiation of the third basic implication is discarded as well.

For the given simple instance r, we do not have to consider basic implications with more than three hypotheses, and thus we finally keep the inference signatures shown in Figure 3. However, due to the cyclic structure of the dependencies in the example, there are basic implications with arbitrarily many hypotheses. So we can extend the basic implications having three hypotheses by a suitable fourth hypothesis, as shown in Figure 4. In fact, e.g., we can

S	D	P
a_S	a_D	b_1
$\mathbf{b_2}$	a_D	b_3
$\mathbf{b_2}$	$\mathbf{b_4}$	b_5
b_6	$\mathbf{b_4}$	a_P
a_S	a_D	a_P

S	D	P
a_S	a_D	b_1
a_S	$\mathbf{b_4}$	b_5
$\mathbf{b_2}$	$\mathbf{b_4}$	b_6
$\mathbf{b_2}$	$\mathbf{b_7}$	a_P
a_S	a_D	a_P

S	D	P
a_S	a_D	b_1
a_S	$\mathbf{b_4}$	b_5
$\mathbf{b_2}$	$\mathbf{b_4}$	b_6
$\mathbf{b_2}$	$\mathbf{b_7}$	b_9
$\mathbf{b_8}$	$\mathbf{b_7}$	b_{11}
$\mathbf{b_8}$	$\mathbf{b_{10}}$	b_{13}
b_{12}	$\mathbf{b_{10}}$	a_P
a_S	a_D	a_P

Fig. 4. Basic implications of Σ having four hypotheses and an example of a basic implication of Σ having "many" hypotheses

generalize the structure of the fourth dependency shown in Figure 3 by extending the present "path" a_S,a_S,b_4,b_4 by $b_2,b_2,b_7,b_7,b_8,b_8,b_{10},b_{10}$, as exhibited by the rightmost dependency shown in Figure 4. However, the instance r lacks sufficient diversity to instantiate such a long path with different constants. But we could employ a single element of the instance for instantiating several hypotheses and would then obtain instantiated signatures that we already got before.

4   Compiling and Monitoring Signatures

Generalizing the example, we now present the new signature-based enforcement mechanism as a two phase protocol.

The *compiling phase* takes the dependencies Σ declared in the schema, the confidentiality policy *psec*, and the instance r as inputs, and proceeds as follows to generate the set *psig* of all inference signatures:

1. It successively, with an increasing number of hypotheses, generates all basic template dependencies implied by Σ, i.e.,
   $\Sigma^{+basic} := \{td \mid \Sigma \models td, td$ is hypothesis-minimal and conclusion-maximal$\}$,
   until no further ones can exist or further ones would not lead to nonredundant instantiations for the given instance r.
2. It determines all pairs (Ψ, td) with $\Psi \in psec$ and $td \in \Sigma^{+basic}$ as generated so far, whose components match in the sense that the scheme of Ψ is a subset of the scheme of td, i.e., of the conclusion's aggregated agree set $\cup_{j=1}^{l} ag(c, h_j)$, where $td = TD[h_1, \ldots, h_l | c]$.
3. For each such pair, the (distinguished) symbols (seen as free variables) in the scheme of td are instantiated with the respective constants appearing in Ψ. Then the instantiation is propagated from the conclusion to the hypotheses. The result is denoted by $td[\Psi]$.
4. For each $td[\Psi]$ obtained so far, the instance r is searched for a minimal set of tuples $r_{td[\Psi]}$ that "uniformly covers" all hypotheses: (i) all equalities required by $td[\Psi]$ are satisfied and, (ii) the tuple values equal the respective already propagated instantiations. For each such set, the hypotheses are further instantiated on agree sets not captured before with the respective values found in the uniform covering. The remaining symbols are left unchanged. If none of the hypotheses covers any of the potential secrets in *psec*, then the resulting inference signature $Sig(\Psi, td, r_{td[\Psi]})$ is inserted into *psig*.
5. If the hypotheses of a result of step 3 or 4 constitute a superset of the hypotheses of another result of step 3 or 4, respectively, then the former element is discarded, since it is redundant.

Given the fullness of the join dependencies in Σ, step 1 can be based on the chase algorithm [16] together with bounded searching for minimization and subsequent maximization. Though being complex in general, the computation is expected to be feasible in practice, since we only deal with schema items. Moreover, in practice, a database administrator will only admit "minor" deviations from Boyce-Codd normal form having a unique key, for example, to ensure *faithful representation* of all dependencies by relaxing Boyce-Codd normal form to 3NF or to provide support of expected queries by a dedicated *denormalization*.

Similarly, seeing the elements of the confidentiality policy as a declaration of exceptions from the general default rule of permission, we expect that in many applications steps 2 and 3 will produce only a manageable number of templates for inference signatures. Moreover, as far as the constants occurring in these templates achieve a high selectivity regarding the instance considered, step 4 will not substantially increase the number of final inference signature.

The *monitoring phase* takes a query Φ, the confidentiality policy *psec*, the set *psig* of all inference signatures determined in the compiling phase, and the instance r as inputs, and proceeds as follows:

1. It checks whether some $\Psi \in psec$ is covered by Φ (equivalently, Ψ is entailed by Φ, see Section 2), and if this is the case, the answer is immediately refused.
2. Otherwise, it determines all hypotheses Π occurring in *psig* and not marked before such that Π is covered by Φ, and it tentatively marks them. If now for some signature in *psig* all hypotheses are marked, then the answer is refused and the tentative marking is aborted. Otherwise, the correct answer is determined from r and then returned, and the tentative marking is committed if a positive answer Φ is returned; otherwise, if $\neg\Phi$ is returned, the tentative marking is aborted.

A straightforward implementation of the monitoring phase keeps the potential secrets and the suitably tagged hypotheses of inference signatures in two dedicated relations. Given a query, these relations are searched for covered tuples and inspected for an inference signature becoming fully marked. Approximating the computational costs of these actions for one instantiated hypothesis by a constant, the overall runtime complexity of an execution of the monitoring phase is at most linear in the size of the dedicated relations. In the next section we will present how this rough design has been converted into an SQL-based prototype.

Theorem 2. *Assume the Situation 4 sketched in Section 1 and analyzed in [8]: the a priori knowledge is restricted to comprise only a schema declaration with functional dependencies and join dependencies, and confidentiality policies and queries are restricted to select-project sentences. Then the signature-based enforcement mechanism preserves confidentiality (in the sense of Section 1).*

Sketch of Proof. "Negative" answers of the form $\neg\Phi_i$ do not contribute to a harmful inference of a potential secret Ψ: on the one hand, the confidentiality policy contains only positive sentences and, on the other hand, the dependencies only generate positive conclusions from positive assumptions.

So let us assume indirectly that there is a harmful inference based on some minimal set of positive answers $\{\Phi_1, \ldots, \Phi_m\}$ to derive Ψ. Then, by Theorem 1 (Theorem 2 of [8]), there exists a corresponding nontrivial template dependency that witnesses such an inference. In step 1 of the compiling phase, a hypothesis-minimal and conclusion-maximal version td of this dependency is added to Σ^{+basic}; this version then witnesses the inference considered as well. In the subsequent steps 2 and 3 of the compiling phase, td together with Ψ is further processed to set up a generic signature of the form $td[\Psi]$. Furthermore, in step 4 of the compiling phase, the tuples in the instance r leading to the harmful positive answers $\Phi_1, \dots, \Phi_m$ or these answers themselves, respectively, contribute to generate an (instantiated) inference signature of the form $Sig(\Psi, td, r_{td[\Psi]})$.

Finally, in the monitoring phase, when the last of these positive answers is controlled, the tentative marking of this inference signature results in a marking of all its hypotheses, and thus the answer is refused. This contradicts the assumption that the positive answers $\Phi_1, \dots, \Phi_m$ are all returned to the client. □

5   A Prototype for Oracle/SQL

We implemented[1] the signature-based enforcement mechanism as part of a larger project to realize a *general prototype* for inference-usability confinement of reactions generated by the server of a relational database management system (see Section 9 of [3]).

This prototype has been designed as a *frontend* to an Oracle/(SQL)-system: the *administration interface* enables officers to declare and manage client-specific data like the postulated a priori knowledge, a required confidentiality policy, a permitted interaction language, and the kind of distortion; the *interaction interface* enables registered clients to send requests like queries and receive reactions. The *interaction language* is uniformly based on the relational calculus as a specific version of first-order logic, which provides the foundation of the semantics of relational databases.

Accordingly, if a client should be permitted to issue queries under a schema $RS = \langle R, \mathcal{U}, \Sigma \rangle$ but be confined by the signature-based enforcement mechanism, then the following has to happen: the client is granted a permission to query the current instance r; the dependencies in Σ are added to that client's a priori knowledge; the language $\mathcal{L}$ is made available to the client to submit queries; a confidentiality policy is declared to confine the client's permission; and refusals are specified as the wanted kind of distortion. Additionally, a compatible enforcement mechanism is selected, either automatically by an optimizer or explicitly by an administrator. In the remainder of this section, we assume that inference signatures are both applicable and necessary as described in Section 1.

In a first attempt, considering the general prototype to mediate access to the underlying Oracle-system would suggest to let a wrapper translate a query $\Phi \in \mathcal{L}$ into an *SQL-query* during the *monitoring phase*. In our case, for instance, assuming that R denotes the Oracle-table for the instance r, a closed (yes/no)-query $(\exists X_1)\dots(\exists X_l)R(X_1, \dots, X_l, c_{l+1}, \dots, c_n)$ would be converted into

SELECT $A_{l+1}, \dots, A_n$ FROM R WHERE $A_{l+1} = c_{l+1}$ AND $\dots$ AND $A_n = c_n$,

---

[1] The following exposition only outlines the implementation and slightly differs from the presently employed version of the code, which is under ongoing development for both improved usability and further optimization.

R	Sym	Dia	Pat
	Fever	Cancer	Smith
	Fever	Fraction	Smith
	Fever	Cancer	Miller
	Fever	Fraction	Miller

SEC	Sym	Dia	Pat
	*	Cancer	Smith

QUE	Sym	Dia	Pat	Rea
	Fever	Cancer	X	

SIG	Sym	Dia	Pat	Id	Imp	Old
	Fever	Cancer	b_1	1		
	Fever	b_2	Smith	1		
	Fever	Cancer	b_1	2		
	Fever	Fraction	b_5	2		
	b_2	Fraction	Smith	2		

Fig. 5. Oracle-tables for signature-based enforcement applied to hospital database

which returns either the empty set or a singleton with a tuple μ of the form $(A_{l+1} : c_{l+1}, \ldots, A_n : c_n)$ over the attribute set $\{A_{l+1}, \ldots, A_n\}$.

However, while forwarding this query to the Oracle-system, we want the server not only to evaluate the query but to perform the further actions on the inference signatures described in Section 4 as well. In principle, this goal can be accomplished by the features of Oracle for *active databases*, i.e., by *triggers*. Since Oracle does not provide means to define a trigger on a query directly, we instead employ a suitable *update command* to an auxiliary Oracle-table *QUE(RY)*, which together with two further tables (which will be described below) has already been created during the *compiling phase*.

- Basically, the table *QUE(RY)* has the attributes in $\mathcal{U}$ specified in the schema for the table *R* and a further attribute *Rea(ction)*, which has a three-valued type $\{ref(used), pos(itive), neg(ative)\}$: a tuple of *QUE(RY)* denotes a query as a generalized tuple combined with an indicator how to react.

Regarding Oracle-*privileges*, any access right the client might have before on the Oracle-table *R* must be revoked, and instead the client is only granted the INSERT-right on the auxiliary table *QUE(RY)*. The needed trigger *CQE* is declared for insertions into the table *QUE(RY)*, and this trigger is executed with the access rights of the owner (administrator) of the table *R*. Accordingly, we employ some kind of "right amplification", as for example offered by the operating system UNIX by means of setting the suid-flag for an executable file: the client only receives a privilege to *initiate* the query-and-control activities, as predefined by the trigger, without being permitted to perform such activities at his own discretion.

The trigger *CQE* operates on the auxiliary Oracle-table *QUE(RY)* and the two further Oracle-tables *(POT)SEC* and *SIG(NATURE)* that already have been created separated from the Oracle-table *R* during the compiling phase.

- The table *(POT)SEC* has the same attribute set $\mathcal{U}$ as *R* such that a declared potential secret $\Psi \in \mathcal{L}$ can be represented as a generalized tuple; however, each (originally existentially quantified) variable is uniformly replaced by a special placeholder " * ".
- The table *SIG(NATURE)* has the attributes in $\mathcal{U}$ as well such that a hypothesis of an inference signature can again be represented as a generalized

tuple, and three further attributes to be used as follows: the attribute *Id* specifies an inference signature the represented hypothesis belongs to; the Boolean attribute *(Flag)Imp* refers to a tentative marking during a current monitoring phase; and the Boolean attribute *(Flag)Old* refers to a marking already committed while controlling a preceding query.

Unfortunately, it turned out that we need the two flags, since we could not employ standard transaction functionality to freshly mark hypotheses tentatively and finally either commit the fresh markings or abort them by a rollback instruction: Oracle does not offer to direct the needed transaction functionality within trigger executions.

Figure 5 shows the Oracle-tables for the small hospital database introduced as Example 2 in Section 3 with the instance inserted into table *R*, after filling the table *(POT)SEC* with the declared confidentiality policy, populating the table *SIG(NATURE)* with the compiled inference signatures, and forwarding the query $(\exists X)R(Fever, Cancer, X)$ to the table *QUE(RY)*.

Activated by the insertion of the query into the Oracle-table *QUE(RY)*, the trigger *CQE* basically proceeds as follows:

1. The trigger extracts the query submitted by the client from *QUE*, constructs an SQL-query to determine whether or not the extracted query covers an element in *SEC*, executes the constructed SQL-query, and then checks the result for emptiness: if the result is nonempty, i.e., a covering has been detected, the trigger modifies the attribute *Rea(ction)* of the single tuple in *QUE* into *ref(used)* (which the frontend retrieves subsequently), and the trigger exits. The following SQL-query is constructed for the example:

   SELECT Sym,Dia,Pat FROM SEC WHERE $(Sym = Fever$ OR $Sym = *)$ AND $(Dia = Cancer$ OR $Sym = *)$ AND $Pat = *$

2. Otherwise, the trigger continues to perform the actions already described in Section 4 by suitably employing the Oracle-tables in a similar way as in the first step; see Figure 6 for the rough design.

6   Experimental Evaluation

To determine the runtime overhead inherently caused by the signature-based enforcement mechanism, we measured the query processing times of the implementation described in Section 5. We started with the following observation. Given the dependencies declared in the schema and the confidentiality policy, a generic inference signature signifies a typical "forbidden structure", and thus all of them together represent all possibilities for harmfulness. Thus, we aimed at constructing the instances to be used for runtime evaluations by varying the following parameters: (1) the included forbidden structures; (2) the instantiations of included forbidden structures; and (3) the fraction of confined "exceptions".

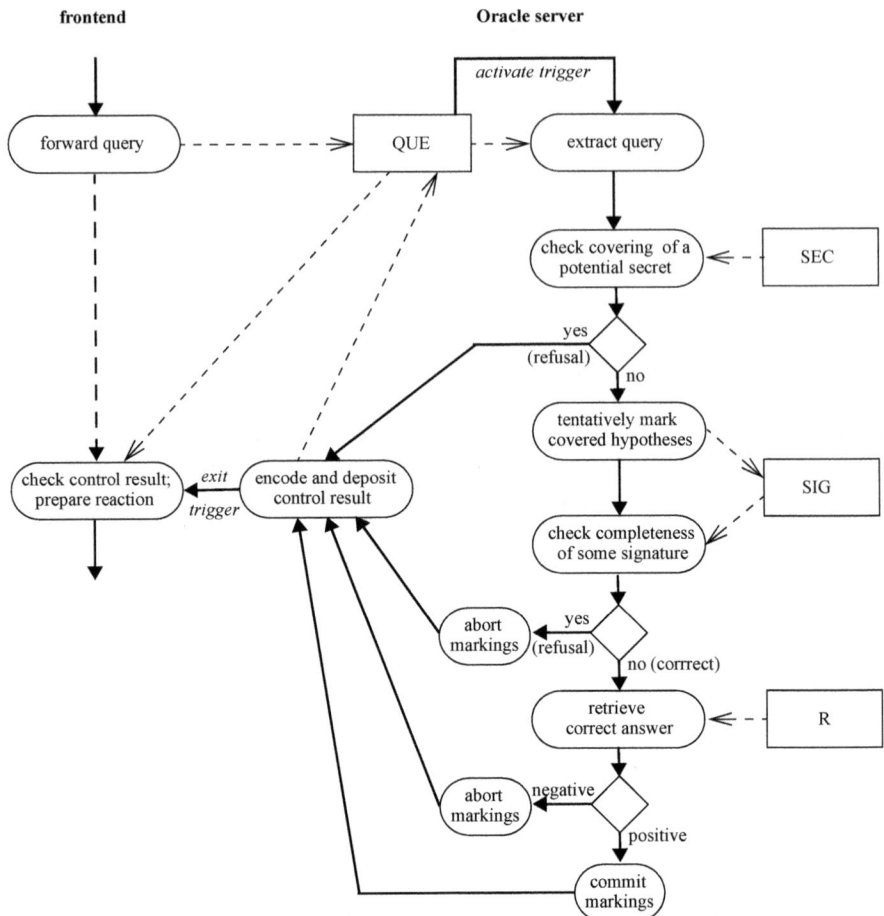

Fig. 6. Design of the trigger *CQE* to control a submitted query

Moreover, we expected an impact of the length of query sequences, since with increasing length more markings of hypotheses will be found.

Accordingly, we measured the following query processing times: the *maximum time* that occurred up to the last query of a sequence for the *whole frontend* and the *trigger alone*, respectively; and the *average time* over a sequence for the *whole frontend* and the *trigger alone*, respectively.

Discarding exceptional measurements caused by external factors and applying suitable roundings, we depict the results for Example 2 in Table 1: We restricted to the forbidden structures shown in Figure 2 and 4; varied the *number of instantiations* from 1 over 10 and 100 up to 1000, in this way getting instances (by applying the chase algorithm to satisfy the dependencies) of size from 79 tuples up to 79000 tuples; declared for each instantiated forbidden structure just one potential secret (as an "exception"); and formed query sequences of length from 100 up to 100000, suitably covering all relevant cases in a random way.

Table 1. Maximum and average query processing times experienced for Example 2

Instantiations	Instance size	Queries	Processing time pro query in msec			
			whole frontend		trigger alone	
			maximum	average	maximum	average
1	79	100	22	15	10	2
10	790	1000	54	17	10	3
100	7900	10000	421	23	150	7
1000	79000	100000	711	78	580	61

The results for the particular example suggest the practical feasibility of our approach, including scalability: a human user acting as a client will basically not realize a query processing time in the range of a few milliseconds up to around half a second. Of course, general practicality still has to be justified by statistically evaluating more advanced experiments for "real-world" applications using a more mature implementation with enhanced functionality and further optimizations.

7   Conclusions

Summarizing and concluding, we presented a signature-based enforcement mechanism that satisfies confidentiality requirements that are very general and have been considered in many contexts before, as summarized and further investigated by Halpern/O'Neill [13] and suitably extended to include policies by Biskup/Tadros [10]. The mechanism can be seen as a variation of a security automaton for the run time enforcement of security properties in the sense of Ligatti/Reddy [15] and others. We demonstrated the effectiveness of the mechanism for relational databases that are constrained by the large class of functional dependencies and join dependencies, which capture a wide range of applications, see, e.g., Abiteboul/Hull/Vianu [1].

Our mechanism differs from previously considered monitoring systems by taking advantage of the particular properties of functional dependencies and join dependencies, without imposing any further restrictions on these dependencies. We provide a proactive control functionality avoiding any confidentiality breach, in contrast to auditing approaches as described by, e.g., Kaushik/Ramamurthy [14], which can only detect violations after the fact.

There are several lines of further research and development, dealing with the following issues: tools for the compiling phase, the distribution of functionality between the two phases, the complete integration into a database management system like Oracle, more advanced interactions like open queries, updates and transactions, and experimental evaluations with "real-world" applications.

Regarding tools for the compiling phase, a major open problem is to design a generally applicable algorithm to effectively and efficiently determine all basic implications up to a suitably chosen number of hypotheses for any set of functional dependencies and join dependencies. We conjecture that properties regarding the occurrence of cyclic structures in the hypergraph of the dependencies has a major impact. The computational complexity of the compiling phase should also be investigated.

Regarding distribution of functionality, we already designed a more dynamic version of the signature-based enforcement mechanism. In this version, we initially keep the inference signatures generic, without instantiating them with specific values from the instance already at compile time. Rather, instantiations are dynamically generated at run time only employing instance tuples actually returned as responses to the client. This more dynamic version can be derived from the static version detailed in this article but some subtle optimization problems still have to be solved in a satisfactory way.

Regarding a complete integration, we first of all face problems of modifying proprietary software, but we would also be challenged to make the added functionality fully compatible with all the many services already offered. Of course, from the point of view of both administrators and clients, in general a full integration would be advantageous: conceptually for employing uniform interfaces, and algorithmically for avoiding the overhead raised by the communication of a separate frontend with a server and for including the security functionality into the scope of the server's optimizer.

Regarding advanced interactions, on the one hand we have to suitably adapt previous theoretical results [5,7] and, again, to exploit the features of the underlying database management system as far possible. On the other hand, in general an update of the database instance will require to update the (instantiated) inference signatures as well. Clearly, both aspects would have to be suitably combined, while also considering the optimization problems mentioned above.

Finally, regarding "real-world" applications, we would have to identify suitable classes of applications, clarify in detail how far the assumptions underlying the signature-based approach are actually satisfied by such applications, and then overcome essential mismatches by additional mechanisms. However, as pointed out in the introduction, there is an inevitable tradeoff between conceptual expressiveness and computational complexity: any extension of the work presented in this article will be challenged to maintain an appropriate balance between the conflicting goals.

Acknowledgments. We would like to sincerely thank Martin Bring and Jaouad Zarouali for improving the implementation and conducting the experiments.

References

1. Abiteboul, S., Hull, R., Vianu, V.: Foundations of Databases. Addison-Wesley, Reading (1995)
2. Biskup, J.: History-Dependent Inference Control of Queries by Dynamic Policy Adaption. In: Li, Y. (ed.) DBSec 2011. LNCS, vol. 6818, pp. 106–121. Springer, Heidelberg (2011)
3. Biskup, J.: Inference-usability confinement by maintaining inference-proof views of an information system. International Journal of Computational Science and Engineering 7(1), 17–37 (2012)
4. Biskup, J., Bonatti, P.A.: Lying versus refusal for known potential secrets. Data Knowl. Eng. 38(2), 199–222 (2001)

5. Biskup, J., Bonatti, P.A.: Controlled query evaluation with open queries for a decidable relational submodel. Ann. Math. Artif. Intell. 50(1-2), 39–77 (2007)
6. Biskup, J., Embley, D.W., Lochner, J.-H.: Reducing inference control to access control for normalized database schemas. Inf. Process. Lett. 106(1), 8–12 (2008)
7. Biskup, J., Gogolin, C., Seiler, J., Weibert, T.: Inference-proof view update transactions with forwarded refreshments. Journal of Computer Security 19, 487–529 (2011)
8. Biskup, J., Hartmann, S., Link, S., Lochner, J.-H.: Chasing after secrets in relational databases. In: Laender, A.H.F., Lakshmanan, L.V.S. (eds.) Alberto Mendelzon International Workshop on Foundations of Data Management, AMW 2010. CEUR, vol. 619, pp. 13.1–13.12 (2010)
9. Biskup, J., Lochner, J.-H., Sonntag, S.: Optimization of the Controlled Evaluation of Closed Relational Queries. In: Gritzalis, D., Lopez, J. (eds.) SEC 2009. IFIP AICT, vol. 297, pp. 214–225. Springer, Heidelberg (2009)
10. Biskup, J., Tadros, C.: Policy-based secrecy in the Runs & Systems Framework and controlled query evaluation. In: Echizen, I., Kunihiro, N., Sasaki, R. (eds.) Advances in Information and Computer Security – International Workshop on Security, IWSEC 2010, Short Papers, pp. 60–77. Information Processing Society of Japan (2010)
11. Biskup, J., Wiese, L.: A sound and complete model-generation procedure for consistent and confidentiality-preserving databases. Theoretical Computer Science 412, 4044–4072 (2011)
12. Fagin, R., Maier, D., Ullman, J.D., Yannakakis, M.: Tools for template dependencies. SIAM J. Comput. 12(1), 36–59 (1983)
13. Halpern, J.Y., O'Neill, K.R.: Secrecy in multiagent systems. ACM Trans. Inf. Syst. Secur. 12(1), 5.1–5.47 (2008)
14. Kaushik, R., Ramamurthy, R.: Efficient auditing for complex SQL queries. In: Sellis, T.K., Miller, R.J., Kementsietsidis, A., Velegrakis, Y. (eds.) ACM SIGMOD International Conference on Management of Data, SIGMOD 2011, pp. 697–708. ACM (2011)
15. Ligatti, J., Reddy, S.: A Theory of Runtime Enforcement, with Results. In: Gritzalis, D., Preneel, B., Theoharidou, M. (eds.) ESORICS 2010. LNCS, vol. 6345, pp. 87–100. Springer, Heidelberg (2010)
16. Sadri, F., Ullman, J.D.: Template dependencies: A large class of dependencies in relational databases and its complete axiomatization. J. ACM 29(2), 363–372 (1982)

Privacy Consensus in Anonymization Systems via Game Theory

Rosa Karimi Adl, Mina Askari, Ken Barker, and Reihaneh Safavi-Naini

Department of Computer Science, University of Calgary, Calgary, AB, Canada
{rkarimia,maskari,kbarker,rei}@ucalgary.ca

Abstract. Privacy protection appears as a fundamental concern when personal data is collected, stored, and published. Several anonymization methods have been proposed to address privacy issues in private datasets. Every anonymization method has at least one parameter to adjust the level of privacy protection considering some utility for the collected data. Choosing a desirable level of privacy protection is a crucial decision and so far no systematic mechanism exists to provide directions on how to set the privacy parameter. In this paper, we model this challenge in a game theoretic framework to find *consensual* privacy protection levels and recognize the characteristics of each anonymization method. Our model can potentially be used to compare different anonymization methods and distinguish the settings that make one anonymization method more appealing than the others. We describe the general approach to solve such games and elaborate the procedure using k-anonymity as a sample anonymization method. Our simulations of the game results in the case of k-anonymity reveals how the equilibrium values of k depend on the number of quasi-identifiers, maximum number of repetitive records, anonymization cost, and public's privacy behaviour.

Keywords: Privacy Protection, Data Anonymization, Privacy/Utility Trade-off, Privacy Parameter Setting, Game Theory, k-Anonymity.

1 Introduction

Massive data collection about individuals on the Web raises the fundamental issue of privacy protection. A common approach to address privacy concerns is to use data anonymization methods [1–5]. During data anonymization identifiers are removed and data perturbation, generalization, and/or suppression methods are applied to data records.

Data anonymization promises privacy up to a certain level specified by some privacy parameter(s). In setting the privacy parameter, usually the amount of the expected data utility is considered and hence the level of privacy offered by an anonymization method is never set to the maximum. Since the risk to privacy is not completely removed, we postulate that data providers must be informed about the amount of privacy risk involved (represented as the privacy parameter's value) before deciding to provide their personal data to data collectors.

To bring data providers' privacy opinion into the cycle of data anonymization, we propose a game theoretic model that finds *consensual* privacy and utility levels by considering preferences of data providers as well as data collectors and data users. More

N. Cuppens-Boulahia et al. (Eds.): DBSec 2012, LNCS 7371, pp. 74–89, 2012.

specifically, we analyze the privacy/utility trade-off from the perspective of three different parties: a *data user* who wants to perform data analysis on a dataset and is willing to pay for it; a *data collector* who collects and provides privacy protected data to the data user; and *data providers* who can choose to participate in data collection if they see it as *worthwhile*. As these parties try to maximize their "profit" (payoff), the collective outcome of the game produces the equilibria [6] in our trade-off system. In an equilibrium state, no single player can achieve higher profits by changing their actions. Therefore, equilibria represent shared agreements (hence the term *consensus*) in which none of the players would attempt to behave differently. Using these equilibria, we are able to examine privacy trade-offs and analyze different characteristics of an anonymization technique such as the expected amount of privacy, precision, database size, and each party's profit. We believe that features of an anonymization technique must be inspected at equilibrium stages to provide more reliable evaluation results. The proposed model can be used as an evaluation framework to compare various anonymization methods from different perspectives. This is the first attempt to use game theory to analyze trade-offs in a private data collection system by considering preferences of data providers.

Paper Organization: The remainder of this paper is organized as follows: Section 2 discusses the related work. Section 3 describes basic definitions in game theory. Section 4 explains our game model and its ingredients. Section 5 provides a general solution to the game. Section 6 demonstrates a sample application of our model for the case of k-anonymity. Section 7 provides conclusions and suggests future directions.

2   Related Work

The issue of protecting individual's privacy while collecting personal information has motivated several research projects in literature. Our work mostly relates to anonymization techniques such as k-anonymity [1, 2], l-diversity [3], t-closeness [4], and differential privacy [5]. Anonymization techniques provide data privacy at the cost of losing some information. Several methods [7–11] have been proposed to evaluate the trade-off of privacy/utility. When data usage is unspecified, similarity between the original data and the privacy protected data is considered as information loss. The average size of equivalence classes [7] and discernibility [8] in k-anonymity are two examples of such generic metrics. However, most scholars have noticed that more reliable utility measures must be defined in the context of data application (*e.g.,*data mining and queries). Various measures of utility such as information-gain-privacy-loss ratio [9] and clustering and partitioning based measure [12] have been proposed to determine the next generalization step within anonymization algorithms. Sramka *et al.* [10] developed a data mining framework that examins the privacy/utility trade-off after the anonymization has been done using a mining utility. Machanavajjhala *et al.* [11] defines an accuracy metric for differential privacy in the context of social recommendation systems and analyzes the trade-off between accuracy and privacy. The existing privacy/utility trade-off methods all assume that a dataset already exists before choosing the privacy protection level for it. These methods do not consider the effect of privacy protection level on data providers' decision and hence the volume of the collected information.

In this work we use game theory to investigate steady levels of privacy protection by adopting a broader view of affecting parameters. Game theory has been successfully applied to analyze privacy issues from legal [13] and economic perspectives [14–17]. Kleinberg *et al.* [15] describe three scenarios modeled as coalition games [6] and use core and shapely values to find a "fair" reward allocation method in exchange for private information. The underlying assumption in these scenarios is that *any* amount of reward compensates for the loss of privacy protection. We believe this assumption oversimplifies the nature of privacy concerns and is not compatible with our perception of privacy. Calzolari and Pavan [16] use game theory to explore the optimum flow of customers' private information between two interested firms. The perspective of their work is possibly closest to ours but their model is substantially different from our work since they define a privacy policy as probability of revealing detailed customers' information to another party. Game theory has also been used as a means to address more technical aspects of privacy such as attacks on private location data [18], implementation of dynamic privacy [17], and questioning the assumption of honest behavior in multiparty privacy preserving data mining [19]. Our work builds on a commonly accepted definition of privacy among computer and social science scholars and adopts a game theoretical approach to find steady privacy levels. The novelty of our research lies on bringing the economic perspective to data anonymization issues and utilizing game theory for the first time to address privacy/utility trade-offs in a more realistic setting.

3   Preliminaries and Assumptions

In this paper we propose a game-theoretic framework to find steady level(s) of privacy protection for any arbitrary anonymization technique. We assume that the data providers are informed about having their personal information collected and the data collector is trustworthy in the sense that he fulfills his promises. Every instance of the game is modeled according to a chosen anonymization technique. A common factor between these techniques is a privacy parameter such as k in k-anonymity, l in l-diversity, and $1/\epsilon$ in differential privacy that indicates the level of privacy protection guaranteed by the corresponding privacy mechanism. To provide a generic game model, we use the letter δ to denote the privacy parameter. For any chosen anonymization technique, larger values for δ lead to higher privacy protection and lower data utility. The exact meaning of δ has to be interpreted according to the privacy definition chosen for the game. In this section we provide a brief overview of the game theoretic definitions used in this paper.

3.1   Sequential Game Model

Game theory is a mathematical approach to study interdependencies between individual's decisions in strategic situations (games). A game is explained by a set of *players* (decision makers), their *strategies* (available moves), and *payoffs* to each player for every possible strategy combination of all players (*strategy profile*). A strategy profile is a *Nash equilibrium* if none of the players can do better by changing their strategy assuming that other players adhere to theirs. Nash equilibrium is commonly used to predict stable outcomes of games and since it represents a steady state of a game [6], we use the term "stable" through the rest of the paper to denote the strategies found in the equilibrium. To capture a pre-specified order for players' turn to move, a *game tree* is used to represent a *sequential game*. In this tree each node is a point of choice for a player and the branches correspond to possible actions. A sequence of actions from the root to any intermediate node or to a leaf node is called a *history* or a *terminal history*, respectively [6]. Payoff functions define Preferences of players over each terminal history. A player's strategy explains his decision at any point in the game that he has to move.

Since the sequential structure of extensive form games is not considered in the concept of Nash equilibrium, the notion of "subgame-perfect Nash equilibrium" [6] is normally used to determine the robust steady states of such games. Every sub-tree of the original game tree represents a subgame. A strategy profile is a subgame perfect equilibrium if it induces a Nash equilibrium in every subgame [6]. The principle of *Backward induction* is a common method to deduce subgame-perfect equilibria of sequential games. Backward induction simply states that when a player has to move, he first deduces the consequences of every available action (how the subsequent player rationally reacts to his decision) and chooses the action that results in the most preferred terminal history.

The challenge of setting a desirable value for privacy parameter δ defines strategic situation with some ordering on players' turn to move. As a result, we model the problem as a sequential game.

4   Game Description

To define a game-theoretic model for the challenge of finding a balanced value of δ, we must specify the decision makers (players), their preferences, and the rules of the sequential game. The following sections explain the details of our model.

4.1   Players

Players of the game are the following three parties:

Data Providers. Data providers are individuals that decide whether to provide their personal information at a specific privacy level δ and use the service offered by the data collector or to reject the offer. For example the service could be a discount on some online purchase activity or a software application offered for free. Since privacy preferences of each data provider is affected by several demographic and socioeconomic factors [20–22], it is practically infeasible to determine how much utility is gained by each data provider for each combination of δ and incentive. In an alternative approach, we rely on the assumption that data providers' behavior is captured by a model based on some observation rather than a game theoretic analysis. Our assumed model is a regression model which captures how the number of data providers increases as the values of δ and incentive increase. Although this specific model has not been developed yet, similar studies have been conducted to explore the effects of other parameters (such as knowledge of privacy risks, trust, age, income level, *etc.*) on public's privacy behavior [20, 22, 23]. A regression model that explains the effects of δ and incentive seems to be a natural extension to those studies. The assumed model generally considers data providers who are interested in both privacy and incentive and is defined as:

$$N = n(\delta, I) = \beta_0 + \beta_1\, h_1(\delta) + \beta_2\, h_2(I) \tag{1}$$

where N represents total number of individuals who accept the offer as a function of δ and incentive I (in terms of a monetary value). h_1 and h_2 are functions of δ and I. Parameters β_0, β_1, and β_2 are the intercept and marginal effects of $h_1(\delta)$ and $h_2(I)$ on individual's decision to participate in the data collection procedure. The functions h_1 and h_2 can be any non-decreasing functions of δ and I. This regression model does not assume accurate knowledge about privacy risks for data providers and as this knowledge increases, we expect to have larger β_1 to reflect a higher level of privacy concerns.

By assuming a regression model, we mostly *observe* data providers' behavior rather than directly *analyzing* it. This assumption *trims* the game tree by removing the data providers from the analysis of the game. Nevertheless, the effect of data providers' decisions is reflected in other players' payoff functions and paying specific attention to their impact on the final level of privacy is one of the distinctive strengths of our work.

Data Collector. A data collector is the entity who collects a dataset of personal data and provides it to some data users. The data collector receives offers from the data users, and based on their needs and the expected cardinality of the collected dataset announces a privacy level and some incentive to collect data from individuals. Once a data collector collects a dataset of personal information, he protects the privacy of the data providers at the consented level δ and provides the private dataset to the data user.

The data collector generally prefers to receive more money from the data user and spend less money on the amount of incentive he pays the data providers. Consequently, cardinality of the dataset (number of data providers) affects the payoff to the data collector. A detailed formulation of data collector's payoff is provided in Sect. 4.3.

Data User. A data user is an entity interested in accessing personal information for some data analysis purposes. A data user prefers a dataset with higher quality (more accurate query results) and higher cardinality (results with higher statistical significance). Privacy parameter δ affects these requirements in positive and negative ways. Therefore a data user chooses a value δ that balances the needs and initiates the game by offering some value for parameter δ and some price, p, for each data record. We give the detailed analysis for games with a single data user. The approach to model multiple data users and data reuse is explained elsewhere [24].

4.2   Game Rules

We model interactions between the data collector and the data user as a sequential game with *perfect* (players are aware of the actions taken by previous players) and *complete* (players are aware of the available strategies and payoff functions of all other players) information. More specifically, both players know data the provider's behavior model. The data user also knows the data collector's available actions and preferences[1].

The game starts with an offer from the data user to the data collector. In the offer, the required value for privacy parameter δ and the price p (per each record) must be

---

[1] Our assumption of complete information does not mean that the data collector and the data user know privacy/incentive trade-off functions of each data provider because individual data providers are not directly modeled as players in the trimmed game tree.

specified. We denote an offer by $Of = \langle \delta, p \rangle$. Once the data collector receives the offer he can either reject or accept it. In case of a rejection, the game terminates with payoff zero to both the data user and the data collector. If the data collector decides to accept then he needs to announce an incentive in exchange for collecting personal information. Here, we assume that I represents monetary value of the incentive and its domain is $\mathbb{R}_{\geq 0}$. The terminal histories of this game are either of the form (Of, I) or

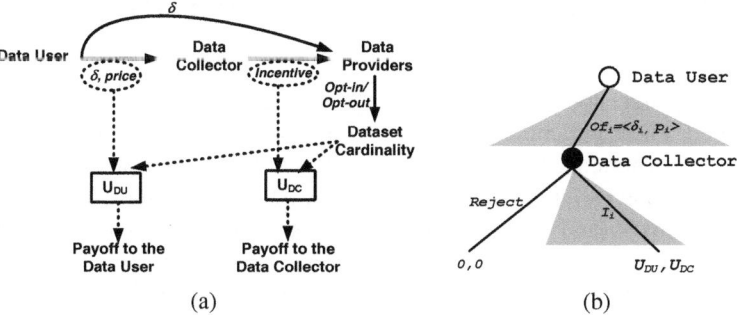

(a)   (b)

Fig. 1. (a) The dynamics of setting a stable level for privacy protection. (b) Trimmed game tree.

$(Of, Reject)$. At any terminal history, the number of data providers who will opt-in is determined by plugging the values of δ and I into Eq(1). Consequently, preferences of the data user and the data collector over all terminal histories are determined based on the payoff function defined over cardinality of dataset and values of δ, p, and I.

The interactions and mutual effects of players' decision are captured in Fig. 1(a). Based on the game's dynamics, Fig. 1(b) illustrates the game tree (triangles represent ranges of possible offers and incentives).

4.3   Payoffs

Payoff to the Data Collector: The data collector receives some money, p, from the data user for each data record. The total number of data records in the dataset is the same as the number of data providers who participate in the data collection procedure and is defined by N in Eq(1). Consequently, the income of the data collector is:

$$income_{DC} = p\,N \qquad (2)$$

Data collection procedure, data anonymization, and storing the dataset are costly and we denote these costs by C. Moreover, the data collector has to pay some incentive, I, to each data provider. As a result, the expenses to the data collector can be defined as:

$$expenditure_{DC} = I\,N + C \qquad (3)$$

For simplicity of analysis we have assumed a fixed cost C for data collector. This assumption can be dropped easily by defining cost as a function of the size of the dataset and privacy level δ without any significant modification to our analysis. The payoff to the data collector is therefore defined as:

$$U_{DC} = income_{DC} - expenditure_{DC} = (p - I)\,N - C \qquad (4)$$

Payoff to the Data User: The data user wants to run some data analysis on the privacy protected dataset T^*. As the cardinality of this dataset increases, the dataset will have higher value to the data user. Let a denote the economic value of each record to the data user, *i.e.*, a represents the net revenue of a data record if the data user gets the record for free. If the number of data records collected from individuals is denoted by N we can initially define the data user's income as $a * N$. However, after anonymization the utility of data drops due to imprecision introduced to results of the queries. We use parameter $0 \leq Precision \leq 1$ as a coefficient of the data user's income to show how the value of the dataset decreases as data become less precise. The income of the data user is:

$$income_{DU} = a\ N\ Precision \tag{5}$$

To estimate the precision of query results on a private dataset, various parameters must be considered. These parameters include the semantics of the query, the anonymization method and algorithm used, database schema, level of privacy protection δ, number of data records N, and *etc.*. For each instance of the game, all of these parameters except for δ and N are fixed (and assumed to be a common knowledge of the game). Therefore, $Precision = prec(\delta, N)$ is defined as a function of two variables δ and N. The main characteristic of the $Precision$ function is that for any **fixed** number of data records N, $Precision$ is a decreasing function of δ [2].

If the data user pays price p per record, his expenditure is $p\ N$ and therefore his payoff can be defined as:

$$U_{DU} = a\ N\ Precision - p\ N \tag{6}$$

5   General Approach to Find Subgame Perfect Equilibria

In this section we explain the steps involved in the process of finding the game's subgame perfect equilibria using backward induction [6]. In the next section, we show the details of this process for k-anonymity as an example.

5.1   Equilibrium Strategies of Data Collector

The first step to find subgame perfect equilibria is to find the optimal actions of the data collector in each subgame of length 1. Subgames of length 1 are represented by subtrees at which the data collector has to move based on a history of the form (Of). Where $Of = \langle \delta, p \rangle$ is an offer made by the data user.

The data collector can estimate the expected cardinality of the dataset for each δ and I based on Eq(1). If we plug this equation into the U_{DC} formula from Eq(4), the data collector's payoff after accepting $Of = \langle \delta, p \rangle$ will be:

$$U_{DC} = (p - I)(\beta_0 + \beta_1 h_1(\delta) + \beta_2 h_2(I)) - C \tag{7}$$

For each offer $Of = \langle \delta, p \rangle$, the values of δ and p are fixed. The data collector needs to find the optimum I (denoted by $\hat{I}$) for which the function U_{DC} attains its maximum value. To find $\hat{I}$ we must find the argument of the maximum:

$$\hat{I} = \arg\max_I U_{DC} = \arg\max_I (p - I)(\beta_0 + \beta_1 h_1(\delta) + \beta_2 h_2(I)) - C \tag{8}$$

---

[2] Notice that N is also an increasing function of δ (see Eq(1)) and therefore $\frac{\partial\ prec}{\partial\ \delta}$ is not always greater than or equal to zero.

Subject to the constraint that $\hat{I} \geq 0$.

If the maximum U_{DC}, $\hat{U}_{DC}$, is greater than zero the data collector accepts the offer. If $\hat{U}_{DC} = 0$ then the data collector will be indifferent between accepting and rejecting and in the case where $\hat{U}_{DC} < 0$ the data collector rejects. Therefore, the data collector's best response, BR_{DC}, to an offer $Of = \langle \delta, p \rangle$ is:

$$BR_{DC}(\delta, p) = \begin{cases} Reject & if \; (p - \hat{I})(\beta_0 + \beta_1 h_1(\delta) + \beta_2 h_2(\hat{I})) - C \leq 0 \\ Accept \; with \; \hat{I} \; if \; (p - \hat{I})(\beta_0 + \beta_1 h_1(\delta) + \beta_2 h_2(\hat{I})) - C \geq 0 \end{cases}$$

(9)

The optimum incentive $\hat{I}$ must only be calculated when the data collector accepts the offer. This means $\hat{I} \leq p$, otherwise $\hat{U}_{DC} < 0$. Since U_{DC} is continuous in the closed and bounded interval $[0, p]$ (the domain of I), according to the Extreme value theorem [25], U_{DC} reaches its maximum at least once and therefore $\hat{I}$ is guaranteed to exist.

5.2   Equilibrium Strategies of Data User

The next step to find the subgame perfect equilibria is to find the most profitable action of the data user; Knowing the data collector's best response (Sect. 5.1) to each $Of = \langle \delta, p \rangle$, what combination of δ and p maximizes the data user's payoff? When the data collector accepts an offer $Of = \langle \delta, p \rangle$, he chooses the optimum incentive $\hat{I}$. Depending on the exact function definitions used in Eq(8), if $\hat{I}$ is unique for every combination of δ and p, then $\hat{I}$ can be defined as a function of δ and p (i.e., $\hat{I} = \hat{i}(\delta, p)$). Without loss of generality, we assume that this is the case. If multiple values of I maximize U_{DC}, the one that also maximizes the data user's payoff is in the equilibria of the game.

According to Sect. 5.1, if the data collector accepts the offer he starts collecting personal information at privacy level δ with incentive $\hat{I} = \hat{i}(\delta, p)$. Otherwise, no dataset will be provided to the data user. As a result, the anticipated number of records N can be determined as:

$$N = n(\delta, \hat{I}) = \begin{cases} \beta_0 + \beta_1 h_1(\delta) + \beta_2 h_2(\hat{I}) & if \; \hat{U}_{DC} \geq 0 \\ 0 & Otherwise \end{cases}$$

(10)

Plugging the function definition of $\hat{I} = \hat{i}(\delta, p)$ into Eq(10), $N = n_2(\delta, p)$ becomes a function of δ and p as well. Recall that $Precision = prec(\delta, N)$ is defined as a function of δ and N. Since N is a function of δ and p, we can define $Precision = prec_2(\delta, p)$ as a function of δ and p as well. After substituting N and $Precision$ with $n_2(\delta, p)$ and $prec_2(\delta, p)$, the U_{DU} function from Eq(6) becomes a function of two variables δ and p. The most profitable strategy for the data user is to choose values of δ and p that maximize his payoff:

$$\langle \hat{\delta}, \hat{p} \rangle = \arg \max_{\delta, p} U_{DU} = \arg \max_{\delta, p} \left(a \; prec_2(\delta, p) - p \right) \left(n_2(\delta, p) \right)$$

(11)

By definition, the lower bounds on p and δ is zero, i.e., $p \geq 0$ and $\delta \geq 0$. Moreover, since $Precision \leq 1$ then $(a * prec_2(\delta, p)) \leq a$. Choosing a value $p > a$ leads to a negative payoff to the data user and he can always do better by choosing $p = 0$

(which leads to payoff zero). Therefore, the upper bound for p is a. Parameter δ is not necessarily bounded from above. Consequently, we cannot use the Extreme value theorem to guarantee an equilibrium.

If U_{DU} has an absolute maximum subject to the bounds defined on δ and p, the game has subgame perfect equilibria of the forms $((\hat{\delta}, \hat{p}), reject)$ or $((\hat{\delta}, \hat{p}), \hat{I})$. The first form occurs when the data collector cannot find any profitable amount of incentive (regardless of δ and p chosen by the data user) and the negotiation is unsuccessful. The second format occurs in games where there are at least one combination of δ and p of which the data collector can make profit. The two types of equilibria provide a means to determine whether an anonymization technique is *practical* or *impractical* given other problem settings. If the cost of implementing an anonymization technique is too high and the public's trust in the method is not high enough, the game might become an instance of unsuccessful negotiations and we have a case of impractical anonymization.

6   Game Theoretic Analysis for k-Anonymity

To demonstrate the details of the steps explained in Sect. 5, we use k-anonymity as the anonymization technique and provide a $Precision$ function for it. The game solution is described and a simulation of the results is provided at the end of this section.

6.1   k-Anonymity Overview

A dataset to be released contains some sensitive attributes, identifying attributes, and *quasi-identifying* attributes. Even after removing the identifying attributes, the values of quasi-identifying attributes can be used to uniquely identify at least a single individual in the dataset via linking attacks. Every subset of tuples in dataset that share the same values for quasi-identifiers is often referred to as an *equivalence class*. A released dataset is said to satisfy k-anonymity, if for each existing combination of quasi-identifier attribute values in the dataset, there are at least $k - 1$ other records in the database that contain such a combination.

There are several methods to achieve k-anonymity. Our work is built on Mondrian algorithm [26]. This greedy algorithm implements *multidimensional* recoding (with no cell suppression) which allows finer-grained search and thus often leads to a better data quality. In Mondrian algorithm all the identifying attributes are suppressed first. Then records are recursively partitioned into $d-$dimensional rectangular boxes (equivalence classes), where d is the number of quasi-identifiers. To partition each box, a quasi-identifier attribute (a dimension) is selected and the *median* value along this attribute is used as a binary cut to split the box into two smaller boxes. Once partitioning is done, records in each partition are generalized so that they all share the same quasi-identifier value, to form an equivalence class. A copy of this algorithm is provided in Fig. 3(b).

6.2   Data Providers' Privacy Model

Based on Sect. 4.1, we assume a regression model to explain data providers' reaction (at an aggregate level) to each combination of privacy protection levels and incentives.

This model is explained in Eq(1). In k-anonymity, privacy parameter is k. Here, we consider the identity function for the incentive (because of its simplicity) and logarithmic function for parameter k. In other words :

$$N = n(k, I) = \beta_0 + \beta_1 log_2(k) + \beta_2 I \qquad (12)$$

To understand our choice of log function for h_1, notice that when k-anonymity is used, it is assumed that the probability of re-identifying an individual is $\frac{1}{k}$. For example, when k is 1, the probability of re-identification is 1 and the guaranteed privacy is 0. When k becomes 2, the probability of re-identification becomes $\frac{1}{2}$ and the amount of uncertainty about the identity of the individual increases from 0 ($log1$) to 1 ($log2$). However, this increase in uncertainty about the identity of individuals (privacy) is not the same as k changes from 99 to 100 because the probability changes from $\frac{1}{99}$ to $\frac{1}{100}$. For this reason we use entropy ($logk$) of this uniform probability distribution ($p = \frac{1}{k}$) as the indicator for privacy protection.

6.3   Precision Estimate

To determine the payoff to the data user (see Eq(6)) we need a metric to calculate *Precision*. A reasonable estimate on the amount of imprecision caused by anonymization depends on the data application. We have briefly discussed the nature of imprecision that can be introduced to the results of any SELECT query executed against an anonymized dataset elsewhere [24] . In this paper we provide the precision estimates for a specific SELECT query type and consider this query as the data analysis purpose. Our SELECT query is of the following form:

   Q_i ≡ SELECT sensitiveAtt FROM T* WHERE q = v$_i$

In this query sensitiveAtt represents the value of sensitive attribute, T^* is the anonymized dataset, q is one of the quasi-identifiers, and v_i is the ith possible value for attribute q. For example, a query Q_{20} can be the following:

   Q_{20} ≡ SELECT disease FROM T* WHERE age = 20

Let $|Q_i(T)|$ denote cardinality of the result set of query Q_i on dataset T. When Q_i is run against T^*, the result set $Q_i(T^*)$ contains two groups of records: a subset of them satisfy the condition q = v$_i$ and the rest of them are just included in the result because they are partitioned into the same equivalence class as the points with q = v$_i$. The latter introduce some quantity imprecision in the result. LeFevre *et al.* [27] introduce an imprecision metric to find the best cuts while running the Mondrian algorithm [26] on experimental datasets. After normalizing this metric, we define *Precision* as:

$$Precision(Q_i, T^*, T) = \frac{|Q_i(T)|}{|Q_i(T^*)|} \quad \text{(where } |Q_i(T^*)| > 0) \qquad (13)$$

As a result, to calculate *Precision* we first need to estimate $|Q_i(T)|$ and $|Q_i(T^*)|$. Let Pr_i denote the portion of the records in the dataset that have value v$_i$ for quasi-identifier q. Then the expected value of $|Q_i(T)|$ is:

$$|Q_i(T)| = Pr_i \, N \qquad (14)$$

Through Theorems 1 and 2 we provide an estimate for $|Q_i(T^*)|$. In Mondrian algorithm the minimum and maximum number of records in each equivalence class are k

and $2d(k - 1) + m$, where m denotes the maximum number of records with identical values for all quasi-identifiers [26]. Since the distribution of equivalence class sizes are not known *a priori*, with a simplifying assumption of uniform distribution, we can estimate the average number of records in each equivalence class, ec_{AVG}, as:

$$ec_{AVG} = \frac{2d(k - 1) + m + k}{2} \tag{15}$$

Theorem 1. *If the average size of each equivalence class is determined by Eq(15), then the depth of the recursive calls, l, in Mondrian algorithm [26] can be estimated as:*

$$l = log_2(\frac{2N}{2d(k - 1) + m + k}) \tag{16}$$

Proof. (*sketch*) Mondrian algorithm starts with the original dataset as a single equivalence class and chooses the *median* value of one of the dimensions to recursively cut each equivalence class into two smaller ones. It stops when there is no more possible cuts for any of the equivalence classes. For this estimate, we assume that the algorithm stops at the point where the size of each class reaches ec_{AVG} from Eq(15). By solving the recursive definition, we get Eq(16). A complete proof is available [24].

Theorem 2. *If N denotes the number of records in a dataset T, the cardinality of the result set of query Q_i on T^* can be estimated as:*

$$|Q_i(T^*)| = (1 - \frac{1}{2d})^l \, N \tag{17}$$

where d is the number of quasi-identifiers and l is the depth of recursive calls estimated in Theorem 1.

Proof. (*sketch*) The core idea of this proof is to note that during the partitioning process, for each equivalence class if the dimension q is chosen as the cutting dimension then half of the records in the class will be partitioned into a new class that will not be included in the result set of Q_i. Otherwise the cut does not reduce the size of the result set. A complete proof is available [24].

Consequently, *Precision* is defined as:

$$Precision = \frac{pr_i \, N}{(1 - \frac{1}{2d})^l \, N} = \frac{pr_i}{(1 - \frac{1}{2d})^l} \tag{18}$$

We can also use Theorem 2 to define pr_i based on the parameters. In real instances of the problem pr_i is independent of any specific algorithm and estimates; it is a property of the dataset. However, since we have made some simplifying assumptions for other estimates the assumptions should also be applied to pr_i to produce a meaningful estimate. Theorem 2 provides an estimate on $|Q_i(T^*)|$. When $k = 1$, there are no irrelevant records in the result set. Therefore, $|Q_i(T^*_{k=1})|$ provides an estimate on the number of records that satisfy the condition q $= v_i$ and $|Q_i(T^*_{k=1})|/N$ can be used as an estimate for pr_i.

Consequently, we can refine Equation(18) as:

$$Precision = \frac{(1 - \frac{1}{2d})^{log_2 \frac{2N}{m+1}}}{(1 - \frac{1}{2d})^l} \tag{19}$$

6.4   Subgame Perfect Equilibria

As explained in Sect. 5.1, the first step to find the game's subgame perfect equilibria is to determine the optimum incentive $\hat{I}$ from Eq(8). If the data collector accepts the offer $Of = \langle k, p \rangle$ with incentive I, his payoff will be:

$$U_{DC} = (p - I)(\beta_0 + \beta_1 log_2(k) + \beta_2 I) - C \tag{20}$$

Calculating the derivative of U_{DC} with respect to I and setting it to zero reveals the maximizing I:

$$\frac{dU_{DC}}{dI} = -(\beta_0 + \beta_1 log_2(k) + \beta_2 I) + \beta_2(p - I) = 0 \implies \hat{I} = \frac{\beta_2 p - \beta_1 log_2(k) - \beta_0}{2\beta_2} \tag{21}$$

$\hat{I}$ is the local maximum since the second derivative of the function is negative. The restriction here is $I \geq 0$. If $\hat{I} < 0$, the maximizing I will be zero. The lower bound on I leads us to consider two separate cases:

Case 1: $\beta_2 p \geq \beta_1 log_2(k) + \beta_0$- In this case the amount of incentive that maximizes U_{DC} is $\hat{I} = \frac{\beta_2 p - \beta_1 log_2(k) - \beta_0}{2\beta_2}$. Plugging $\hat{I}$ into Eq(20) gives us the maximum payoff to the data collector for Case 1 (denoted as $\hat{U}_{DC}^1$):

$$\hat{U}_{DC}^1 = \frac{\beta_2}{4}(p + \frac{\beta_1 log_2(k) + \beta_0}{\beta_2})^2 - C \tag{22}$$

The data collector will accept the offer $Of = \langle k, p \rangle$ if $\hat{U}_{DC}^1 \geq 0$. In other words, the data collector accepts if:

$$p + \frac{\beta_1 log_2(k)}{\beta_2} \geq \sqrt{\frac{4C}{\beta_2}} - \frac{\beta_0}{\beta_2} \tag{23}$$

Case 2: $\beta_2 p < \beta_1 log_2(k) + \beta_0$- The optimum incentive in this case would be $\hat{I} = 0$. With this incentive the maximum payoff to the data collector (denoted as $\hat{U}_{DC}^2$) is:

$$\hat{U}_{DC}^2 = p(\beta_0 + \beta_1 log_2(k)) - C \tag{24}$$

The data collector will accept this offer if $\hat{U}_{DC}^2 \geq 0$. More precisely, the data collector accepts the offer if:

$$p(\beta_0 + \beta_1 log_2(k)) \geq C \tag{25}$$

If the values of $\hat{I}$ (from the two cases) are plugged into Eq(10), we can define the cardinality of the private dataset as a piecewise function of k and p:

$$N = \begin{cases} \frac{\beta_0 + \beta_1 log_2(k) + \beta_2 p}{2} & if \ \beta_2 p \geq \beta_1 log_2(k) + \beta_0 \ \wedge p + \frac{\beta_1 log_2(k)}{\beta_2} \geq \sqrt{\frac{4C}{\beta_2}} - \frac{\beta_0}{\beta_2} \\ \\ \beta_0 + \beta_1 log_2(k) & if \ \beta_2 p < \beta_1 log_2(k) + \beta_0 \ \wedge p(\beta_0 + \beta_1 log_2(k)) \geq C \\ \\ 0 & Otherwise \end{cases} \tag{26}$$

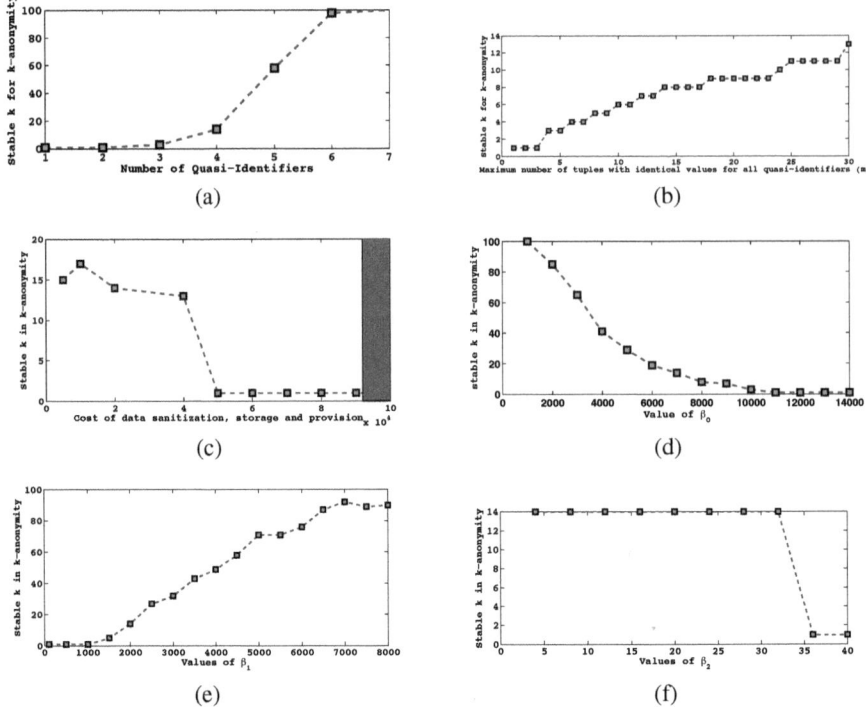

Fig. 2. Changes to the stable k due to an increase in: (a) the number of quasi-identifiers d; (b) the maximum number of data providers with identical values for their quasi-identifiers m; (c) the cost of data anonymization and storage C; (d) the number of privacy unconcerned data providers β_0; (e) the effect of privacy protection level on data providers' decision β_1; (f) the effect of incentive on data providers' decision β_2.

If the new definition of N is plugged into the $Precision$ function, precision becomes a function of k and p. As a result, U_{DU} from Eq(6) becomes a function of k and p. The best strategy for the data user is to compute $\hat{k}$ and $\hat{p}$ according to Eq(11). The optimum offer is $Of = \langle \hat{\delta}, \hat{p} \rangle$ and this completes the process of finding perfect equilibria.

6.5   Simulation Results

If the players of the game are rational and have the required information, the equilibria of the game would always conform to what Sect. 6.4 suggests because we used an analytical method to find the game's equilibria. In our proposed method, a dataset does not exist before the game is complete and the specifications of the collected dataset depend on the parameters chosen while the game is played. Therefore, running experiments on real databases does not provide meaningful results for this work. Alternatively, we choose to simulate the game and visualize the results by testing multiple parameter settings using MATLAB R2008a. In every setting, the effect of one of the parameters a, C, d, m, and β is examined on the stable values of k (while the values of the rest of the parameters are fixed to $\beta_0 = 7000$, $\beta_1 = 2000$, $\beta_2 = 20$, $a = \$10$, $C = \$20,000$, $m = 5$, and $d = 4$). The results are shown in Fig. 2.

The values for a and C are randomly selected as an estimate of reasonable values commonly used in real instances of the problem. We assumed a population size of 55,000 potential data providers and the values selected for parameters β_0, β_1, and β_2 are chosen to reflect Westin's privacy indexes [28]. Based on the maximum values of k ($k = 100$) and p ($p = a$), β_1 and β_2 are chosen such that the effect of maximum privacy is almost the same as maximum incentive. The value of β_0 is chosen such that 17% of the data providers fall in the *privacy unconcerned* category [28].

Figure 2(a) shows how stable values of k increase as the number of quasi-identifiers increase. To understand the reason, we have provided another diagram in Fig. 3(a) which illustrates the precision curves for different values of d. According to this figure, with fewer quasi-identifiers the precision curve decreases at a higher rate. Therefore, as the number of quasi-identifiers increase, offering larger values for k becomes a better option for the data user since it can increase the size of the dataset without severely affecting data quality.

In Fig. 2(b) we can see the effect of m (maximum number of data providers with identical quasi-identifier values) on the stable values of k. We have chosen the values of m from $\{1, ..., 30\}$. As the value of m increases the stable value of k increases. To understand this counter-intuitive result, notice that as m increases less generalization will be needed to group the tuples in equivalence classes of size k. Therefore, compared to the cases with smaller m, the same precision can be achieved with higher values of k. Larger values of k attract more data providers without largely affecting the precision of query results and consequently, the data user can make more profit in this case.

The effects of anonymization, and maintenance cost (C) on stable values of k are illustrated in Fig. 2(c). Based on the settings chosen for other parameters, after a certain point the cost becomes too high for condition of the Eq(25) to be satisfied and case 1 (from Sect. 6.4) happens. In this case, the data collector is receiving a payment high enough to announce non-zero incentives. This incentive convinces several privacy concerned data providers to participate even with a low privacy protection level. As a result, the data user simply asks for no privacy protection since he is confident that enough data providers will participate to receive the incentive. Finally, after a certain value for C, the game reaches a point (demonstrated by a shaded rectangle) where no combination of $\langle k, p \rangle$ can be found that is acceptable by the data collector and $U_{DU} \geq 0$. This situation represents an instance of *impractical* anonymization.

Figures 2(d), 2(e), and 2(f) represent the effects of data providers' privacy attitude on stable values of k. According to Fig. 2(d) as the number of *privacy unconcerned* group (data providers who provide their personal information without any privacy or incentive) increase, the data user can receive larger volume of data without asking for sanitized dataset. By increasing the value of β_1 we model a privacy aware population. As can be seen in Fig. 2(e), when privacy has more significant impact on data providers' decisions, data will be sanitized with larger values of k. In Fig. 2(f) we showed how the value of β_2 impacts stable values of k. If β_2 is less than a certain level then it mostly affects the price of information and not the level of privacy protection. However if the weight of incentive on data providers' privacy decisions becomes heavier than a certain point, case 1 (refer to Sect. 6.4) happens and the data user can maximize his benefit by just increasing the price and asking for no privacy. These diagrams show how public's privacy awareness can force the firms to protect privacy of data providers.

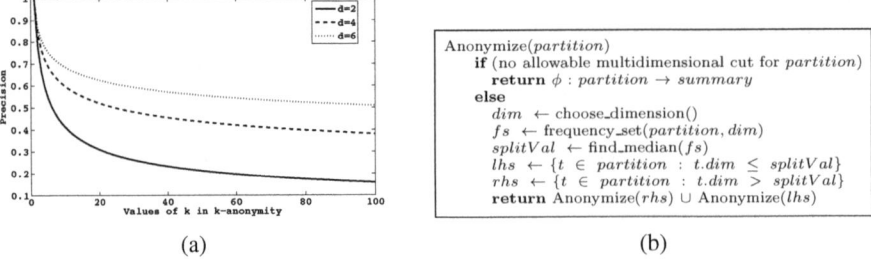

(a)   (b)

Fig. 3. (a) Precision curves for different number of quasi-identifiers d. The value of m is fixed by 5. (b) Mondrian Algorithm.

7   Conclusions and Future Work

In this paper we modeled the process of private data collection as a sequential game to achieve consensus on the level of privacy protection. We explained the general approach to solve the game and as an example provided the details of game analysis for k-anonymity. Players of the game are a data user, a data collector, and a group of data providers. We use the method of backward induction to explore the game's subgame perfect equilibria. Equilibria of the game suggest stable values of the privacy parameter that are unlikely to be changed when other parties move according to their equilibria strategies. For the k-anonymity case, we found the stable values of k and showed that these values are related to number of quasi-identifiers, maximum number of identical tuples (in their quasi-identifier values), cost of data sanitization and storage, and coefficients of public's privacy behavior model. Our results illustrate the significant impact of the number of quasi-identifiers on the decision about the value of k.

We are plannig to analyze other privacy definitions such as l-diversity [3] and differential privacy [5] and for each privacy definition, distinguish the settings which make it the most profitable option to the players of the game. We are also planning to improve the model by dropping the assumption about the amount of information available to the data collector and data user. Our goal is to design a new evaluation framework that uses our game theoretic model to compare different anonymization methods and distinguish the settings that make one anonymization method more appealing than another.

References

1. Samarati, P., Sweeney, L.: Generalizing data to provide anonymity when disclosing information (abstract). In: PODS, p. 188. ACM Press (1998)
2. Sweeney, L.: k-anonymity: a model for protecting privacy. International Journal on Uncertainty, Fuzziness and Knowledge-Based Systems 10(5), 557–570 (2002)
3. Machanavajjhala, A., Kifer, D., Gehrke, J., Venkitasubramaniam, M.: L-diversity: Privacy beyond k-anonymity. ACM Trans. Knowl. Discov. Data 1(1), 24 pages (2007)
4. Li, N., Li, T., Venkatasubramanian, S.: t-closeness: Privacy beyond k-anonymity and l-diversity. In: ICDE 2007, pp. 106–115 (2007)
5. Dwork, C.: Differential Privacy. In: Bugliesi, M., Preneel, B., Sassone, V., Wegener, I. (eds.) ICALP 2006, Part II. LNCS, vol. 4052, pp. 1–12. Springer, Heidelberg (2006)

6. Osborne, M.J.: 8,9,16. In: An Introduction to Game Theory. Oxford University Press, USA (2003)
7. LeFevre, K., DeWitt, D.J., Ramakrishnan, R.: Workload-aware anonymization. In: KDD, pp. 277–286 (2006)
8. Bayardo Jr., R.J., Agrawal, R.: Data privacy through optimal k-anonymization. In: ICDE, pp. 217–228 (2005)
9. Fung, B.C.M., Wang, K., Yu, P.S.: Top-down specialization for information and privacy preservation. In: ICDE, pp. 205–216 (2005)
10. Sramka, M., Safavi-Naini, R., Denzinger, J., Askari, M.: A practice-oriented framework for measuring privacy and utility in data sanitization systems. In: EDBT/ICDT Workshops (2010)
11. Machanavajjhala, A., Korolova, A., Sarma, A.D.: Personalized social recommendations - accurate or private? CoRR abs/1105.4254 (2011)
12. Loukides, G., Shao, J.: Data utility and privacy protection trade-off in k-anonymisation. In: PAIS 2008, pp. 36–45. ACM (2008)
13. Anderson, H.E.: The privacy gambit: Toward a game theoretic approach to international data protection. bepress Legal Series (2006)
14. Böhme, R., Koble, S., Dresden, T.U.: On the viability of privacy-enhancing technologies in a self-regulated business-to-consumer market: Will privacy remain a luxury good? In: WEIS 2007 (2007)
15. Kleinberg, J., Papadimitriou, C.H., Raghavan, P.: On the value of private information. In: TARK 2001, pp. 249–257. Morgan Kaufmann Publishers Inc. (2001)
16. Calzolari, G., Pavan, A.: Optimal design of privacy policies. Technical report, Gremaq, University of Toulouse (2001)
17. Preibusch, S.: Implementing Privacy Negotiations in E-Commerce. In: Zhou, X., Li, J., Shen, H.T., Kitsuregawa, M., Zhang, Y. (eds.) APWeb 2006. LNCS, vol. 3841, pp. 604–615. Springer, Heidelberg (2006)
18. Gianini, G., Damiani, E.: A Game-Theoretical Approach to Data-Privacy Protection from Context-Based Inference Attacks: A Location-Privacy Protection Case Study. In: Jonker, W., Petković, M. (eds.) SDM 2008. LNCS, vol. 5159, pp. 133–150. Springer, Heidelberg (2008)
19. Kargupta, H., Das, K., Liu, K.: Multi-party, Privacy-Preserving Distributed Data Mining Using a Game Theoretic Framework. In: Kok, J.N., Koronacki, J., Lopez de Mantaras, R., Matwin, S., Mladenič, D., Skowron, A. (eds.) PKDD 2007. LNCS (LNAI), vol. 4702, pp. 523–531. Springer, Heidelberg (2007)
20. Acquisti, A., Grossklags, J.: Privacy and rationality in individual decision making. IEEE Security & Privacy 3(1), 26–33 (2005)
21. Culnan, M.J., Armstrong, P.K.: Information privacy concerns, procedural fairness, and impersonal trust: An empirical investigation. Organization Science 10, 104–115 (1999)
22. Singer, E., Mathiowetz, N.A., Couper, M.P.: The impact of privacy and confidentiality concerns on survey participation: The case of the 1990 U.S. census. The Public Opinion Quarterly 57(4), 465–482 (1993)
23. Milne, G.R., Gordon, M.E.: Direct mail privacy-efficiency trade-offs within an implied social contract framework. Journal of Public Policy & Marketing 12(2), 206–215 (1993)
24. Adl, R.K., Askari, M., Barker, K., Safavi-Naini, R.: Privacy consensus in anonymization systems via game theory. Technical Report 2012-1021-04, University of Calgary (2012)
25. Sydsaeter, K., Hammond, P.: Mathematics for economic analysis. Prentice-Hall International (1995)
26. LeFevre, K., DeWitt, D.J., Ramakrishnan, R.: Mondrian multidimensional k-anonymity. In: ICDE 2006, p. 25. IEEE Computer Society (2006)
27. LeFevre, K., DeWitt, D.J., Ramakrishnan, R.: Workload-aware anonymization techniques for large-scale datasets. ACM Trans. Database Syst. 33, 17:1–17:47 (2008)
28. Kumaraguru, P., Cranor, L.F.: Privacy indexes: A survey of westin's studies. ISRI Technical Report (2005)

Uniform Obfuscation for Location Privacy

Gianluca Dini and Pericle Perazzo

University of Pisa,
Department of Information Engineering, via Diotisalvi 2,
56122 Pisa, Italy
{g.dini,p.perazzo}@iet.unipi.it

Abstract. As location-based services emerge, many people feel exposed to high privacy threats. Privacy protection is a major challenge for such applications. A broadly used approach is *perturbation*, which adds an artificial noise to positions and returns an obfuscated measurement to the requester. Our main finding is that, unless the noise is chosen properly, these methods do not withstand attacks based on probabilistic analysis. In this paper, we define a strong adversary model that uses probability calculus to de-obfuscate the location measurements. Such a model has general applicability and can evaluate the resistance of a generic location-obfuscation technique. We then propose UNILO, an obfuscation operator which resists to such an adversary. We prove the resistance through formal analysis. We finally compare the resistance of UNILO with respect to other noise-based obfuscation operators.

Keywords: location-based services, privacy, obfuscation, perturbation, uniformity.

1 Introduction

Recent years have seen the widespread diffusion of very precise localization technologies and techniques. The most known is GPS, but there are many other examples, like Wi-Fi fingerprinting, GSM trilateration, etc. The emergence of such technologies has brought to the development of *location-based services* (*LBS*) [3,6,10], which rely on the knowledge of location of people or things. The retrieval of people's location raises several privacy concerns, as it is personal, often sensitive, information. The indiscriminate disclosure of such data could have highly negative effects, from undesired location-based advertising to personal safety attempts.

A classic approach to the problem is to introduce strict access-control policies in the system [9,16]. Only some trusted (human or software) entities will be authorized to access personal data. This access-control-based approach has a main drawback: if the entity does not need complete (or exact) information, it is a useless exposure of personal data. The "permit-or-deny" approach of access control is often too rigid. Some services require more flexible techniques which can be tailored to different user preferences.

N. Cuppens-Boulahia et al. (Eds.): DBSec 2012, LNCS 7371, pp. 90–105, 2012.

Samarati and Sweeney [18] introduced the simple concept of *k-anonymity*: a system offers a *k*-anonymity to a subject if his identity is undistinguishable from (at least) $k - 1$ other subjects. *K*-anonymity is usually reached by obfuscating data with some form of generalization. The methods based on *k*-anonymity [4,10,12] offer generally a high level of privacy, because they protect both the personal data and the subject's identity. However, they have some limitations:

- They do not permit the authentication of the subject and the customization of the service. Since they cannot identify the subject, some identity-based services like social applications or pay-services could not work.
- They are usually more complex and inefficient than methods based only on data obfuscation. This happens because their behavior must depend on a set of (at least) *k* subjects and not on a single subject only.
- They need a centralized and trusted obfuscator. In distributed architectures, such an entity may be either not present or not trusted by all the subjects.
- They are not applicable when the density of the subjects is too low. Obviously, if there are only 5 subjects in a system, they will never reach a 10-anonymity.

A simpler approach is *data-only obfuscation* [1,15], whose aim is not to guarantee a given level of anonymity, but simply to protect the personal data. This is done by obfuscating data before disclosing it, in a way that it is still possible for the service provider to offer his service. Data obfuscation adds some artificial imperfection to information. The nature of such imperfection can fall into two categories [8]: *inaccuracy*, and *imprecision*. Inaccuracy concerns a lack of correspondence with reality, whereas imprecision concerns a lack of specificity in information. Deliberately introducing inaccuracy requires the obfuscation system to "lie" about the observed values. This can reduce significantly the number of assumptions the service can trust on. For this reason, the majority of obfuscation methods operates by adding imprecision, both by means of *generalization* or *perturbation* [5]. Generalization replaces the information with a value range which contains it, whereas perturbation adds random noise to it. We focus on the perturbation method. This method is both simple and efficient, and is often used to obfuscate data [14]. In spite of its simplicity, it requires to choose a suitable noise to effectively perturb data. In case of location data - and non-scalar data in general - such a problem is not trivial and should not be underrated. We found that if the noise is not chosen properly, perturbation will not resist to attacks based on statistical analysis. In particular, an obfuscation operator must offer a spacial *uniformity* of probability.

We present an analytical adversary model, which performs attacks based on statistical analysis. We show how such attacks can be neutralized by the property of uniformity. We present a metric for quantifying uniformity of an obfuscation system, called *uniformity index*. We further propose UNILO, an obfuscation operator for location data that introduces imprecision while maintaining accuracy. UNILO is simple and $\mathcal{O}(1)$-complex. It does not require a centralized and trusted obfuscator and can be seamlessly added to a distributed architecture as a building block. We show how UNILO offers a better uniformity with respect to other noise-based obfuscation operators. To the best of our knowledge, UNILO is the first obfuscation operator which offers guarantees on uniformity.

The rest of the paper is organized as follows. Section 2 introduces some basic concepts concerning the system model and the terminology. Section 3 formally describes the adversary model. Section 4 presents the UNILO operator in detail and its properties. Section 5 evaluates UNILO resistance by means of experimental results, and compares it to other obfuscation operators. Section 6 presents some examples of location-based services that can be built on UNILO operator. Section 7 explains some related works and analyzes differences and similarities with UNILO techniques. Finally, the paper is concluded in Section 8.

2 System Model

In our system, a *subject* is an entity whose location is measured by a *sensor*. A *service provider* is an entity that receives the subject's location in order to provide him with a *location-based service*. The subject applies an *obfuscation operator* to location information, prior to releasing it to the service provider. The obfuscation operator purposefully reduces the precision to guarantee a certain privacy level. Such a precision is defined by the subject and reflects his requirements in terms of privacy. The more privacy the subject requires, the less precision the obfuscation operator returns.

The subject is usually a person who has agreed to reveal - with some level of privacy - his location to one or more service providers. The service provider can be a human or a piece of software, depending on the kind of location-based service. For instance, a security service in an airport or in a train station often requires a human service provider. In contrast, in a customer-oriented service, for example, returning the nearest restaurant to the subject, the service provider may be a piece of software. The obfuscation operator can be applied to the data directly by the subject. Alternatively, a central obfuscator could be provided as well, serving several subjects at once.

For the sake of simplicity, the arguments and results we present in this paper refer to the two-dimensional case. However, they can be extended to the three-dimensional case in a straightforward way.

In the most general case, a *location measurement* is affected by an intrinsic error that limits its precision. Such error depends on several factors including the localization technology, the quality of the sensor, the environment conditions. Different technologies have different degrees of precision. For instance, the 68-th percentile of the error on a Garmin professional GPS receiver is 1.1 meters, on the iPhone's GPS is 8.6 meters, and on the iPhone's Wi-Fi localization system is 88 meters [21]. This implies that the location cannot be expressed as a geographical point but rather as a neighborhood of the actual location. We assume that locations are always represented as *planar circular areas* [1,21], because it is a good approximation for many location techniques [17]. A location measurement (Fig. 1) can be defined as follows:

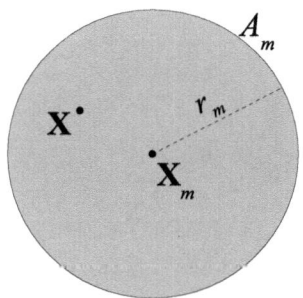

Fig. 1. Location measurement

Definition 1 (Location measurement). *Let* $\mathbf{X}$ *be the* actual *position of the subject. A* location measurement *is a circular area* $A_m = \langle \mathbf{X}_m, r_m \rangle \subseteq \mathbb{R}^2$, *where* $\mathbf{X}_m$ *is the center of* A_m *and* r_m *is the radius, such that* $P\{\mathbf{X} \in A_m\} = 1$ *(Accuracy Property).*

The Accuracy Property guarantees that the location measurement actually contains the subject, or, equivalently, that the distance $\overline{\mathbf{X}\mathbf{X}_m}$ does not exceed r_m. The radius r_m specifies the precision of the localization technology, and we call it *precision radius*. Different technologies have different values for the precision radius. If a technology has a precision radius r_m, then a subject cannot be located with a precision better than r_m. We assume that r_m is constant over time. This means either that the precision does not change over time, or that r_m represents the worst-case precision.

A subject can specify his privacy preference in terms of *privacy radius* (r_p). If the subject specifies r_p, $r_p > r_m$, as his privacy radius, then he means that he wishes to be located with a precision not better than r_p. The task of an obfuscation operator is just to produce an obfuscated position $\mathbf{X}_p$, appearing to the provider as a measurement with precision r_p, worse than r_m. More formally, the obfuscation operator has to solve the following problem:

Problem 1 (Obfuscation). Let $\mathbf{X}$ be the actual position of a subject, $A_m = \langle \mathbf{X}_m, r_m \rangle$, be the location measurement and, finally, $r_p, r_p > r_m$, be his desired privacy radius. Transform, A_m into an *obfuscated measurement* (also called *privacy area*) $A_p = \langle \mathbf{X}_p, r_p \rangle$ such that the following properties hold:

1. (Accuracy) $P\{\mathbf{X} \in A_p\} = 1$
2. (Uniformity) $pdf(\mathbf{X}) : \mathbb{R}^2 \to \mathbb{R}$ (probability density function) as uniform as possible over A_p.

Property 1 guarantees that the obfuscated measurement actually contains the subject. Property 2 guarantees that the subject can be located everywhere in A_p with an almost-uniform probability. This property is particularly important because it prevents an adversary from determining areas that more likely contain the subject, and thus jeopardize the user privacy requirements. We will show how to quantify such a uniformity in Section 3.

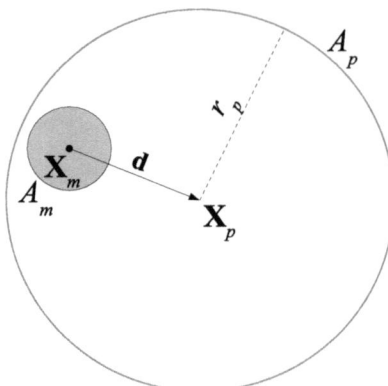

Fig. 2. Obfuscation and shift vector

With reference to Fig. 2, in order to produce an obfuscated measurement A_p, the obfuscation operator applies both an enlargement and a translation to the location measurement A_m. Intuitively, the operator enlarges the location measurement in order to decrease its precision and thus achieve the desired privacy level r_p. However, if A_m and A_p were concentric, determining the former from the latter would be trivial once the precision radius r_m is known. Therefore, the operator randomly selects a *shift vector* $\mathbf{d}$ and translates the enlarged measurement by $\mathbf{d}$, i.e., $\mathbf{X}_m + \mathbf{d} = \mathbf{X}_p$. Of course, the system has to keep the shift vector secret.

The enlargement and translation operations must be such that, when composed, the resulting obfuscation satisfies the Accuracy and Uniformity Properties. Whereas the enlargement operation is straightforward, the translation operation is instead more subtle. As to the Accuracy Property, we state the following:

Proposition 1. *Given a location measurement A_m and an obfuscation $(r_p, \mathbf{d})$, the resulting obfuscated measurement A_p fulfills the Accuracy Property iff:*

$$\|\mathbf{d}\| \leq (r_p - r_m)$$

Proof. In order to guarantee the Accuracy Property, it is necessary and sufficient that $A_m \subset A_p$. Thus, with reference to Fig. 2, the distance between $\mathbf{X}_m$ and $\mathbf{X}_p$ must not exceed the difference between the precision radius and the privacy radius, i.e., $\|\mathbf{d}\| \leq (r_p - r_m)$.

3   Adversary Model and Uniformity Index

We assume the adversary knows the obfuscated measurement $\mathbf{X}_p$, the privacy radius r_p, and the precision radius r_m. She aims at discovering the actual subject's position $\mathbf{X}$. Since $\mathbf{X}$ cannot be known with infinite precision, the result of the attack will have a probabilistic nature.

Three kinds of information could help the adversary: (i) the probability density of the *measurement error*, which depends on the sensor's characteristics, (ii) the probability density of the *shift vector*, which depends on the obfuscation operator, and (iii) the probability density of the *population*, which depends on the map's characteristics. In the following, we will consider the population's density as *irrelevant* or, equivalently, *uniform*. This is a broadly used hypothesis in obfuscation systems [1]. In fact, landscape non-neutrality can be faced by means of complementary techniques, such as enlarging the privacy radius [2].

Basing on the measurement error's density and the shift vector's density, the adversary computes the *pdf* $f_{\mathbf{X}}(x, y)$ of the subject's position. After that, she defines a *confidence goal* $c \in (0, 1]$ and computes the smallest area which contains the subject with a probability c:

Definition 2 (Smallest c-confidence area).

$$\hat{A}^c = \arg \min_{A \in \mathcal{A}^c} \{|A|\}$$

where:

$$\mathcal{A}^c = \left\{ A | A \subseteq \mathbb{R}^2, P\{\mathbf{X} \in A\} = c \right\}$$

$$P\{\mathbf{X} \in A\} = \iint_A f_{\mathbf{X}}(x, y) \, \mathrm{d}x\mathrm{d}y$$

and $|A|$ indicates the size of A.

The adversary can find the smallest c-confidence area either analytically, by algebraic calculus, or statistically, by simulating many obfuscated measurements. $\hat{A}^c$ will cover the zones where $f_{\mathbf{X}}(x, y)$ is more concentrated. It is the result of the attack, and the adversary's most precise c-confidence estimation of the position. The smaller $\hat{A}^c$ is, the more precise is the adversary in locating the subject. A good obfuscation operator should keep $\hat{A}^c$ as larger as possible for every value of c. This is done by fulfilling the Uniformity Property. The best case occurs when the Perfect Uniformity Property is fulfilled, defined as follows:

Definition 3 (Perfect Uniformity Property). *An obfuscation operator fulfills the Perfect Uniformity Property iff $f_{\mathbf{X}}(x, y)$ is perfectly uniform over A_p.*

An obfuscation operator which fulfills such a property is *ideal*. It serves only for comparisons with real operators, and it is not realizable in the general case. This is because we cannot force a particular *pdf* inside A_p if we cannot control the *pdf* inside A_m, which depends on the measurement error.

Another way to state the Perfect Uniformity is the following:

Proposition 2. *A privacy area A_p fulfills the Perfect Uniformity Property iff:*

$$\forall A \subseteq A_p, P\{\mathbf{X} \in A\} = \frac{|A|}{|A_p|} \tag{1}$$

That is, each sub-area of A_p contains the subject with a probability proportional to its size. In such a case:

$$\left|\hat{A}^c\right| = c \cdot |A_p| \qquad (2)$$

Otherwise:

$$\left|\hat{A}^c\right| < c \cdot |A_p| \qquad (3)$$

The uniformity can be quantified by means of Eq. 3, by measuring how much, for a given c, $\left|\hat{A}^c\right|$ gets close to $c \cdot |A_p|$. We define the following *uniformity index* by fixing $c = 90\%$:

Definition 4 (Uniformity index).

$$\text{unif}\,(A_p) = \frac{\left|\hat{A}^{90\%}\right|}{90\% \cdot |A_p|}$$

The constant factor in the denominator is for normalization purposes. The uniformity index ranges from 0% (worst case), if the subject's position is perfectly predictable, to 100% (best case), if the subject's position is perfectly uniform. A uniformity index of 100% is necessary and sufficient for the Perfect Uniformity.

The uniformity index has a direct practical application. For example, if an obfuscation operator produces a privacy area of $400\,\mathrm{m}^2$ with a uniformity index of 80%, the subject will be sure that an adversary cannot find his position (with 90% confidence) with more precision than $80\% \cdot 90\% \cdot 400 = 288\,\mathrm{m}^2$. In other words, the uniformity index is proportional to the lack of precision of the attack.

4   UNILO

UNILO operator adds to $\mathbf{X}_m$ a shift vector $\mathbf{d} = (\mu\cos\phi, \mu\sin\phi)$ with the following probability densities (Fig. 3):

$$f\,(\phi) = \begin{cases} \frac{1}{2\pi} & \phi \in [0, 2\pi) \\ 0 & \text{otherwise} \end{cases} \qquad (4)$$

$$f\,(\mu) = \begin{cases} 2\mu/(r_p - r_m)^2 & \mu \in [0, r_p - r_m] \\ 0 & \text{otherwise} \end{cases} \qquad (5)$$

These equations aim at producing shift vectors with uniform spacial probability density, and magnitude less than or equal to $r_p - r_m$. This will greatly improve the uniformity of $f_{\mathbf{X}}\,(x, y)$. However, remind that $f_{\mathbf{X}}\,(x, y)$ depends even on the measurement error's density, over which we have no control. So it will not be perfectly uniform in the general case. UNILO fulfills the following properties:

Accuracy Property. The privacy area always contains the subject. We give a formal proof of this.

Uniformity Property. For $r_p/r_m \geq 10$, the uniformity index is above 81%. We will prove this by simulations, in Section 5.

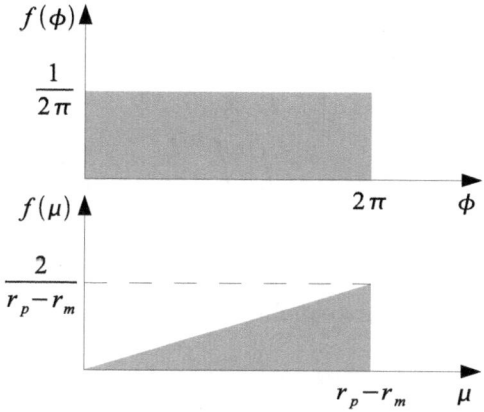

Fig. 3. ϕ and μ *pdf*s of a UNILO vector

Perfect Uniformity Property as $r_m \to 0$. With highly precise sensors, UNILO tends to be an ideal obfuscation operator. We give a formal proof of this.

Theorem 1. UNILO *fulfills Accuracy Property.*

Proof. By construction, $\|\mathbf{d}\| \le r_p - r_m$. Hence, from Prop. 1, Accuracy holds.

Theorem 2. *As* $r_m \to 0$, UNILO *fulfills Perfect Uniformity Property.*

Proof. If $r_m \to 0$, A_m will narrow to a point, with $\mathbf{X} \equiv \mathbf{X}_m$, and the probability density of the magnitude in Eq. 5 will become:

$$f(\mu) = \begin{cases} 2\mu/r_p^2 & \mu \in [0, r_p] \\ 0 & \text{otherwise} \end{cases} \tag{6}$$

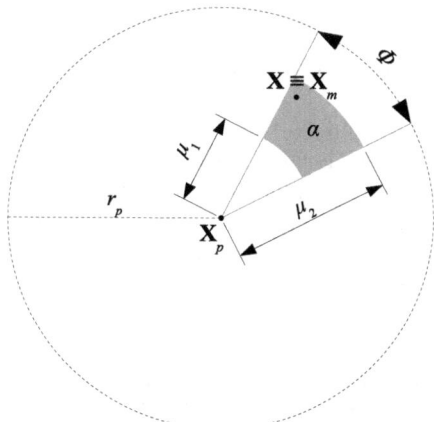

Fig. 4. Generic annular sector

Initially we prove that the hypothesis of Prop. 2 is satisfied for a generic annular sector α (Fig. 4). From Eqq. 4 and 6, and since $\mathbf{X} \equiv \mathbf{X}_m$:

$$P\{\mathbf{X} \in \alpha\} = P\{\mathbf{X}_m \in \alpha\}$$

$$= \int_0^\Phi \int_{\mu_1}^{\mu_2} \frac{2\mu}{r_p^2}\, \mathrm{d}\mu\, \frac{1}{2\pi} \mathrm{d}\phi$$

$$= \frac{\Phi}{2} \frac{(\mu_2^2 - \mu_1^2)}{\pi r_p^2}$$

Since the sizes of α and A_p are equal to:

$$|\alpha| = \frac{\Phi}{2}(\mu_2^2 - \mu_1^2)$$

$$|A_p| = \pi r_p^2$$

then:

$$P\{\mathbf{X} \in \alpha\} = \frac{|\alpha|}{|A_p|}$$

If the hypothesis of Prop. 2 holds for a generic annular sector, it holds even for a composition of annular sectors, because the total size is the sum of the sizes, and the total probability is the sum of the probabilities. Since a generic $A \subseteq A_p$ can be partitioned in a set of infinitesimal annular sectors, the hypothesis of Prop. 2 holds for each $A \subseteq A_p$. Hence, Perfect Uniformity is satisfied.

It is worth remarking that UNILO operator protects a *single* obfuscated position. If the adversary can access many obfuscated positions at different times, as it happens in tracking systems, additional protection mechanisms must be deployed. In fact, if the subject does not move or moves slowly, the adversary could overlap the different privacy areas, thus reducing the uncertainty. A common countermeasure is to reuse the same shift vector every time [5]. If the subject does not move, the adversary will receive the same privacy area, and no overlap strategy will be possible.

5   Attack Resistance Analysis

UNILO has been implemented and used to obfuscate simulated location measurements. The error on the location measurements was assumed to follow a Rayleigh distribution, as it is usually done in GPS [13]. We truncated the distribution at $r_m = 3\sigma$, so that no sample falls outside A_m. Such truncated Rayleigh distribution differs from the untruncated one for only 1.1% of samples. The tests aim at evaluating the uniformity of UNILO with respect to the ratio r_p/r_m (*radius ratio*).

Figure 5 shows the statistical distribution of $\mathbf{X}$ in A_p of 2.000 UNILO samples for different values of the radius ratio. They give a first visual impression about the uniformity of UNILO. We note that the distribution tends to be perfectly uniform as $r_p/r_m \to \infty$. The inner areas are $\hat{A}^{90\%}$.

We compared UNILO with other common obfuscation noises:

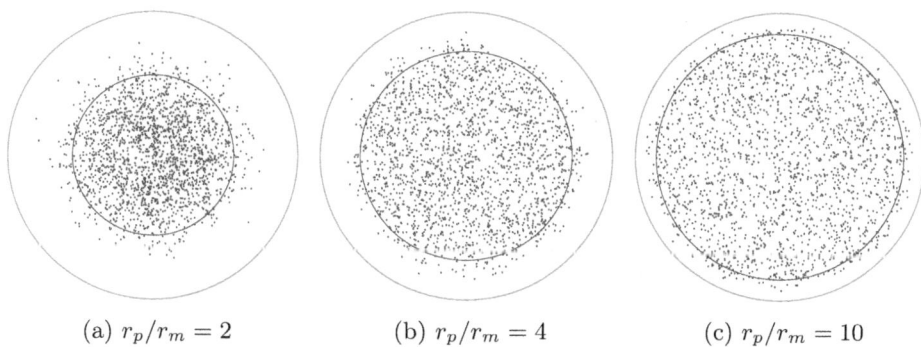

(a) $r_p/r_m = 2$ (b) $r_p/r_m = 4$ (c) $r_p/r_m = 10$

Fig. 5. 2.000-sample simulations

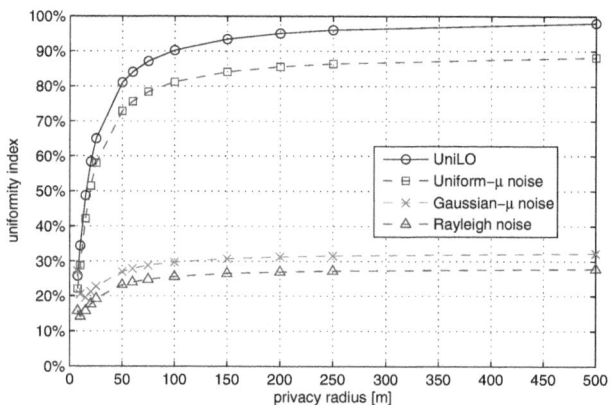

Fig. 6. unif (A_p) with $r_m = 5\,\mathrm{m}$

- A Rayleigh noise (i.e. gaussian X - gaussian Y), used for modeling 2-dimensional measurement errors. The Rayleigh distribution is truncated at $r_p - r_m$, in order to fulfill Accuracy Property. The σ parameter is fixed at $(r_p - r_m)/3$.
- A gaussian-μ noise (i.e. uniform angle - gaussian magnitude), used by Krumm to perturb GPS data [14]. The gaussian distribution is truncated at $r_p - r_m$, in order to fulfill Accuracy Property. The σ parameter is fixed at $(r_p - r_m)/3$.
- A uniform-μ noise (i.e. uniform angle - uniform magnitude). This is the simplest two-dimensional noise.

Figure 6 shows the uniformity indexes of the noises. Each uniformity index estimation was obtained by means of 50 million samples. As we told in Section 4, UNILO offers a uniformity index above 81% for $r_p/r_m \geq 10$. We can see how UNILO performs better than all the other noises for all the radius ratii. In

Fig. 7. Employee localizer screenshot

particular, gaussian-magnitude and Rayleigh-magnitude noises are particularly bad for obfuscating. We believe this is the reason why Krumm needed a surprisingly high quantity of noise ($\sigma = 5\,\mathrm{Km}$) to effectively withstand inference attacks [14].

6   Service Examples

UNILO operator has the advantage to be transparent to the service provider, in the sense that a privacy area has the same properties as an ordinary measurement area. A software service provider designed for receiving non-obfuscated inputs can be seamlessly adapted for receiving UNILO-obfuscated inputs. The following subsections describe some examples of services which can be deployed over UNILO operator.

6.1   Employee Localizer

The aim is to retrieve the instantaneous locations of a set of employees to better coordinate work operations. Before giving their consensus, employees specify their privacy radii. A software service provider displays the locations on the monitor of a human operator, in the form of circles on a map. Each circle is larger or smaller depending on the privacy radius. The privacy radius may depend on context-based rules. For example, an employee may require a high privacy radius when standing in some zones of the map and a small one when standing in others. Figure 7 shows a screenshot of such a service, taken from a practical implementation.

6.2   Find the Near Friends

This is a social application, in which the users share their obfuscated positions with their friends. Alice wants to find out which of her friends are in her

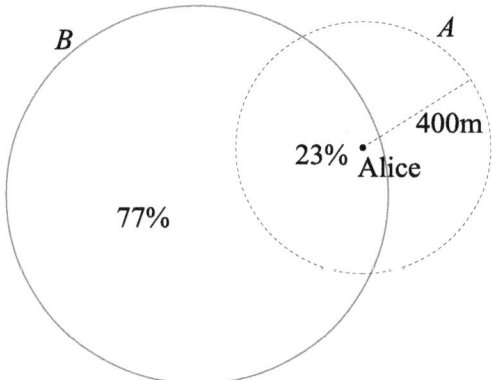

Fig. 8. Find the near friends

proximity. We define "being in the proximity of Alice" as "being at a distance of 400 meters or less from Alice". In this case Alice is the service provider and her friends are the subjects. While Alice knows its own position, the locations of her friends are obfuscated. Suppose Bob is one of Alice's friends. Since Alice does not know his exact location, the question "is Bob in my proximity?" will necessarily have a probabilistic answer, like "60% yes, 40% no".

The problem can be modeled as depicted in Fig. 8. Alice builds a circle centered on its position and with 400 meters of radius (*proximity circle*, A), and computes the intersection between that circle and the privacy circle of Bob (B). If Bob is inside this intersection, he will be in Alice's proximity. The probability that such an event happens is:

$$P\{\text{Bob is in Alice's proximity}\} = \iint_{A \cap B} f(x, y) \, \mathrm{d}x\mathrm{d}y \qquad (7)$$

Alice can numerically compute such an integral to find out the probability. If the privacy area of Bob can be assumed as perfectly uniform, the Eq. 7 will become:

$$P\{\text{Bob is in Alice's proximity}\} = \frac{|A \cap B|}{|B|}$$

In the figure, such probability is 23%. The service provider performs this calculus only for each friend whose X_p is nearer than $r_p + 400\,$m. The others have no intersection, and thus 0% probability. Alice finds an answer like the following:

- Bob is in the proximity with 23% probability.
- Carol with 10% probability.
- Dave with 100% probability.
- All the others with 0% probability.

6.3   Find the Nearest Taxi

Alice calls a taxi and releases her obfuscated GPS position in order to speed-up the procedure. The taxi company knows the positions of the available taxis.

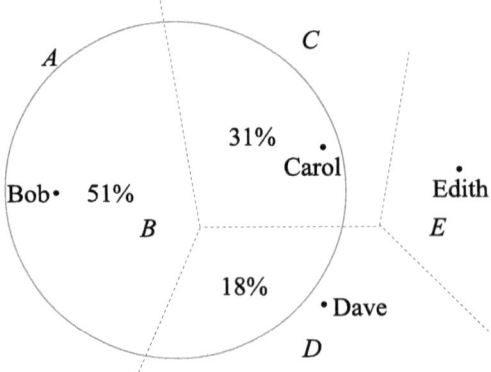

Fig. 9. Find the nearest taxi

Then, it finds the one which is probabilistically the nearest to Alice, and forwards the request to it. In this way, only the taxi driver needs to know Alice's exact position.

The problem can be modeled as depicted in Fig. 9, by means of a Voronoi diagram. Each region of the diagram corresponds to a taxi. Let us call the taxi drivers Bob (region B), Carol (C), Dave (D) and Edith (E). If Alice is inside B, Bob's will be the nearest taxi, and so on. Fortune's algorithm [11] can compute the Voronoi diagram in $\mathcal{O}(n \log n)$ time, where n is the number of taxis. The taxi company obtains the probabilities by simply integrating $f(x,y)$ over the intersections between the privacy area and the Voronoi regions. If the privacy area of Alice can be assumed as perfectly uniform, the integral becomes a simple area ratio, like in Subsection 6.2. In the figure, the taxi company will obtain the following probabilities:

- Bob's taxi is the nearest with 51% probability.
- Carol's taxi with 31% probability.
- Dave's taxi with 18% probability.
- All the others with 0% probability.

The taxi company will then forward the request to Bob.

7   Related Works

Conway and Strip published a seminal work about general-purpose database-oriented obfuscation methods [5]. The authors introduced two obfuscation approaches that, with some generalization, have been used until today: *value distortion*, which perturbs value with a random noise; and *value-class member-ship*, which partitions the whole value domain in classes, and discloses only the class where the value is in.

Gruteser and Grunwald first approached k-anonymity problem in location-based services. The proposed solution involves the subdivision of the map in static quadrants with different granularities [12]. Mascetti et al. proposed an obfuscation method that divides the map in quadrants like [12], but it does not aim at k-anonymity [15]. It focuses only on data obfuscation and proximity services.

Duckham and Kulik took a radically different approach, that models a map as an adjacency graph, where the vertices represent zones of the map and the edges the adjacency between two zones [7]. A graph modelization is more powerful in some applications, because it can model obstacles, unreachable zones and hardly viable passages through edge costs. The obfuscation method reveals a set of nodes where the subject could be. Proximity services are realized by means of Dijkstra-like algorithms. Shokri et al. took a similar approach, and involves also the anonymization of the subjects [20]. A drawback is that a graph-based description of the map must be available, and shared between the subjects and the service providers. Calculating a graph model of a geographic map that is both simple and accurate may be not trivial. Another drawback is that the proximity services are not based on simple and efficient Voronoi diagrams (cfr. Section 6), but they have to involve more complex Dijkstra-like algorithms.

Ardagna et al. proposed a set of obfuscation operators that perturb the location: radius enlargement, radius restriction, center shift [1]. These operators transform a measurement area into an obfuscated one. To the best of our knowledge, this is the most similar work to our approach, but it contains relevant differences with respect to UNILO in the initial requirements and the final results:

- The subject's actual location could be outside the obfuscated area. This happens in case of radius reduction or center shift operators. Thus, the obfuscation introduces inaccuracy which does not allow the service provider to offer some services, like those described in Section 6. In contrast, UNILO always guarantees that the obfuscated area contains the subject.
- The quantity of privacy is measured by a parameter, called *relevance*, which is quite unintuitive. Final users prefer parameters they can easily understand such as the *privacy radius* used by UNILO. If a user specifies a privacy radius of 100 m, then he means that he wishes to be located with a precision not better than 100 m. Relevance has not a 1-to-1 relationship with the privacy radius: the same relevance corresponds to a small privacy radius if the location technology is precise, or to a larger one if is imprecise.
- The resistance against attacks relies on the fact that the system chooses the obfuscation operators at random. However, the adversary can make probabilistic hypothesis on them. This possibility is not investigated. De facto, the adversary is assumed to be unaware of the obfuscation method. This is an optimistic assumption, which features a form of security by obscurity that should be avoided [19].

8 Conclusions and Future Works

We have proposed UNILO, an obfuscation operator for location data, which adds a special random noise which maximizes probability uniformity. UNILO is simple and $\mathcal{O}\,(1)$-complex. We have presented an adversary model which performs statistical-based attacks. We have shown that the property of uniformity neutralizes such attacks. We have proved the resistance of UNILO in terms of uniformity, through both formal analysis and experimental results. To the best of our knowledge, UNILO is the first obfuscation operator which offers guarantees on uniformity.

The work leaves space for extensions to noncircular or nonplanar location measurements, extensions for tracking systems, and extensions to offer multiple contemporaneous levels of privacy.

Acknowledgment. This work has been supported by the EU-funded Integrated Project PLANET "PLAtform for the deployment and operation of heterogeneous NETworked cooperating objects," and the Network of Excellence CONET "Cooperating Objects Network of Excellence."

References

1. Ardagna, C.A., Cremonini, M., De Capitani di Vimercati, S., Samarati, P.: An obfuscation-based approach for protecting location privacy. IEEE Transactions on Dependable and Secure Computing 8(1), 13–27 (2011)
2. Ardagna, C.A., Cremonini, M., Gianini, G.: Landscape-aware location-privacy protection in location-based services. Journal of Systems Architecture 55(4), 243–254 (2009)
3. Barkuus, L., Dey, A.: Location-based services for mobile telephony: a study of users privacy concerns. In: Proceedings of the INTERACT 2003, 9th IFIP TC13 International Conference on Human-Computer Interaction, pp. 709–712 (July 2003)
4. Beresford, A.R., Stajano, F.: Location privacy in pervasive computing. IEEE Pervasive Computing 2(1), 46–55 (2003)
5. Conway, R., Strip, D.: Selective Partial Access to a Database. In: Proceedings of the 1976 Annual Conference, pp. 85–89. ACM (1976)
6. D'Roza, T., Bilchev, G.: An overview of location-based services. BT Technology Journal 21(1), 20–27 (2003)
7. Duckham, M., Kulik, L.: A Formal Model of Obfuscation and Negotiation for Location Privacy. In: Gellersen, H.-W., Want, R., Schmidt, A. (eds.) PERVASIVE 2005. LNCS, vol. 3468, pp. 152–170. Springer, Heidelberg (2005)
8. Duckham, M., Mason, K., Stell, J., Worboys, M.: A formal approach to imperfection in geographic information. Computer, Environment and Urban Systems 25, 89–103 (1999)
9. Duri, S., Gruteser, M., Liu, X., Moskowitz, P., Perez, R., Singh, M., Tang, J.M.: Framework for security and privacy in automotive telematics. In: Proceedings of the 2nd International Workshop on Mobile Commerce, pp. 25–32. ACM (2002)
10. Espinoza, F., Persson, P., Sandin, A., Nyström, H., Cacciatore, E., Bylund, M.: GeoNotes: Social and navigational aspects of location-based information systems. Tech. Rep. T2001/08, Swedish Institute of Computer Science (SICS) (May 2001)
11. Fortune, S.: A sweepline algorithm for voronoi diagrams. In: Proceedings of the Second Annual ACM SIGACT/SIGGRAPH Symposium on Computational Geometry, SCG 1986, pp. 313–322. ACM (1986)

12. Gruteser, M., Grunwald, D.: Anonymous Usage of Location-Based Services Through Spatial and Temporal Cloaking. In: Proceedings of the MobiSys 2003: 1st International Conference on Mobile Systems, Applications and Services, pp. 31–42 (2003)
13. Hofmann-Wellenhof, B., Lichtenegger, H., Collins, J.: Global Positioning System: Theory and Practice. Springer (2001)
14. Krumm, J.: A survey of computational location privacy. Personal and Ubiquitous Computing 13(6), 391–399 (2008)
15. Mascetti, S., Bettini, C., Freni, D., Wang, X.S., Jajodia, S.: Privacy-Aware Proximity Based Services. In: Proceedings of the MDM 2009: 10th International Conference on Mobile Data Management: Systems, Services and Middleware, pp. 31–40. IEEE (2009)
16. Myles, G., Friday, A., Davies, N.: Preserving privacy in environments with location-based applications. IEEE Pervasive Computing 2(1), 56–64 (2003)
17. Pal, A.: Localization algorithms in wireless sensor networks: Current approaches and future challenges. Network Protocols and Algorithms 2(1), 45–74 (2010)
18. Samarati, P., Sweeney, L.: Protecting privacy when disclosing information: k-anonymity and its enforcement through generalization and suppression. Tech. rep., Computer Science Laboratory SRI International (1998)
19. Schneier, B.: Secrecy, security, and obscurity (May 2002), http://www.schneier.com/crypto-gram-0205.html
20. Shokri, R., Freudiger, J., Jadliwala, M., Hubaux, J.P.: A distortion-based metric for location privacy. In: Proceedings of the 8th ACM Workshop on Privacy in the Electronic Society, WPES 2009, pp. 21–30. ACM (2009)
21. Zandbergen, P.A.: Accuracy of iPhone locations: A comparison of assisted GPS, WiFi and cellular positioning. Transactions in GIS 13(s1), 5–26 (2009)

Security Vulnerabilities of User Authentication Scheme Using Smart Card

Ravi Singh Pippal[1], Jaidhar C.D.[2], and Shashikala Tapaswi[1],[*]

[1] ABV-Indian Institute of Information Technology and Management,
Gwalior-474015, India
{ravi,stapaswi}@iiitm.ac.in
[2] Defence Institute of Advanced Technology, Girinagar,
Pune-411025, India
jaidharcd@diat.ac.in

Abstract. With the exponential growth of Internet users, various business transactions take place over an insecure channel. To secure these transactions, authentication is the primary step that needs to be passed. To overcome the problems associated with traditional password based authentication methods, smart card authentication schemes have been widely used. However, most of these schemes are vulnerable to one or the other possible attack. Recently, Yang, Jiang and Yang proposed RSA based smart card authentication scheme. They claimed that their scheme provides security against replay attack, password guessing attack, insider attack and impersonation attack. This paper demonstrates that Yang et al.'s scheme is vulnerable to impersonation attack and fails to provide essential features to satisfy the needs of a user. Further, comparative study of existing schemes is also presented on the basis of various security features provided and vulnerabilities present in these schemes.

Keywords: Authentication, Cryptanalysis, Impersonation, Password, Smart card.

1 Introduction

Remote user authentication is used to verify the legitimacy of a remote user and it is mandatory for most of the applications like online banking, ID verification, medical services, access control and e-commerce. One among various authentication schemes is password based authentication scheme. In traditional password based authentication schemes, server keeps verification table securely to verify the legitimacy of a user. However, this method is insecure since an attacker may access the contents of the verification table to break down the entire system. Lamport [1] proposed password authentication scheme to authenticate remote users by storing the passwords in a hashed format. Nevertheless, this scheme has a security drawback as an intruder can go through the server and modify the contents of the verification table. To resist all possible attacks on the

---
[*] Corresponding author.

N. Cuppens-Boulahia et al. (Eds.): DBSec 2012, LNCS 7371, pp. 106–113, 2012.

verification tables, smart card based password authentication scheme has been proposed. This scheme eliminates the use of verification table.

Today, authentication based on smart card is employed continuously in several applications like cloud computing, healthcare, key exchange in IPTV broadcasting, wireless networks, authentication in multi-server environment, wireless sensor networks and many more. Hence, it is necessary that the authentication scheme must be efficient as well as secure enough so that it can be utilized for practical applications.

1.1 Contribution of This Paper

Recently, Yang et al. [20] proposed an access control scheme using smart card. This paper demonstrates that Yang et al.'s scheme has following weaknesses: (i) unauthorized user can easily forge a valid login request. (ii) user is not able to choose and change the password freely. (iii) it does not provide mutual authentication, session key generation and early wrong password detection. (iv) it fails to solve time synchronization problem. Further, comparative study of existing schemes is also done on the basis of various security features provided and vulnerabilities present in these schemes.

The remainder of this paper is organized as follows. The existing literature related to smart card authentication schemes is explored in section 2. Section 3 describes a brief review of Yang et al.'s access control scheme using smart card. Security flaws of Yang et al.'s scheme along with comparison of existing schemes based on various security features and attacks are presented in section 4. Finally, section 5 concludes the paper.

2 Literature Review

Throughout the last two decades, various smart card authentication schemes have been proposed [2, 4, 6–9, 11–13, 15, 17, 20]. However, most of these schemes fail to fulfill the essential requirements of users. Hwang and Li [2] presented a remote user authentication scheme using ElGamal's cryptosystem and claimed that their scheme is free from maintaining verification table and able to resist replay attack. Chan and Cheng [3] found that Hwang-Li's scheme is vulnerable to impersonation attack. To improve efficiency, Sun [4] suggested a remote user authentication scheme using one way hash function. However, Hsu [5] proved that Sun's scheme is insecure against offline and online password guessing attacks. To handle these flaws, Chien et al. [6] proposed remote user authentication scheme using one-way hash function. Nevertheless, it exhibits parallel session attack [5]. To defend against insider attack and reflection attack over [6], Ku and Chen [7] presented an improved scheme which also provides the facility to change the password freely. But, it is found that the scheme is weak against parallel session attack and has insecure password change phase [8]. Further improvement has also been suggested by Yoon et al. [8]. However, the improved scheme remains vulnerable to guessing attack, Denial-of-Service attack and impersonation attack

[9]. To remedy these drawbacks, Wang et al. [9] proposed an enhanced scheme. Though, the scheme is weak against guessing attack, denning sacco attack and does not offer perfect forward secrecy [10].

Das et al. [11] offered a dynamic ID based remote user authentication scheme using one way hash function. They claimed that their scheme is secure against ID theft and able to withstand replay attack, forgery attack, guessing attack, insider attack and stolen verifier attack. Though, the scheme is weak against guessing attack [12] and insider attack [12,13]. Additionally, the scheme is password independent [13] and does not provide mutual authentication [12, 13]. To beat these flaws, Wang et al. [13] suggested an improved scheme. However, Ahmed et al. [14] found that the scheme does not provide security against password guessing attack, masquerade attack and Denial-of-Service attack. An enhanced scheme has also been given to resist password guessing attack, user masquerade attack and server masquerade attack [15]. Nevertheless, the scheme is exposed to password guessing attack, server masquerade attack and lack of password backward security [16]. Song [17] proposed symmetric key cryptography based smart card authentication scheme and claimed that the scheme is able to resist the existing potential attacks. In addition, it provides mutual authentication and shared session key. However, Song's scheme fails to provide early wrong password detection [18] and perfect forward secrecy [18, 19]. Moreover, it does not resist offline password guessing attack and insider attack [19]. All the schemes discussed so far have their pros and cons. Recently, Yang, Jiang and Yang [20] proposed RSA based smart card authentication scheme. The authors claimed that their scheme has the ability to withstand existing attacks. Though, this paper proves that Yang et al.'s scheme is exposed to impersonation attack and does not provide essential features.

3   Review of Yang et al.'s Scheme

This section briefly reviews Yang et al.'s access control scheme using smart card [20]. The notations used throughout this paper are summarized in Table 1. The scheme consists of four phases: Initialization phase, Registration phase, Login phase and Authentication phase. Three phases are shown in Fig. 1.

Table 1. Notations used in this paper

Symbols	Their meaning	Symbols	Their meaning
U_i	Remote user	T_A	Attacker time stamp
S	Authentication server	$\phi(N)$	Euler's totient function
U_A	Attacker	$H(\cdot)$	Collision-resistant hash function
ID_i	Identity of U_i	$\parallel$	Message concatination
PW_i	Password generated by S	--→	Secure channel
T_C	User time stamp	⟶	Insecure channel

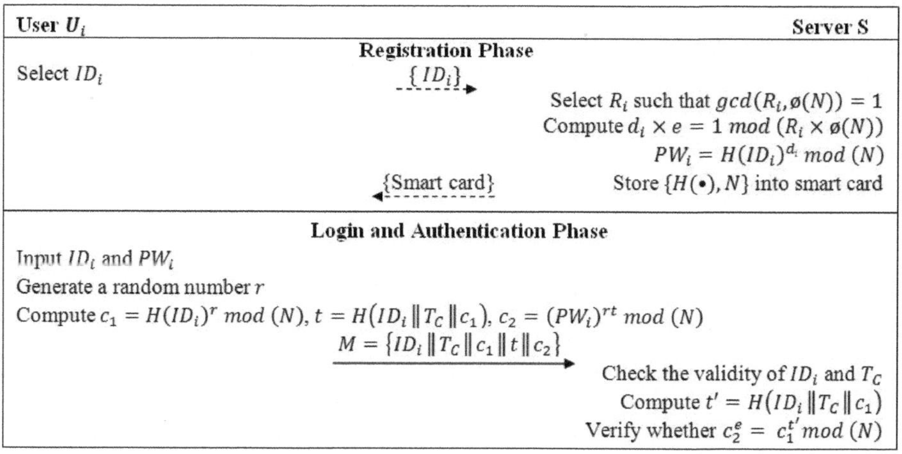

User U_i		Server S
	Registration Phase	
Select ID_i	$\xrightarrow{\{ID_i\}}$	Select R_i such that $gcd(R_i, \emptyset(N)) = 1$
		Compute $d_i \times e = 1 \ mod \ (R_i \times \emptyset(N))$
		$PW_i = H(ID_i)^{d_i} \ mod \ (N)$
	$\xleftarrow{\{Smart \ card\}}$	Store $\{H(\bullet), N\}$ into smart card
	Login and Authentication Phase	
Input ID_i and PW_i		
Generate a random number r		
Compute $c_1 = H(ID_i)^r \ mod \ (N), t = H(ID_i \| T_C \| c_1), c_2 = (PW_i)^{rt} \ mod \ (N)$		
$M = \{ID_i \| T_C \| c_1 \| t \| c_2\}$		
$\xrightarrow{\hspace{4cm}}$		Check the validity of ID_i and T_C
		Compute $t' = H(ID_i \| T_C \| c_1)$
		Verify whether $c_2^e = c_1^{t'} mod \ (N)$

Fig. 1. Yang et al.'s scheme

3.1 Initialization Phase

In this phase, server S generates the following system parameters.
$N : N = p \times q$ such that $p = 2p_1 + 1$, $q = 2q_1 + 1$, where p, q, p_1, q_1 are all primes.
e : Secret key of the system satisfying $gcd(e, \phi(N)) = 1$.

3.2 Registration Phase

In this phase, U_i selects ID_i and submits it to S over a secure channel. Upon receiving the registration request from U_i, S selects R_i such that $gcd(R_i, \phi(N)) = 1$ and computes d_i such that $d_i \times e = 1 \ mod(R_i \times \phi(N))$, U_i's password $PW_i = H(ID_i)^{d_i} mod(N)$ and delivers PW_i as well as smart card over secure channel to U_i by storing $\{H(\cdot), N\}$ into smart card memory.

3.3 Login Phase

U_i inserts the smart card to the card reader and keys in ID_i and PW_i. The card reader generates a random number r, computes $c_1 = H(ID_i)^r mod(N)$, $t = H(ID_i \| T_C \| c_1)$, $c_2 = (PW_i)^{rt} mod(N)$ and sends the login request $M = \{ID_i \| T_C \| c_1 \| t \| c_2\}$ to S.

3.4 Authentication Phase

Upon receiving the login request $M = \{ID_i \| T_C \| c_1 \| t \| c_2\}$; S first checks the validity of ID_i and T_C to accept/reject the login request. If true, S computes $t' = H(ID_i \| T_C \| c_1)$ and checks whether $c_2^e = c_1^{t'} mod(N)$ holds or not. If it holds, S accepts the login request M otherwise rejects it.

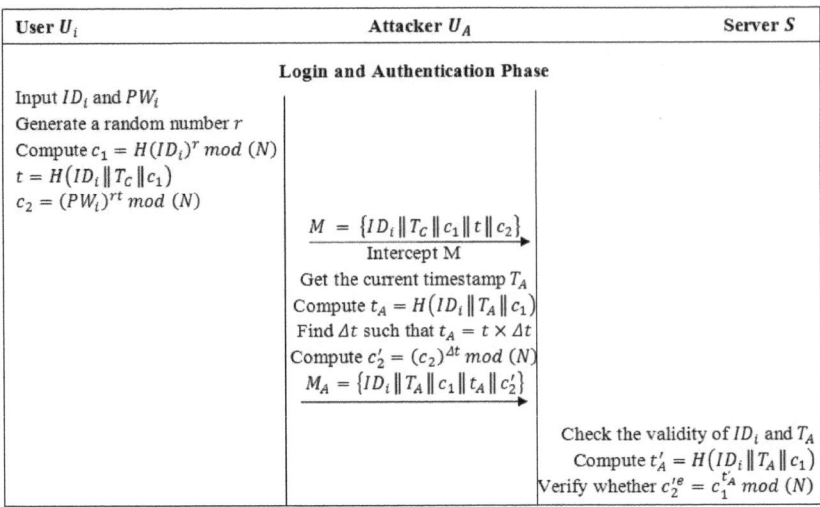

Fig. 2. Impersonation attack on Yang et al.'s scheme

4   Weaknesses Present in Yang et al.'s Scheme

This section demonstrates the security flaws in Yang et al.'s scheme under the assumption that the attacker is able to intercept all the messages exchanged between U_i and S. It is found that this scheme has following weak spots: (i) vulnerable to impersonation attack, (ii) no early wrong password detection, (iii) no mutual authentication and (iv) no session key generation. In addition, this scheme fails to solve time synchronization problem and does not allow users to choose and change the password freely.

4.1   Vulnerable to Impersonation Attack

First, an attacker U_A intercepts the login request $\{ID_i \parallel T_C \parallel c_1 \parallel t \parallel c_2\}$ transmitted from user U_i to server S (as shown in Fig. 2). U_A gets the current timestamp T_A, computes $t_A = H(ID_i \parallel T_A \parallel c_1)$ and finds a value Δt such that $t_A = t \times \Delta t$. After getting Δt, U_A computes $c_2' = (c_2)^{\Delta t} mod(N) = (PW_i)^{r \times t_A} mod(N)$ and sends forged login request $M_A = \{ID_i \parallel T_A \parallel c_1 \parallel t_A \parallel c_2'\}$ to S. Once the request M_A is received, S computes $t_A' = H(ID_i \parallel T_A \parallel c_1) = t_A$ and verifies whether $c_2'^e = c_1^{t_A'} mod(N)$ or not which is obviously true. Hence, U_A is able to impersonate as legitimate user U_i.

4.2   No Early Wrong Password Detection

To prevent Denial-of-Service attack, password needs to be verified at the user side prior to login request creation. In this scheme, adversary can create invalid login request by entering wrong password which will be detected only at the server side not at the user side. Hence, it leads to Denial-of-Service attack.

4.3   No Mutual Authentication

It is necessary that not only server verifies the legal users, but users also need to verify the identity of the legal server to achieve two way secure communication. In this scheme, only the login request is verified at the server side to verify the legitimacy of the user. Hence, this scheme fails to provide mutual authentication. Further, session key is used to secure the entire communication between them and it must be changed from session to session. In this scheme, there is no session key generation.

In timestamp-based authentication schemes, the clock of the server and all registered user systems need to be synchronized. In addition, transmission delay of the login request needs to be limited. However, it is inefficient from the practical point of view specially for a large network where clock synchronization is hard to achieve. Yang et al.'s scheme fails to solve this problem. Moreover, users are not able to choose the password as per their convenience. They must remember the password issued by the server which causes inconvenience. Further, they are not able to change the password whenever they feel.

4.4   Performance Comparison

This paper identifies the possible attacks for existing smart card authentication schemes. These include (i) impersonation attack (SA1), (ii) replay attack (SA2), (iii) password guessing attack (SA3), (iv) reflection attack (SA4), (v) parallel session attack (SA5), (vi) insider attack (SA6) and (vii) attack on password change phase (SA7). A comparison is presented in Table 2 based on the ideas given by different authors.

Further, essential security features that have to be offered by any authentication scheme is also spotted out. These features include (i) user chooses the password (SF1), (ii) user changes the password (SF2), (iii) early wrong password detection (SF3), (iv) mutual authentication (SF4), (v) session key generation (SF5) and (vi) free from time synchronization problem (SF6). A comparative study of existing schemes is given in Table 3 on the basis of these security features.

Table 2. Comparison based on various security attacks

Security Attacks	SA1	SA2	SA3	SA4	SA5	SA6	SA7
Hwang-Li [2]	Insecure[3]	Secure	Secure	NA	NA	NA	NA
H. M. Sun [4]	Secure	Secure	Insecure[5]	NA	NA	NA	NA
Chien et al. [6]	Secure	Secure	Insecure[7]	Insecure[7]	Insecure[5]	Insecure[7]	NA
Ku-Chen [7]	Insecure[9]	Secure	Insecure[9]	Secure	Insecure[8]	Secure	Insecure[8]
Yoon et al. [8]	Insecure[9]	Secure	Insecure[9]	Secure	Secure	Secure	Secure
Wang et al. [9]	Secure	Secure	Insecure[10]	Secure	Secure	Secure	Secure
Das et al. [11]	Secure	Secure	Insecure[12]	NA	NA	Insecure[12],[13]	Insecure[14]
Wang et al. [13]	Insecure[14]	Secure	Insecure[14]	Secure	Secure	NA	Insecure[14]
Hao-Yu [15]	Secure	Secure	Insecure[16]	Secure	Secure	NA	Secure
R. Song [17]	Secure	Secure	Insecure[19]	Secure	Secure	Insecure[19]	Secure
Yang et al. [20]	Insecure	Secure	Secure	NA	NA	NA	NA

[x] As per the reference [x]

Table 3. Comparison based on various security features provided

Security Features	SF1	SF2	SF3	SF4	SF5	SF6
Hwang-Li [2]	No	No	No	No	No	No
H. M. Sun [4]	No	No	No	No	No	No
Chien et al. [6]	Yes	No	No	Yes	No	No
Ku-Chen [7]	Yes	Yes	No	Yes	No	No
Yoon et al. [8]	Yes	Yes	No	Yes	No	No
Wang et al. [9]	Yes	Yes	Yes	Yes	Yes	No
Das et al. [11]	Yes	Yes	No	No	No	No
Wang et al. [13]	No	Yes	No	Yes	No	No
Hao-Yu [15]	No	Yes	No	Yes	No	No
R. Song [17]	Yes	Yes	No	Yes	Yes	No
Yang et al. [20]	No	No	No	No	No	No

From both of these tables, it is clear that none of these schemes offer protection against identified attacks and fulfill the needs of a user.

5   Conclusion

Authentication is the imperative factor for any scheme that deals with the transmission of secret information over a public network. This paper pointed out that Yang et al.'s scheme has security flaws as an intruder can easily impersonate legal users to pass the authentication phase. Moreover, it does not allow users to choose and change the password freely which results inconvenience from the user's point of view. In addition, it does not provide mutual authentication, session key generation, early wrong password detection and fails to solve time synchronization problem. Hence, the scheme is computationally inefficient as well as insecure for practical applications.

Furthermore, performance comparison of existing smart card authentication schemes is also presented which shows that a lot of work has to be done in this field to provide secure and efficient authentication scheme. Before designing any authentication scheme, the identified security attacks must be taken into consideration along with the requirements desired by the end users.

Acknowledgments. The authors would like to thank ABV-Indian Institute of Information Technology and Management, Gwalior, India for providing the academic support.

References

1. Lamport, L.: Password authentication with insecure communication. Communications of the ACM 24, 770–772 (1981)
2. Hwang, M.S., Li, L.H.: A new remote user authentication scheme using smart cards. IEEE Transactions on Consumer Electronics 46, 28–30 (2000)
3. Chan, C.K., Cheng, L.M.: Cryptanalysis of a remote user authentication scheme using smart cards. IEEE Transactions on Consumer Electronics 46, 992–993 (2000)

4. Sun, H.M.: An efficient remote user authentication scheme using smart cards. IEEE Transactions on Consumer Electronics 46, 958–961 (2000)
5. Hsu, C.L.: Security of two remote user authentication schemes using smart cards. IEEE Transactions on Consumer Electronics 49, 1196–1198 (2003)
6. Chien, H.Y., Jan, J.K., Tseng, Y.M.: An efficient and practical solution to remote authentication: smart card. Computers and Security 21, 372–375 (2002)
7. Ku, W.C., Chen, S.M.: Weaknesses and improvements of an efficient password based remote user authentication scheme using smart cards. IEEE Transactions on Consumer Electronics 50, 204–207 (2004)
8. Yoon, E.J., Ryu, E.K., Yoo, K.Y.: Further improvement of an efficient password based remote user authentication scheme using smart cards. IEEE Transactions on Consumer Electronics 50, 612–614 (2004)
9. Wang, X.M., Zhang, W.F., Zhang, J.S., Khan, M.K.: Cryptanalysis and improvement on two efficient remote user authentication scheme using smart cards. Computer Standards and Interfaces 29, 507–512 (2007)
10. Yoon, E.J., Lee, E.J., Yoo, K.Y.: Cryptanalysis of Wang et al.'s remote user authentication scheme using smart cards. In: 5th International Conference on Information Technology: New Generations, Las Vegas, USA, pp. 575–580 (2008)
11. Das, M.L., Saxena, A., Gulati, V.P.: A dynamic ID-based remote user authentication scheme. IEEE Transactions on Consumer Electronics 50, 629–631 (2004)
12. Liao, I.E., Lee, C.C., Hwang, M.S.: Security enhancement for a dynamic ID-based remote user authentication scheme. In: International Conference on Next Generation Web Services Practices, Seoul, Korea, pp. 437–440 (2005)
13. Wang, Y.Y., Liu, J.Y., Xiao, F.X., Dan, J.: A more efficient and secure dynamic ID-based remote user authentication scheme. Computer Communications 32, 583–585 (2009)
14. Ahmed, M.A., Lakshmi, D.R., Sattar, S.A.: Cryptanalysis of a more efficient and secure dynamic id-based remote user authentication scheme. International Journal of Network Security and its Applications 1, 32–37 (2009)
15. Hao, Z., Yu, N.: A security enhanced remote password authentication scheme using smart card. In: 2nd International Symposium on Data, Privacy and E-Commerce, Buffalo, USA, pp. 56–60 (2010)
16. Zhang, H., Li, M.: Security vulnerabilities of an remote password authentication scheme with smart card. In: 2011 International Conference on Consumer Electronics, Communications and Networks, XianNing, China, pp. 698–701 (2011)
17. Song, R.: Advanced smart card based password authentication protocol. Computer Standards and Interfaces 32, 321–325 (2010)
18. Pippal, R.S., Jaidhar, C.D., Tapaswi, S.: Comments on symmetric key encryption based smart card authentication scheme. In: 2nd International Conference on Computer Technology and Development, Cairo, Egypt, pp. 482–484 (2010)
19. Horng, W.B., Lee, C.P., Peng, J.W.: Security weaknesses of Song's advanced smart card based password authentication protocol. In: 2010 IEEE International Conference on Progress in Informatics and Computing, Shanghai, China, pp. 477–480 (2010)
20. Yang, C., Jiang, Z., Yang, J.: Novel access control scheme with user authentication using smart cards. In: 3rd International Joint Conference on Computational Science and Optimization, Huangshan, China, pp. 387–389 (2010)

Secure Password-Based Remote User Authentication Scheme with Non-tamper Resistant Smart Cards

Ding Wang[1,2,*], Chun-guang Ma[1,**], and Peng Wu[1]

[1] Harbin Engineering University, Harbin City 150001, China
[2] Automobile Management Institute of PLA, Bengbu City 233011, China
wangdingg@mail.nankai.edu.cn, chunguangma@hrbeu.edu.cn

Abstract. In DBSec'11, Li et al. showed that Kim and Chung's password-based remote user authentication scheme is vulnerable to various attacks if the smart card is non-tamper resistant. Consequently, an improved version was proposed and claimed that it is secure against smart card security breach attacks. In this paper, however, we will show that Li et al.'s scheme still cannot withstand offline password guessing attack under the non-tamper resistance assumption of the smart card. In addition, their scheme is also prone to denial of service attack and fails to provide user anonymity and forward secrecy. Therefore, a robust scheme with a brief analysis is presented to overcome the identified drawbacks.

Keywords: Cryptanalysis, Network security, Authentication protocol, Smart card, Non-tamper resistant, User anonymity.

1 Introduction

Password-based authentication is widely used for systems that control remote access to computer networks. In order to address some of the security and management problems that occur in traditional password authentication protocols, research in recent decades has focused on smart card based password authentication. Since Chang and Wu [1] introduced the first remote user authentication scheme using smart cards in 1993, there have been many smart card based authentication schemes proposed. In most of the previous authentication schemes, the smart card is assumed to be tamper-resistant, i.e., the secret information stored in the smart card cannot be revealed. However, recent research results have shown that the secret data stored in the smart card could be extracted by some means, such as monitoring the power consumption [2,3] or analyzing the leaked information [4]. Therefore, such schemes based on the tamper resistance assumption of the smart card are vulnerable to some types of attacks, such as impersonation attacks, offline password guessing attacks, etc., once an adversary

---

* This is the extended abstract and a full version [7] of this paper is available at http://machunguang.hrbeu.edu.cn/Research/
** Corresponding author.

N. Cuppens-Boulahia et al. (Eds.): DBSec 2012, LNCS 7371, pp. 114–121, 2012.
© IFIP International Federation for Information Processing 2012

has obtained the secret information stored in a user's smart card and/or just some intermediate computational results in the smart card.

In DBSec'11, Li et al. [5] identified that Kim and Chung's scheme [6] cannot withstand various attacks and further proposed an enhanced remote authentication scheme. They claimed their scheme is secure and can overcome all the identified security flaws of Kim and Chung's scheme even if the smart card is non-tamper resistant. In this work, however, we will demonstrate that Li et al.'s scheme cannot withstand denial of service attack, and it is still vulnerable to off-line password guessing attack under their assumption. In addition, their scheme does not provide forward secrecy and user anonymity. To conquer the identified weaknesses, a robust authentication scheme based on the secure one-way hash function and the well-known discrete logarithm problem is presented.

2   Review of Li et al.'s Scheme

In this section, we briefly illustrate the remote user authentication scheme proposed by Li et al. [5] in DBSec 2011. Their scheme consists of four phases: registration, login, verification and password update. For ease of presentation, we employ some intuitive abbreviations and notations listed in Table 1.

Table 1. Notations

Symbol	Description	Symbol	Description
U_i	i^{th} user	x	the secret key of remote server S
S	remote server	$\|$	the string concatenation operation
ID_i	identity of user U_i	$h(\cdot)$	collision free one-way hash function
P_i	password of user U_i	$\to$	a common channel
$\oplus$	the bitwise XOR operation	$\Rightarrow$	a secure channel

2.1   Registration Phase

The registration phase involves the following operations:

1) User U_i chooses his/her identity ID_i, password P_i, and then generates a random number RN_1.

2) $U_i \Rightarrow S : \{ID_i, h(h(P_i \oplus RN_1))\}$.

3) On receiving the registration message from U_i, the server S creates an entry $\{ID_i, N, h(h(P_i \oplus RN_1))\}$ in the verification table,where $N = 0$ if it is U_i's initial registration, otherwise S set $N = N + 1$. Then, server S computes $C_1 = h(ID_i \parallel x \parallel N) \oplus h(h(P_i \oplus RN_1))$

4) $S \Rightarrow U_i$: A smart card containing security parameters $\{ID_i, C_1, h(\cdot)\}$.

5) Upon receiving the smart card,user U_i stores RN_1 into his/her smart card.

2.2   Login Phase

When U_i wants to login to S, the following operations will be performed:

1) U_i inserts his/her smart card into the card reader, and inputs ID_i,P_i and a random number RN_2.

2) The smart card generates a random number RC and then computes $C_2 = h(P_i \oplus RN_1)$,$C_3 = C_1 \oplus h(C_2)$,$C_4 = C_3 \oplus C_2$,$C_5 = h(h(P_i \oplus RN_2))$ and $C_6 = E_{KU_i}(C_5, RC)$,where $K_{U_i} = h(C_2 \| C_3)$.

3) $U_i \to S : \{ID_i, C_4, C_6\}$.

2.3 Verification Phase

After receiving the login request from U_i, S performs the following operations:

1) The server S first checks the validity of identity ID_i and then computes $C_7 = h(ID_i \| x \| N)$, $C_8 = C_4 \oplus C_7$, $C_9 = h(C_8)$, and compares C_9 with the third field of the entry corresponding to ID_i in its verification table. If it equals, S successfully authenticates U_i and computes symmetric key $K'_{U_i} = h(C_8 \| C_7)$, and obtains (C_5,RC) by decrypting C_6. Then, S replaces the third field $h(h(P_i \oplus RN_i))$ of the entry corresponding to ID_i with $C_5 = h((P_i \oplus RN_2))$, generates a random RS and computes $K_5 = h(C_7 \| C_8)$.

2) $S \to U_i : \{E_{K_5}(RC, RS, C_5)\}$.

3) On receiving the response from server S, the smart card computes the symmetric key $K'_s = h(C_3 \| C_2)$ and obtains (RC', C'_5) by decrypting the received message using K_s. Then, the smart card checks whether (RC', C'_5) equals to (RC, C_5) generated in the login phase. This equivalency authenticates the legitimacy of the server S, and smart card replaces original RN_1 and C_1 with new RN_2 and $C_3 \oplus C_5$, respectively.

4) $U_i \to S : \{h(RS)\}$

5) On receiving $h(RS)'$, the serve S compares the computed $h(RS)$ with the received value of $h(RS)'$. If they are not equal, the connection is terminated.

6) The user U_i and the server S agree on the session key $SK = h(RC \oplus RS)$ for securing future data communications.

2.4 Password Change Phase

The password change phase is provided to allow users to change their passwords freely. Since the password change phase has little to do with our discussion, we omit it here and detailed information is referred to Ref. [5].

3   Cryptanalysis of Li et al.'s Scheme

In this section we will show that Li et al.'s scheme is vulnerable to offline password guessing attack and denial of service attack. In addition, their scheme fails to preserve user anonymity and forward secrecy. Although tamper resistant smart card is widely assumed in the literature, such an assumption is difficult in practice. Many researchers have shown that the secret information stored in a smartcard can be breached [2–4]. Be aware of this threat, Li et al. intentionally based their scheme on the assumption of non-tamper resistance of the smart card. However, Li et al.'s scheme fails to serve its purposes.

3.1   Offline Password Guessing Attack

Let us consider the following scenarios. In case a legitimate user U_i's smart card is stolen by an adversary $\mathcal{A}$ just before U_i's jth login, and the stored secret values such as C_1 and RN_j can be revealed. Then, $\mathcal{A}$ returns the smart card to U_i and eavesdrops on the insecure channel. Because U_i's identity is transmitted in plaintext within the login request, it is not difficult for $\mathcal{A}$ to identify the login request message from U_i. Once the jth login request message $\{ID_i, C_i^j - h(ID_i \parallel x \parallel N) \oplus h(P_i \oplus RN_i), C_6^j\}$ is intercepted by $\mathcal{A}$, an offline password guessing attack can be launched in the following steps:

Step 1. Guesses the value of P_i to be P_i^* from the password dictionary.

Step 2. Computes $T = h(h(P_i^* \oplus RN_j)) \oplus h(P_i^* \oplus RN_j)$, as RN_j is known. .

Step 3. Computes $T' = C_1 \oplus C_4^j$, as C_1 has been extracted and C_4^j has been intercepted, where $C_1 = h(ID_i \parallel x \parallel N) \oplus h(h(P_i \oplus RN_j)), C_4^j = h(ID_i \parallel x \parallel N) \oplus h(P_i \oplus RN_j)$.

Step 4. Verifies the correctness of P_i^* by checking if T is equal to T'.

Step 5. Repeats Steps 1, 2, 3, and 4 until the correct value of P_i is found.

After guessing the correct value of P_i, the adversary $\mathcal{A}$ can compute $C_3^j = C_1 \oplus h(h(P_i \oplus RN_j)), C_2^j = h(P_i \oplus RN_j)$ and $K_{U_i}^j = h(C_2^j \parallel C_3^j)$. Then the adversary can obtain RC_j by decrypting C_6^j using $K_{U_i}^j$, and gets RS_j in a similar way. Hence the malicious user can successfully compute the session key $SK_j = h(RC_j \oplus RS_j)$ and renders the jth session between U_i and S completely insecure.

3.2   Denial of Service Attack

A denial of service attack is an offensive action whereby the adversary could use some methods to work upon the server so that the login requests issued by the legitimate user will be denied by the server. In Li et at.'s scheme, an adversary can easily launch a denial of service attack in the following steps:

Step 1. Eavesdrops over the channel, intercepts a login request $\{ID_i, C_4^j, C_6^j\}$ from U_i and blocks it, supposing it is U_i's jth login.

Step 2. Replaces C_6^j with an equal-sized random number R, while ID_i and C_4^j are left unchanged.

Step 3. Sends $\{ID_i, C_4^j, R\}$ instead of $\{ID_i, C_4^j, C_6^j\}$ to the remote server S.

After receiving this modified message, S will perform Step $V1$ and $V2$ of the verification phase without observing any abnormality, as a result, the verifier corresponding to ID_i in the verification table will be updated and the response $E_{K_S}(RC_j^*, RS_j, C_5^{j*})$ will be sent to U_i. On receiving the response from S, U_i decrypts $E_{K_S}(RC_j^*, RS_j, C_5^{j*})$ and will find (RC_j^*, C_5^{j*}) unequal to (RC, C_5) , thus the session will be terminated. Thereafter, U_i's succeeding login requests will be denied unless he/she re-registers to S again. That is, the adversary can easily lock the account of any legitimate user without using any cryptographic techniques. Thus, Li et al.'s protocol is vulnerable to denial of service attack.

3.3   Failure to Achieve Forward Secrecy

Let us consider the following scenarios. Supposing the server S's long time private key x is leaked out by accident or intentionally stolen by an adversary $\mathcal{A}$. Once the value of x is obtained, with previously intercepted C_4^j, C_6^j and $E_{K_S}(RC, RS, C_5)$ transmitted in the legitimate user U_i's jth authentication process, $\mathcal{A}$ can compute the session key of S and U_i's jth encrypted communication through the following method:

Step 1. Assumes $N = 0$.
Step 2. Computes $C_7^* = h(ID_i \parallel x \parallel N)$ and $C_8^* = C_7^* \oplus C_4^j$, where ID_i is previously obtained by eavesdropping on the insecure channel.
Step 3. Computes $K_{U_i}^* = h(C_8^* \parallel C_7^*)$ and $K_S^* = h(C_7^* \parallel C_8^*)$.
Step 4. Decrypts C_6^j with $K_{U_i}^*$ to obtain RC_i^*.
Step 5. Decrypts $E_{K_S}(RC, RS, C_5)$ with K_S^* to obtain RC_i^{**}.
Step 6. Verifies the correctness of N by checking if RC_i^* is equal to RC_i^{**}. If they are unequal, sets $N = N + 1$ and goes back to Setp 2.
Step 7. Decrypts $E_{K_S}(RC, RS, C_5)$ to obtain RS_i using K_S^*.
Step 8. Computes $SK_i = h(RC_i \oplus RS_i)$.

Note that the value of N should not be very big, since the re-registration phase is not performed frequently in practice, and thus the above procedure can be completed in polynomial time, which results in the breach of forward secrecy.

3.4   Failure to Preserve User Anonymity

In Li et al.'s scheme, user's identity ID is static and in plaintext form in all the transaction sessions, an adversary can easily obtain the plaintext identity of this communicating client once the login messages were eavesdropped. Hence, different login request messages belonging to the same user can be traced out and may be interlinked to derive some secret information related to the user [8]. Consequently, user anonymity is not preserved in their scheme.

4   Our Proposed Scheme

According to our analysis, three principles for designing a sound password-based remote user authentication scheme are presented. First, user anonymity, especially in some application scenarios, (e.g., e-commence), should be preserved, because from the identity ID_i, some personal secret information may be leaked about the user. Second, a nonce based mechanism is often a better choice than the timestamp based design to resist replay attacks, since clock synchronization is difficult and expensive in existing network environment, especially in wide area networks, and these schemes employing timestamp may still suffer from replay attacks as the transmission delay is unpredictable in real networks. Finally, the password change process should be performed locally without the hassle of interaction with the remote authentication server for the sake of security, user friendliness and efficiency. In this section, we present an improved remote user authentication scheme against smart card security breach.

4.1   Registration Phase

Let $(x, y = g^x \bmod n)$ denote the server S's private key and its corresponding public key, where x is kept secret by the server and y is stored inside each user's smart card. The registration phase involves the following operations:

Step R1. U_i chooses his/her identity ID_i, password P_i and a random number b.

Step R2. $U_i \Rightarrow S : \{ID_i, h(b \parallel P_i)\}$.

Step R3. On receiving the registration message from U_i, the server S computes $N_i = h(b \parallel P_i) \oplus h(x \parallel ID_i)$ and $A_i - h(ID_i \parallel h(b \parallel P_i))$.

Step R4. $S \Rightarrow U_i :$ A smart card containing security parameters $\{N_i, A_i, n, g, y, h(\cdot)\}$.

Step R5. Upon receiving the smart card, U_i enters b into his smart card.

4.2   Login Phase

When U_i wants to login the system, the following operations will be performed:

Step L1. U_i inserts his/her smart card into the card reader and inputs ID_i^*, P_i^*.

Step L2. The smart card computes $A_i^* = h(ID_i^* \parallel h(b \parallel P_i^*))$ and verifies the validity of A_i^* by checking whether A_i^* equals to the stored A_i. If the verification holds, it implies $ID_i^* = ID_i$ and $P_i^* = P_i$. Otherwise, the session is terminated.

Step L3. The smart card chose a random number u and computes $C_1 = g^u \bmod n$, $Y_1 = y^u \bmod n$, $h(x \parallel ID_i) = N_i \oplus h(b \parallel P_i)$, $CID_i = ID_i \oplus h(C_1 \parallel Y_1)$ and $M_i = h(CID_i \parallel C_1 \parallel h(x \parallel ID_i))$.

Step L4. $U_i \rightarrow S : \{C_1, CID_i, M_i\}$.

4.3   Verification Phase

After receiving the login request, the server S performs the following operations:

Step V1. The server S computes $Y_2 = (C_1)^x \bmod n$ using its private key x, and derives $ID_i = CID_i \oplus h(C_1 \parallel Y_2)$ and $M_i^* = h(CID_i \parallel C_1 \parallel h(x \parallel ID_i))$. S compares M_i^* with the received value of M_i. If they are not equal, the request is rejected. Otherwise, server S generates a random number v and computes the session key $SK = (C_1)^v \bmod n$, $C_2 = g^v \bmod n$ and $C_3 = h(SK \parallel C_2 \parallel h(x \parallel ID_i))$.

Step V2. $S \rightarrow U_i : \{C_2, C_3\}$.

Step V3. On receiving the reply message from the server S, U_i computes $SK = (C_2)^u \bmod n$, $C_3^* = h(SK \parallel C_2 \parallel h(x \parallel ID_i))$, and compares C_3^* with the received C_3. This equivalency authenticates the legitimacy of the server S, and U_i goes on to compute $C_4 = h(C_3 \parallel h(x \parallel ID_i) \parallel SK)$.

Step V4. $U_i \rightarrow S : \{C_4\}$

Step V5. Upon receiving $\{C_4\}$ from U_i, the server S first computes $C_4^* = h(C_3 \parallel h(x \parallel ID_i) \parallel SK)$ and then checks if C_4^* is equal to the received value of C_4. If this verification holds, the server S authenticates the user U_i and the login request is accepted else the connection is terminated.

Step V6. The user U_i and the server S agree on the common session key SK for securing future data communications.

4.4   Password Change Phase

In this phase, we argue that the user's smart card must have the ability to detect the failure times. Once the number of login failure exceeds a predefined system value, the smart card must be locked immediately to prevent the exhaustive password guessing behavior. This phase involves the following local operations:

Step P1. U_i inserts his/her smart card into the card reader and inputs the identity ID_i and the original password P_i. The smart card computes $A_i^* = h(ID_i \parallel h(b \parallel P_i))$ and verifies the validity of A_i^* by checking whether A_i^* equals to the stored A_i. If the verification holds, it implies the input ID_i and P_i are valid. Otherwise, the smart card rejects.

Step P2.   The smart card asks the cardholder to resubmit a new password P_i^{new} and computes $N_i^{new} = N_i \oplus h(b \parallel P_i) \oplus h(b \parallel P_i^{new})$, $A_i^{new} = h(ID_i \parallel h(b \parallel P_i^{new}))$. Thereafter, smart card updates the values of N_i and A_i stored in its memory with N_i^{new} and A_i^{new}.

5   Security Analysis

In the following, we briefly analyze the enhanced security of the proposed scheme under the assumption that the secret information stored in the smart card can be revealed, i.e., the security parameters N_i, A_i and y can be obtained by a malicious privileged user. A comprehensive analysis is available in [7].

(1) **User anonymity:** Suppose that the attacker has intercepted U_i's authentication messages $\{CID_i, M_i, C_1, C_2, C_3, C_4\}$. Then, the adversary may try to retrieve any static parameter from these messages, but these messages are all session-variant and indeed random strings due to the randomness of u and/or v. Accordingly, without knowing the random number u, the adversary will face to solve the discrete logarithm problem to retrieve the correct value of ID_i from CID_i,, while ID_i is the only static element corresponding to U_i in the transmitted messages. Hence, the proposed scheme can preserve user anonymity.

(2) **Offline password guessing attack:** Suppose that a malicious privileged user U_i has got U_k's smart card, and the secret information b, N_k, A_k and y can also be revealed under our assumption of the non-tamper resistant smart card. Even after gathering this information, the attacker has to at least guess both ID_i and P_i correctly at the same time, because it has been demonstrated that our scheme can provide identity protection. It is impossible to guess these two parameters correctly at the same time in polynomial time, and thus the proposed scheme can resist offline password guessing attack with smart card security breach.

(3) **Denial of service attack:** Assume that an adversary $\mathcal{A}$ has got the legitimate user U_i's smart card. However, in our scheme, the smart card computes $A_i^* = h(ID_i \parallel h(b \parallel Pi))$ and compares it with the stored value of A_i in its memory to checks the validity of submitted ID_i and P_i before

the password update procedure. It is not possible for $\mathcal{A}$ to guess out U_i's identity ID_i and password P_i correctly at the same time in polynomial time. Moreover, once the number of login failure exceeds a predefined system value, the smart card will be locked immediately. Therefore, the proposed protocol is secure against denial of service attack.

(4) Forward secrecy: Following our scheme, the client and the server can establish the same session key $SK = (C_1)^v = (C_2)^u = g^{uv} \bmod n$. Based on the difficulty of the computational Diffie-Hellman problem, any previously generated session keys cannot be revealed without knowledge of the ephemeral u and v. As a result, our scheme provides forward secrecy.

6   Conclusion

In this paper, we have demonstrated several attacks on Li et al.'s scheme and a robust authentication scheme is thus proposed to remedy these identified flaws. The security analysis demonstrates our scheme eliminates several hard security threats that are difficult to be solved at the same time in previous scholarship.

References

1. Chang, C.C., Wu, T.C.: Remote password authentication with smart cards. IEE Proceedings-E 138(3), 165–168 (1993)
2. Kocher, P., Jaffe, J., Jun, B.: Differential Power Analysis. In: Wiener, M. (ed.) CRYPTO 1999. LNCS, vol. 1666, pp. 388–397. Springer, Heidelberg (1999)
3. Messerges, T.S., Dabbish, E.A., Sloan, R.H.: Examining Smart-Card Security under the Threat of Power Analysis Attacks. IEEE Transactions on Computers 51(5), 541–552 (2002)
4. Kasper, T., Oswald, D., Paar, C.: Side-Channel Analysis of Cryptographic RFIDs with Analog Demodulation. In: Juels, A., Paar, C. (eds.) RFIDSec 2011. LNCS, vol. 7055, pp. 61–77. Springer, Heidelberg (2012)
5. Li, C.T., Lee, C.C., Liu, C.J., Lee, C.W.: A Robust Remote User Authentication Scheme against Smart Card Security Breach. In: Li, Y. (ed.) DBSec 2011. LNCS, vol. 6818, pp. 231–238. Springer, Heidelberg (2011)
6. Kim, S.K., Chung, M.G.: More secure remote user authentication scheme. Computer Communications 32(6), 1018–1021 (2009)
7. Wang, D., Ma, C.G., Wu, P.: Secure Password-based Remote User Authentication Scheme with Non-tamper Resistant Smart Cards. NSR Technical Report 2012/011 (2012), http://machunguang.hrbeu.edu.cn/Research/
8. Ma, C.G., Wang, D., Zhang, Q.M.: Cryptanalysis and Improvement of Sood et al.'s Dynamic ID-Based Authentication Scheme. In: Ramanujam, R., Ramaswamy, S. (eds.) ICDCIT 2012. LNCS, vol. 7154, pp. 141–152. Springer, Heidelberg (2012)

A Friendly Framework for Hidding *fault enabled virus* for Java Based Smartcard

Tiana Razafindralambo, Guillaume Bouffard, and Jean-Louis Lanet

Secure Smart Devices (SSD) Team
XLIM/Université de Limoges – 123 Avenue Albert Thomas, 87060 Limoges, France
aina.razafindralambo@etu.unilim.fr,
{guillaume.bouffard,jean-louis.lanet}@xlim.fr

Abstract. Smart cards are the safer device to execute cryptographic algorithms. Applications are verified before being loaded into the card. Recently, the idea of combined attacks to bypass byte code verification has emerged. Indeed, correct and legitimate Java Card applications can be dynamically modified on-card using a laser beam to become mutant applications or *fault enabled viruses*. We propose a framework for manipulating binary applications to design viruses for smart cards. We present development, experimentation and an example of this kind of virus.

Keywords: Java Card, Virus, Logical Attack, Hidding Code.

1 Introduction

Nowadays, a new deployment model has been developed which has the ability to load third tier application in the SIM card through an application store controlled by the network operator. Unfortunately, these applications are being subjected to fault attacks as it is possible to design inoffensive applications, made hostile once hit by a laser beam. We call them *fault enabled viruses*. Our contribution is twofold, first we propose an architecture as tool and we provide a set of constraints to choose an instruction which will be subjected to a laser attack.

2 Context

Software attacks against smart card can be classified into two categories: ill-typed applications or well-typed applications. But the second category is again divided into permanent well-typed applications or transient well-typed applications. In ill-typed applications [9,4] the input file has been modified in order to illegally obtain information. Permanent well-typed application [8], relies on some weakness of the specification. Transient well-typed applications is a new research field [3,16,4] where an application mutes when a fault occurs. In this way, we have *fault enabled viruses*. Ill-typed applications and transient well-typed applications need to apply byte code transformation engineering at the CAP file level.

N. Cuppens-Boulahia et al. (Eds.): DBSec 2012, LNCS 7371, pp. 122–128, 2012.

2.1   State of the Art

Physical Attacks. As explained by [2], a modification of the input current may modify the execution flow as the card is not self-powered as described in [1,10]. We also have attacks, explained by S. Skorobogatov and R. Anderson in [15], that use the light (LED, laser, *etc.*) and focus on a specific part of the chip, and the light provides enough energy in the memory-cell to change its value. Electromagnetic attack, presented in [13] and [14], as the inducted current provides a way to modify the memory value, and it also helps in characterizing the chip area used during a critical operation.

Logical Attacks. In E. Hubbers *et al.*'s paper [8], they presented a quick overview of the available classical attacks and gave some counter-measures. There are different way to get the type confusion: CAP file manipulation after the building step to bypass an off-card Byte Code Verifier (BCV); using fault injection to bypass the on-card one (difficult and expensive). There is also the use of the shareable interface mechanism, but on recent cards this attack is no longer possible. And finally, we have the transaction mechanism, that consists of making a set of atomic operations. By definition, the rollback mechanism should also deallocate any objects allocated during an aborted transaction and reset references to such objects as `null`. However, the authors found some cases where the card keeps the reference to the objects allocated during transaction even after a rollback. The idea of EMAN attack [9], explained by J. Iguchi-Cartigny *et al.*, is to abuse the firewall mechanism with the unchecked static instructions (as `getstatic`, `putstatic` and `invokestatic`) to call malicious byte codes. In a malicious CAP file, the parameter of `invokestatic` instruction may redirect the control flow graph (CFG) of another installed applet in the targeted smart card. At CARDIS 2011, G. Bouffard *et al.* described, in [4], two methods to change the Java Card CFG. The EMAN2 attack will be further explained in the subsection 3.1.

2.2   The CAP File

As described by S. Hamadouche in [7], the CAP (`Convert APplet`) file format is based on the notion of interdependent components that contain specific information from the Java Card package. For instance, the `Method` component contains the methods byte code, and the `Class` component contains the information on classes such as references to their super-classes or declared methods.

3   The CapMap

3.1   Modification of a CAP File

CapMap has been developed [12] with the aim of having a handy and a friendly way to parse and modify a CAP file. It is very useful and very convenient while designing a logical attack to test Java Cards security. There are three steps to modify a CAP file using the CapMap: identifying which CAP file's components are located in our target, getting the right set of elements, and then applying changes to the components; thanks to setters provided by the CapMap over each CAP file elements. This is a simple example that makes the use of CapMap more clear: it is a reference to the EMAN2 attack. We are going to use the CapMap to particularly manipulate the instruction `sstore` to perform our attack. First, we need to target our method within the `Method` Component, interdependent to the other components. Element within it are indexed. A method is a set of instructions, and an instruction is a set of byte-values. They both are indexed in structures provided by CapMap. Secondly, to target the `sstore` instruction, we are going to change its operand value. By changing the operand value we can write in return function address as listing 1.1.

```
CapFileEditable capFile = new CapFileEditable ();
capFile.load (MY_CAP_FILE);          // Load the cap file
ArrayList<MethodInfo> methods =      // Get methods
        capFile.getMethodComponent ().getMethods ();
//Set the instruction you want to replace
methods.get (METHOD_INDEX).getBytecodes ().set
        (SSTORE_OPERAND_INDEX, RETURN_ADDRESS_REGISTER);
```

Listing 1.1. CAP File modification with CapMap

3.2   Stack Evaluation

If the byte code of a java program is dedicated to be a *fault enabled virus* it needs to avoid the software counter-measures embedded into the card. This type of verification is performed for each method presented in the package. The type checking ensures that no disallowed type conversion is performed. For example, an integer cannot be converted into an object reference. A downcast can only be performed using the `checkcast` instruction, and the arguments which are given to the methods have to be compatible types. The most complicated step and quite expensive (both time and memory), is to retrieve the type of local variables by analyzing the byte code. It requires computing the type of each variable and stack element for each instruction and each execution path, accepting programs (set of instructions) where each stack element and local variable have the same type whatever the path taken to reach an instruction. This also requires that the stack size is the same for each instruction and for each path that can reach this instruction. Another constraint is that the stack must never reach a maximum size which allows checking, if we are not overflowing or underflowing the stack. So, each time we modify a method we can verify the correctness type of the modification. The most important thing for virus implementation is to define the set of instructions eligible to be added to the byte array: only instructions that are compatible with the previous instruction execution can be added to the method. The type information associated to an instruction corresponds to the type of the local variables and of the runtime stack **before the instruction is executed**. The post conditions generated by the execution of the instruction must be checked as pre-condition for the next instruction. This defines a set of constraints that must be guaranteed by each byte code sequence.

3.3   Constraint Solving

To design a *fault enabled virus* we have to hide the real operation as a part of
the operands of the preceding instruction. Thus, when the preceding instruction
is hit by the laser and transformed as a `NOP` instruction: its operand becomes an
instruction. Within this fault model, we need to find an instruction which needs
one operand and satisfies several constraints, or an instruction which needs two
operands. In such a case, the first operand becomes either the first instruction
of the virus, or an instruction without operand and the second operand becomes
the first instruction of the virus. We need to be able to select an instruction
that satisfies several constraints, hence we will be able to hide viruses in a well-
typed program. We try to build a sequence of instructions `prog`, empty at the
beginning, such that it exists an instruction `ins`, with an operand number greater
than one, for which the consumption of the stack is empty and the production
on the stack is lower than the maximum value of the stack. If such an instruction
exists, we can concatenate the sequence `prog` with the sequence `virus` minus
its head. Executing the new sequence `prog` must lead to an empty stack at the
end of execution. Unfortunately, the resulting program may be a non valid Java
program: not all sequences of byte code can be generated by a compiler. But the
certification scheme proposed by GlobalPlatform [5] do not indicate the source
code. The certification process must be done at the CAP file level.

3.4   Java Card Code Reverser

The complete process of generating a *fault enabled virus* needs four steps using
CapMap. Firstly, finding a sequence of instructions which hides the virus code that
satisfies a set of constraints. The resulting CAP File represents a valid Java program in term of stack typing. Next, to evaluate the resulting cap file using an *off-
card* BCV is the second step. If it is rejected, it means that either stack evaluation
goes wrong, or the constraint solver failed. If the *off-card* BCV evaluation succeeds,
the third step is, using our `Cap2Class` tool to reverse the code. Finally, converting
the class file to `Java` file by means of existing tools, if the generated code is valid.

4   Evaluation of the Threat Capacities

4.1   Building a *fault enabled virus* with the CapMap

The listing 1.2 explains how to build the virus. It's aim is to send a clear text which
has the value of an encrypted key container. Of course any analysis will reject this
code as the secret key is being sent to the external world. This code can be split
into three parts. The first one (`B1`) is mandatory and corresponds to the APDU
reception. The second block (`B2`) corresponds to the code to obfuscate and which
should only be executable once a fault occurs. It decrypts the key container and
put the value in the APDU buffer at offset `0`. The last one (`B3`) sends the content
of the apdu buffer from offset `0` for `16` elements (a 3-DES key) to the reader. If we
can replace the `B2` block by an inoffensive code, it is said to be a *fault enabled smart
card virus*. This code corresponds to the following byte code listed in 1.3.

```
public void process(APDU apdu) {
  short localS; byte localB;
  byte[] apduBuffer = apdu.getBuffer(); //get the APDU buffer
  if (selectingApplet()) { return; }  B1
  byte receivedByte = (byte) apdu.setIncomingAndReceive();

  // any code can be placed here
  DES_keys.getKey(apduBuffer, (short)0);  B2

    apdu.setOutgoingAndSend((short)0,16);  B3
}
```

Listing 1.2. The unwanted code

```
/*00bd*/ L0: aload_1              // apdu
/*00be*/     invokevirtual    8  // getBuffer (APDU class)
/*00c1*/     astore           4  // L4 = apduBuffer
/*00c3*/     aload_0             // this=Applet instance
/*00c4*/     invokevirtual    9  // selectingApplet()
/*00c7*/     ifeq            L1  // rel:+3 (@00CA)
/*00c9*/     return
/*00ca*/ L1: aload_1              // apdu               B1
/*00cb*/     invokevirtual   10
/*00ce*/     s2b                 // redByte
/*00cf*/     sstore           5  // L5 = redByte

/*00d6*/     getfield_a_this  1  // DES_keys
/*00d8*/     aload            4  // L4=>apdubuffer
/*00da*/     sconst_0
/*00db*/     invokeinterface  nargs : 3, index : 0,  B2
                              const : 3, method: 4 //getkey
/*00e0*/     pop                 // returned Le byte

/*00e1*/     aload_1             //L1 apdu
/*00e2*/     sconst_0
/*00e3*/     bspush 0x0F         // DES_keys size
/*00e5*/     invokeinterface  nargs : 1, index : 0,  B3
                              const : 3, meth. : 1
/*00ea*/     invokevirtual   11 // setOutgoingAndSend
/*00ed*/     return
```

Listing 1.3. The virus code at the byte code level

The B1 block is the preamble, a correct code that must be executed. The B2 block corresponds to the code that must be obfuscated, and the last one B3 is the postamble. After the execution of the B1 block the state of the stack is {ref, ref, value}. By obfuscating B2 will insert an instruction before in a such a way that constraints explained in the previous section are verified. But prior to select an instruction, we need to link statically the B2 code fragment. The final linking process is done inside the card and we can not rely on this process to resolve automatically the addresses. For that purpose, we have developed an attack, presented in [6], that provides us the way to retrieve (for most of the current cards) the linking information. For this card, the linked address of the getKey method is 0x023C. Then the code to hide becomes:

```
/*00db*/ invokeinterface nargs: 3, @023c,  method: 4
/*00e0*/ pop     // pop the return byte of the method
```

Listing 1.4. Resolved address of the B2 block

If we consider the single fault model then one of the selectable instructions is ifle (0x65) . It uses a short value and its operand is an offset to the branching instruction. The B2 code fragment to be loaded into the card is given in the listing 1.5. If the byte at the offset 0x00D6 becomes 0x0000 (thanks to the laser hit) the original B2 code will be executed.

```
/*00d6*/ [65]  ifle @0x8D // 0x8D corresponds to invokestatic
/*00d8*/ [03]  sconst_0   // corresponds to the nargs
/*00d9*/ [02]  sconst_m1  // corresponds to the address high
/*00da*/ [3c]  pop2       // corresponds to the address low
/*00db*/ [04]  sconst_1   // corresponds to the method number
/*00dc*/ [3b]  pop        // resynchronized with the original code
```

Listing 1.5. The hiding code

4.2 Detecting a *fault enabled virus* with SmartCM

The starting point of this study was the development of SmartCM [11], a simulator that detects such attack, and aims to analyze the effect of a fault on a Java Card program using different modules like the code mutation engine, the risk analysis tool, and the mutants reducer.

5 Conclusion

We have presented in this paper a complete CAP file engineering tool to modify each component of the CAP file in a coherent way. Within this tool, we have the possibility to design a very efficient attack using ill-typed application but also *fault enabled viruses*. It includes a stack checker to avoid embedded countermeasures and a minimalist constraint solver to generate the hiding sequence. We demonstrated the efficiency of the constraint solver to build a valid program which hides a *fault enabled virus*. We have developed a static analyzer *SmartCM* that is able to detect such a *fault enabled virus*. Recently, it appears that the single fault model is out of date and we must consider the possibility of a dual fault attack as a valid hypothesis. Thus, the CapMap tool is able to build such a second order virus by simply applying twice the process. But the constraints for the second pass must be different, and should not reveal the hidden code. This is a new research direction on which we are working now.

References

1. Agoyan, M., Dutertre, J.-M., Naccache, D., Robisson, B., Tria, A.: When Clocks Fail: On Critical Paths and Clock Faults. In: Gollmann, D., Lanet, J.-L., Iguchi-Cartigny, J. (eds.) CARDIS 2010. LNCS, vol. 6035, pp. 182–193. Springer, Heidelberg (2010)
2. Aumüller, C., Bier, P., Fischer, W., Hofreiter, P., Seifert, J.-P.: Fault Attacks on RSA with CRT: Concrete Results and Practical Countermeasures. In: Kaliski Jr., B.S., Koç, Ç.K., Paar, C. (eds.) CHES 2002. LNCS, vol. 2523, pp. 260–275. Springer, Heidelberg (2003)
3. Barbu, G., Thiebeauld, H., Guerin, V.: Attacks on Java Card 3.0 Combining Fault and Logical Attacks. In: Gollmann, D., Lanet, J.-L., Iguchi-Cartigny, J. (eds.) CARDIS 2010. LNCS, vol. 6035, pp. 148–163. Springer, Heidelberg (2010)
4. Bouffard, G., Iguchi-Cartigny, J., Lanet, J.-L.: Combined Software and Hardware Attacks on the Java Card Control Flow. In: Prouff, E. (ed.) CARDIS 2011. LNCS, vol. 7079, pp. 283–296. Springer, Heidelberg (2011)
5. Global Platform: Composition Model Security Guidelines for Basic Applications (2012)
6. Hamadouche, S., Bouffard, G., Lanet, J.L., Dorsemaine, B., Nouhant, B., Magloire, A., Reygnaud, A.: Subverting Byte Code Linker service to characterize Java Card API. Submitted at SAR-SSI (2012)
7. Hamadouche, S.: Étude de la sécurité d'un vérifieur de Byte Code et génération de tests de vulnérabilité. Master's thesis, Université de Boumerdés (2012)
8. Hubbers, E., Poll, E.: Transactions and non-atomic API calls in Java Card: specification ambiguity and strange implementation behaviours. Tech. rep., University of Nijmegen (2004)
9. Iguchi-Cartigny, J., Lanet, J.: Developing a trojan applets in a smart card. Journal in Computer Virology 6(4), 343–351 (2010)
10. Kömmerling, O., Kuhn, M.: Design principles for tamper-resistant smartcard processors. In: Proceedings of the USENIX Workshop on Smartcard Technology (1999)
11. Machemie, J.B., Mazin, C., Lanet, J.L., Cartigny, J.: SmartCM A Smart Card Fault Injection Simulator. In: IEEE International Workshop on Information Forensics and Security - WIFS (2011)
12. Noubissi, A., Séré, A., Iguchi-Cartigny, J., Lanet, J., Bouffard, G., Boutet, J.: Cartes à puce: Attaques et contremesures. MajecSTIC 16(1112) (November (2009)
13. Quisquater, J., Samyde, D.: Eddy current for magnetic analysis with active sensor. In: Proceedings of Esmart (2002)
14. Schmidt, J., Hutter, M.: Optical and em fault-attacks on crt-based rsa: Concrete results. In: Proceedings of the Austrochip, pp. 61–67. Citeseer (2007)
15. Skorobogatov, S., Anderson, R.: Optical Fault Induction Attacks. In: Kaliski Jr., B.S., Koç, Ç.K., Paar, C. (eds.) CHES 2002. LNCS, vol. 2523, pp. 2–12. Springer, Heidelberg (2003)
16. Vetillard, E., Ferrari, A.: Combined Attacks and Countermeasures. In: Gollmann, D., Lanet, J.-L., Iguchi-Cartigny, J. (eds.) CARDIS 2010. LNCS, vol. 6035, pp. 133–147. Springer, Heidelberg (2010)

Approximate Privacy-Preserving Data Mining on Vertically Partitioned Data

Robert Nix[1], Murat Kantarcioglu[1], and Keesook J. Han[2]

[1] Jonsson School of Engineering and Computer Science
The University of Texas at Dallas
800 West Campbell Road
Richardson, Texas, USA
{rcn062000,muratk}@utdallas.edu
[2] Air Force Research Laboratory
Information Directorate
525 Brooks Road
Rome, New York, USA
Keesook.Han@rl.af.mil

Abstract. In today's ever-increasingly digital world, the concept of data privacy has become more and more important. Researchers have developed many privacy-preserving technologies, particularly in the area of data mining and data sharing. These technologies can compute exact data mining models from private data without revealing private data, but are generally slow. We therefore present a framework for implementing efficient privacy-preserving secure approximations of data mining tasks. In particular, we implement two sketching protocols for the scalar (dot) product of two vectors which can be used as sub-protocols in larger data mining tasks. These protocols can lead to approximations which have high accuracy, low data leakage, and one to two orders of magnitude improvement in efficiency. We show these accuracy and efficiency results through extensive experimentation. We also analyze the security properties of these approximations under a security definition which, in contrast to previous definitions, allows for very efficient approximation protocols.[1]

1 Introduction

Privacy is a growing concern among the world's populace. As social networking and cloud computing become more prevalent in today's world, questions arise about the safety and confidentiality of the data that people provide to such services. In some domains, such as medicine, laws such as HIPAA and the Privacy Act of 1979 step in to make certain that sensitive data remains private. This is great for ordinary consumers, but can cause problems for the holders of the data. These data holders would like to create meaningful information from the data that they have, but privacy laws prevent them from disclosing the data

---

[1] Approved for Public Release; Distribution Unlimited: 88ABW-2011-4946, 16-Sep 2011.

N. Cuppens-Boulahia et al. (Eds.): DBSec 2012, LNCS 7371, pp. 129–144, 2012.
© IFIP International Federation for Information Processing 2012

to others. In order to allow such collaboration between the holders of sensitive data, *privacy-preserving* data mining techniques have been developed.

In privacy-preserving data mining, useful models can be created from sensitive data without revealing the data itself. One way to do this is to perturb the data set using anonymization or noise addition [7] and perform the computation on that data. This approach was first pioneered by Agrawal and Srikant [3]. These methods can suffer from low utility, since the data involved in the computation is not the actual data being modeled. In addition, these protocols can suffer from some security problems[18,13,21], which can lead to the retrieval of private data from the perturbed data given.

The other way to do this is using secure multiparty computation techniques to compute the exact data mining result, on the actual data. Secure computation makes use of encryption schemes to keep the data secret, but relies on other tactics, such as encrypting the function itself, or homomorphic properties of the encryption, to perform the computation. This approach was first used by Lindell and Pinkas [20]. These schemes generally rely on very slow public key encryption, which results in a massive decrease in information output. The exact computation of data mining models can take thousands of times longer when using these public key cryptosystems.

While many functions are very difficult to compute using secure multiparty computation, some of these functions have approximations which are much easier to compute. This is especially true in those data mining tasks that deal with aggregates of the data, since these aggregates can often be easily estimated. Approximating the data mining result, however, can lead to some data leakage if the approximation is not done very carefully. The security of approximations has been analyzed by Feigenbaum, et al., [8], but the results of their analysis showed that to make an approximation fully private, the process of the computation must be substantially more complex. Sometimes, this complexity can make computing the approximation more difficult than computing the function itself!

Here, we present another security analysis that, while allowing some small, parameter defined data leakage, creates the opportunity to use much simpler and less computationally expensive approximations securely. We then use this model of security to show the security of two approximation methods for a sub-protocol of many vertically partitioned data mining tasks: the two-party dot product. The dot product is used in association rule mining, classification, and other types of data mining. We prove that our approximations are secure under our reasonable security definitions. These approximations can provide one to two orders of magnitude improvement in terms of efficiency, while sacrificing very little in accuracy.

1.1   Summary of Contributions

A summary of our contributions are as follows:

- We outline a practical security model for secure approximation which allows simple protocols to be implemented securely.

- We showcase two sketching protocols for the dot product and prove their security under our model.
- Through experimentation, we show the practicality of these protocols in vertically partitioned privacy-preserving data mining tasks. These protocols can lead to a two order of magnitude improvement in efficiency, while sacrificing very little in terms of accuracy.

In section 2, we summarize the current state of work in this area. Section 3 provides the standard definitions of secure approximations, and our minor alteration thereof. Section 4 outlines the approximation protocols we use. Section 5 gives the proof that these simple approximation protocols are secure under our definition of secure approximation. In section 6, we give experimental results for different data mining tasks using the approximations. Finally, we offer our overall conclusions and future directions in section 7.

2   Related Work

Privacy-preserving data mining (PPDM) is a vast field with hundreds of publications in many different areas. The two landmark papers by Agrawal and Srikant [3] and Lindell and Pinkas [20] began the charge, and soon many privacy preserving techniques emerged for computing many data mining models [16,27,5,24]. Other techniques can be found in the survey [2]. For our purposes, we will focus on those works which are quite closely related to the work in our paper.

There are quite a few protocols previously proposed for the secure computation of the dot product. The protocol proposed by [27] is quadratic in the size of the vector (times a security parameter). It does, however, have some privacy concerns accoring to [11]. This same work, along with several others [6,14] propose other protocols which are based on very slow public key cryptography. [26] proposes a sampling-based algorithm for secure dot product computation which relies on secure set intersection as a sub-protocol. However, the secure set intersection problem is also nontrivial. It either relies on a secure dot product protocol [27] (which would lead to a circular dependency with [26]), or a large amount of extremely expensive cryptographic operations [30].

The sketching primitives used in this work have been applied to data mining in several different capacities. [25] uses Bloom filters to do association rule mining. However, the model employed in this framework requires a server hierarchy, in which the association rule mining is done at the top level, and represents transactions, not itemsets, as Bloom filters. The Johnson-Lindenstrauss theorem is employed for data mining by [22], however, they employ the Johnson-Lindenstrauss theorem as the sole means of preserving privacy, whereas we are using it as part of a process. Other works [9,31] use Johnson-Lindenstrauss projection as an approximation tool. These, however, do not make use of the projection in a privacy-preserving context, and are merely concerned with fast approximations.

The work of [17] presents a sketching protocol for the scalar product based on Bloom filters. However, its experimentation and discussion of actual data mining tasks was insufficient. Our protocols perform better on real data mining tasks, especially at high compression ratios.

3   Secure Approximations

Much has been written about secure computation, and the steps one must go through in order to compute functions without revealing anything about the data involved. Securely computing the *approximation* of a function poses another challenge. In addition to not revealing the data through the computation process, we must also assure that the function we use to approximate the actual function must not reveal anything about the data! To this end, we outline a definition of secure approximations given by [8], and then propose an alteration to this framework. This alteration, while allowing a very small amount of data leakage, allows for the use of very efficient approximation protocols, which can improve the efficiency of exact secure computation by orders of magnitude.

3.1   A Secure Approximation Framework

The work of Feigenbaum, et. al. [8] gives a well-constructed and thorough definition of secure approximations. In the paper, they first define a concept called *functional privacy*, then use this definition to define the notion of a secure approximation. First, we examine the definition of functional privacy, as follows:

Definition 1 *Functional Privacy*: Let $f(\mathbf{x})$ be a deterministic, real valued function. Let $\hat{f}(\mathbf{x})$ be a (possibly randomized) function. $\hat{f}$ is *functionally private* with respect to f if there exists a probabilistic, expected polynomial time sampling algorithm $\mathbf{S}$ such that for every input $\mathbf{x} \in X$, the distribution $\mathbf{S}(f(\mathbf{x}))$ is indistinguishable from $\hat{f}(\mathbf{x})$.

Note that the term "indistinguishable" in the definition is left intentionally vague. This could be one of the standard models of perfect indistinguishability, statistical indistinguishability, computational indistinguishability [23], or any other kind of indistinguisability. In these cases, the adjective applied to the indistinguishablity is also applied to the functional privacy (i.e., statistical functional privacy for statistical indistinguishability).

Intuitively, this definition means that the result of $\hat{f}$ yields no more information about the input than the actual result of f would. Note, however, that this does not claim that there is any relation between the two outputs, other than the privacy requirement. This does not require that the function $\hat{f}$ be a good approximation of f. Feigenbaum, et al., therefore, also provide a definition for approximations, which is also used in the final concept of a secure approximation.

Definition 2 *P-approximation*: Let $P(f, \hat{f})$ be a predicate for determining the "closeness" of two functions. A function $\hat{f}$ is a *P-approximation* of f if $P(f, \hat{f})$ is satisfied.

Now, for this definition to be useful, we need to define a predicate P to use for the closeness calculation. The most commonly used predicate P is the $\langle \epsilon, \delta \rangle$

criterion, in which $\langle \epsilon, \delta \rangle \, (f, \hat{f})$ is satisfied if and only if $\forall \mathbf{x}, Pr[(1 - \epsilon)f(\mathbf{x}) \leq \hat{f}(\mathbf{x}) \leq (1 + \epsilon)f(\mathbf{x})] > 1 - \delta$. We do not refer to any other criterion in our work, but the definition is provided with a generic closeness predicate for the sake of completeness.

Finally, we present the liberal definition of secure two party approximations as outlined in Feigenbaum, et al.

Definition 3 *Secure Approximation (2-parties)*: Let $f(\mathbf{x}_1, \mathbf{x}_2)$ be a deterministic function mapping the two inputs $\mathbf{x}_1$ and $\mathbf{x}_2$ to a single output. A protocol p is a secure P-approximation protocol for f if there exists a functionally private P-approx-imation $\hat{f}$ such that the following conditions hold:

Correctness. The outputs of the protocol p for each player are in fact equal to the same $\hat{f}(\mathbf{x}_1, \mathbf{x}_2)$.

Privacy. There exist probabilistic polynomial-time algorithms $\mathbf{S}_1, \mathbf{S}_2$ such that

$$\{(\mathbf{S}_1(\mathbf{x}_1, f(\mathbf{x}_1, \mathbf{x}_2), \hat{f}(\mathbf{x}_1, \mathbf{x}_2)), \hat{f}(\mathbf{x}_1, \mathbf{x}_2))\}_{(\mathbf{x}_1, \mathbf{x}_2) \in X} \overset{c}{\equiv}$$
$$\{(\text{view}_1^p(\mathbf{x}_1, \mathbf{x}_2), \text{output}_2^p(\mathbf{x}_1, \mathbf{x}_2))\}_{(\mathbf{x}_1, \mathbf{x}_2) \in X},$$
$$\{(\hat{f}(\mathbf{x}_1, \mathbf{x}_2), \mathbf{S}_2(\mathbf{x}_1, f(\mathbf{x}_1, \mathbf{x}_2), \hat{f}(\mathbf{x}_1, \mathbf{x}_2)))\}_{(\mathbf{x}_1, \mathbf{x}_2) \in X} \overset{c}{\equiv}$$
$$\{(\text{output}_1^p(\mathbf{x}_1, \mathbf{x}_2), \text{view}_2^p(\mathbf{x}_1, \mathbf{x}_2))\}_{(\mathbf{x}_1, \mathbf{x}_2) \in X}$$

where $A \overset{c}{\equiv} B$ means that A is computationally equivalent to B. Note that in the above definition all instances of $\hat{f}(\mathbf{x}_1, \mathbf{x}_2)$ have the same value, as opposed to being some random value from the distribution of $\hat{f}$. This limits the application of the simulators to a single output. This definition essentially says that we have a functionally private function $\hat{f}$ which is a P-approximation of f which itself is computed in a private manner, such that no player learns anything else about the input data.

3.2 Our Definition

Having defined the essential notions of functional privacy, approximations, and secure approximations, we now define another notion of functional privacy, which, while less secure than the above model, allows for vastly more efficient approximations.

Definition 4 $\langle \epsilon, \delta \rangle$-*functional privacy*: A function $\hat{f}$ is $\langle \epsilon, \delta \rangle$-*functionally private* with respect to f if there exists a polynomial time simulator $\mathbf{S}$ such that $Pr[\|\mathbf{S}(f(\mathbf{x}), R) - \hat{f}(\mathbf{x})\| < \epsilon] > 1 - \delta$, where R is a shared source of randomness involved in the calculation of $\hat{f}$.

Intuitively, this definition allows for a non-negligible but still small acceptable information loss of at most ϵ, while still otherwise retaining security. In practice, the amount of information revealed could be much smaller, but this puts a maximum bound on the privacy of the function. In addition, we allow the simulator access to the randomness function used in computing $\hat{f}$, which allows the simulator to more accurately produce similar results to $\hat{f}$.

The acceptable level of loss ϵ can vary greatly with the task at hand. For example, if the function is to be run on the same data set several times, the leakage from that data set would increase with each computation. Thus, for applications with higher repetition, we would want a much smaller ϵ. The ϵ can be adjusted by using a more accurate approximation.

In their work describing the original definition above, Feigenbaum, et al. [8] dismissed a simple, efficient approximation protocol based on their definition of functional privacy. This approximation was a simple random sampling based method for approximating the hamming distance between two vectors. The claim was that even if the computation was done entirely securely, some information about the randomness used in the computation would be leaked into the final result. Thus, we simply explicitly allow the randomness to be used by the simulator in our model. We feel this is realistic, as the randomness is common knowledge to all parties in the computation.

In short, the previous definition of [8] aims to eliminate data leakage from the approximation result. Our definition simply seeks to quantify it and reduce it to acceptable levels. In return, we can use much simpler approximation protocols securely. For example, the eventual secure hamming distance protocol given by [8] has two separate protocols (one which works for high distance and one for low distance) each of which requires several rounds of oblivious transfers between the two parties. Under our definition, protocols can be used which use only a single round of computation and work for any type of vector, as we will show in the next section.

4 Scalar Product Approximation Techniques for Distributed Data Mining

Data mining is, in essence, the creation of useful models from large amounts of raw data. This is typically done through the application of machine learning based model building algorithms such as association rules mining, naive bayes classification, linear regression, or other model creation algorithms. Distributed data mining, then, is the creation of these models from data which is distributed (partitioned) across multiple owners. The dot product of two vectors has many applications in vertically partitioned data mining. Many data mining algorithms can be reduced to one or more dot products between two vectors in the vertically partitioned case. Vertical partitioning can be defined as follows:

Let X be a data set containing tuples of the form $(a_1, a_2, ..., a_k)$ where each a is an attribute of the tuple. Let S be a subset of $\{1, 2, ..., k\}$. Let X_S be the data set where the tuples contain only those attributes specified by the set S. For example, $X_{\{1,2\}}$ would contain tuples of the form (a_1, a_2). The data set X is said to be vertically partitioned across n parties if each party i has a set S_i, and the associated data X_{S_i}, and

$$\bigcup_{i=1}^{n} S_i = \{1, 2, ..., k\}$$

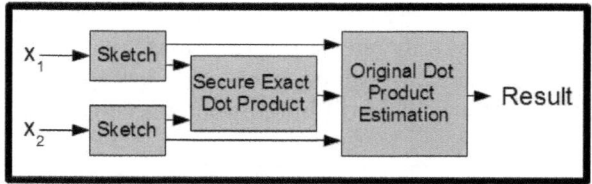

Fig. 1. Dot Product Approximation Concept

In previous work, it has been shown that the three algorithms we test in this paper can in fact be reduced to the dot product of two zero-one vectors in the vertically partioned case. These algorithms are association rules mining[17], naive Bayes classification[28], and C4.5 decision tree classification[29].

We developed two sketching protocols for the approximation of the dot product of two zero-one vectors. These protocols are used to provide smaller input to an exact dot product protocol, which is then used to estimate the overall dot product, as outlined in figure 1. First, we present a protocol based on the Johnson-Lindenstrauss theorem [15] and the work of [1] and [19]. Then, we present a simple sampling algorithm which is also secure under our model. Finally, we present a proof of the security of these approximations in our security model.

4.1 Johnson-Lindenstrauss (JL) Sketching

The Johnson-Lindenstrauss theorem [15] states that for any set of vectors, there is a random projection of these vectors which preserves Euclidean distance within a tolerance of ϵ. More formally, for a given ϵ, there exists a function $f : \mathbb{R}^d \to \mathbb{R}^k$ such that for all u and v in a set of points,

$$(1 - \epsilon)||u - v||^2 \leq ||f(u) - f(v)||^2 \leq (1 + \epsilon)||u - v||^2$$

It is shown in that because of this property, the dot product is also preserved within a tolerance of ϵ. As with any sketching scheme, the probability of being close to the correct answer increases with the size of the sketch.

As outlined in [1] and [19], to do our random projection, we generate a $k \times n$ matrix R, where n is the number of rows in the data set, and k is the number of rows in the resultant sketch. Each value of this matrix has the value 1, 0, or -1, with probabilities set by a sparisity factor s. The value 0 has a probability of $1 - \frac{1}{s}$, and the values 1 and -1 each have a probability of $\frac{1}{2s}$. In order to sketch a vector a of length n, we do $\frac{\sqrt{s}}{\sqrt{k}}Ra$, which will have a length of k. This preserves the dot product to within a certain tolerance. So, to estimate the dot product of two vectors a and b, we merely compute $\frac{\sqrt{s}}{\sqrt{k}}Ra \cdot \frac{\sqrt{s}}{\sqrt{k}}Rb$. Note that this will be equal to $s\frac{Ra \cdot Rb}{k}$, and in practice, we typically omit the $\frac{\sqrt{s}}{\sqrt{k}}$ term from the sketching protocol, and simply divide by the length of the sketch and multiply by the sparsity factor after performing the dot product. This yields the same result. This is shown below as Algorithm 4.1.

According to [19], the sparsity factor s can be as high as $\frac{n}{logn}$ before significant error is introduced, and as s increases, the time and space requirements for the sketch decrease. Nevertheless we still used relatively low sparsity factors, to show that even in the slowest case, we still have an improvement.

Algorithm 4.1. Johnson-Lindenstrauss(JL) Dot Product Protocol

---

RandomMatrixGeneration(n,k):
Matrix R
for $i \leftarrow 1...n$ **do**
  for $j \leftarrow 1...k$ **do**
    $R_{j,i} \overset{\$}{\leftarrow} \{\frac{1}{2s} : -1, 1 - \frac{1}{s} : 0, \frac{1}{2s} : 1\}$
  end for
end for
return R

DotProductApproximation(Vector u,Vector v, k):
Matrix $R \leftarrow$ RandomMatrixGeneration($|u|, k$)
$u' \leftarrow Ru$
$v' \leftarrow Rv$
return $\frac{s \cdot \text{SecureDotProduct}(u',v')}{k}$

4.2   Random Sampling

In addition to the more complicated method above, to estimate the dot product of two vectors, one could simply select a random sample of both vectors, compute the dot product, then multiply by a scaling factor to estimate the total dot product. Note that this works fairly well on vectors where the distribution of values is known, such as zero-one vectors, but can work quite poorly on arbitrary vectors. The sampling algorithm is shown below in Algorithm 4.2.

5   Approximation Protocol Security

We now provide a proof that each of the above protocols provides a secure approximation in the sense outlined above. We first show the $\langle 2\epsilon, \delta^2 \rangle$-functional privacy of the protocols, then show that the protocols are secure under the liberal definition of secure approximations.

Theorem. The protocols outlined in section 4 are both $\langle 2\epsilon, \delta^2 \rangle$-functionally private, and meet the liberal definition for secure approximations (definition 3).

Proof:
Functional Privacy Let ϵ, δ be the approximation guarantees granted by the above protocols. That is, $Pr[|u \cdot v - \text{DotProductApproximation}(u, v)| > \epsilon] < 1 - \delta$. For JL, these bounds are provided by the Johnson-Lindenstrauss theorem itself, as shown by the work of [22]. For sampling, we can use the Hoeffding inequality [12] to establish a bound on the error:

Algorithm 4.2. Sampling Protocol

Sketch(Vector v, $samplingFactor \in [0...1]$):
$sketch \leftarrow []$
for $i \leftarrow 1...n$ **do**
  $r \overset{\$}{\leftarrow} [0...1]$
  if $r < samplingFactor$ **then**
    $sketch.append(v_i)$
  end if
end for
return $sketch$

DotProductApproximation(u,v,$samplingFactor$)
$u' \leftarrow$ Sketch(u)
$v' \leftarrow$ Sketch(v)
return $\frac{\text{SecureDotProduct}(u',v') \cdot |u|}{|u'|}$

$$Pr[|\hat{f}(\mathbf{x}) - f(\mathbf{x})| \geq \epsilon] \leq 2e^{-2\epsilon^2 n^2}$$

Where n is the sample size. As $\hat{f}$ can be taken to be an estimate of the mean of the product of the random variables, the Hoeffding inequality holds for the dot product of the samples. So, we set our δ to $2e^{-2\epsilon^2 n^2}$.

Note that, with both of these approximation protocols, adjusting the size (for JL, the matrix size, and for sampling, the sample size), allows us to adjust the ϵ of the functional privacy requirement. This would allow us to adjust the ϵ value to be as low as we deemed necessary for our purposes.

Now, let our simulator $\mathbf{S}(f(\mathbf{x}), R)$ generate two random zero-one vectors u and v such that $f(u, v) = u \cdot v = f(\mathbf{x})$. We then apply the randomness given to perform a calculation of the dot product approximation $\beta = \hat{f}(u, v)$. Now, the probability that $|f(\mathbf{x}) - \hat{f}(\mathbf{x})| \geq \epsilon$ is $1 - \delta$. The probability that $|f(\mathbf{x}) - \beta| \geq \epsilon$ is also $1 - \delta$, since $f(\mathbf{x}) = f(u, v)$. As these are independent events, the probability that neither occurs is δ^2. In the case this occurs, we have $|f(\mathbf{x}) - \hat{f}(\mathbf{x})| \leq \epsilon$ and $|f(\mathbf{x}) - \beta| \leq \epsilon$, which means that $-\epsilon \leq f(\mathbf{x}) - \hat{f}(\mathbf{x}) \leq \epsilon$ and $-\epsilon \leq f(\mathbf{x}) - \beta \leq \epsilon$. Because of this, the difference between the two quantities $(f(\mathbf{x}) - \hat{f}(\mathbf{x})) - (f(\mathbf{x}) - \beta) = \beta - \hat{f}(\mathbf{x})$ can be no more than 2ϵ. If our simulator returns β, then we have shown that $\hat{f}$ is $\langle 2\epsilon, \delta^2 \rangle$-functionally private with respect to f.

Secure Approximation (under Definition 3). For the approximation to be considered secure, it must compute the same value for both players (which is trivially true for both protocols), and be private with respect to the views of each player. Now, consider, in each case, what each player sees. Player 1 sees his input, a sketch of that input, and the inputs and outputs of a secure dot product protocol. Our simulator can take that input, sketch it, and simulate the secure dot product protocol, altering its output to be $\hat{f}(\mathbf{x})$ to player 1. Since this output is all player 1 sees outside of the secure dot product protocol, it cannot distinguish this from the true output. Player 2 sees the same thing, his input, a sketch of that input, and the operations of a secure dot product protocol on the inputs. Since the subprotocol is secure, neither player can learn anything about the inputs that the sketches would not tell them.

Having shown that the sketching protocols are $\langle \epsilon, \delta \rangle$-functionally private, and that the computation protocol is secure under definition 3, we now claim that the entire protocols are secure under our model.   □

6 Experiments

In order to determine the efficiency and effectiveness of the algorithms proposed, we conducted several experiments. Each of the sketching protocols presented were inserted into the data mining process for three different data mining tasks: association rules mining, naive Bayes classification, and C4.5 decision tree classification. We used three separate sparsity values for JL sketching: $s = 1$, which results in a matrix completely full of 1 and -1, $s = 100$, and $s = 1000$. The efficiency of JL increases with s, and these values are much lower than what is required to achieve good accuracy [19].

For association rules mining, we used the retail data set found at [10], which lists transactions from an anonymous Belgian retail store. We considered three variables in the association rules experiments: the required support, the required confidence, and the compaction ratio of the sketching protocol. For testing the required support, we used 2%, 3%, 4%, 5%, and 6%, while holding the confidence constant at 70% and the compaction ratio constant at 10%. For the confidence, we used 60%, 65%, 70%, 75%, and 80% while holding the support constant at 4% and the compaction ratio constant at 10%. Finally, for the compaction ratio, we used 1%, 5%, 10%, 15%, and 20%, holding the support constant at 4% and the confidence constant at 70%. For naive Bayes and C4.5 decision tree classification, we used the Adult data set from the UC Irvine Machine Learning Repository [4], which consists of data from the 1993 US Census. As there were no paramaters to set for naive Bayes or the decision tree, we varied only the compaction ratio as above. We did, however, discretize each attribute of the data set before performing the data mining, as continuous data would not function under our model. For each task and variable set, we ran ten separate experiments, using different initialization values for the inherent randomness in the sketching protocols. We employed ten-fold cross-validation for the classification tasks. The accuracy results were then averaged over all ten trials to come up with the final result.

6.1 Accuracy

Association Rules Mining. To assess the accuracy of the algorithms on association rules mining, we look at both the number of false positives (that is, the number of invalid associations returned by the algorithm) and false negatives (the number of valid assocations not returned by the algorithm). For the association rules mining, this is a better picture of the accuracy than overall accuracy, since the true positives are so much rarer than the true negatives. Figure

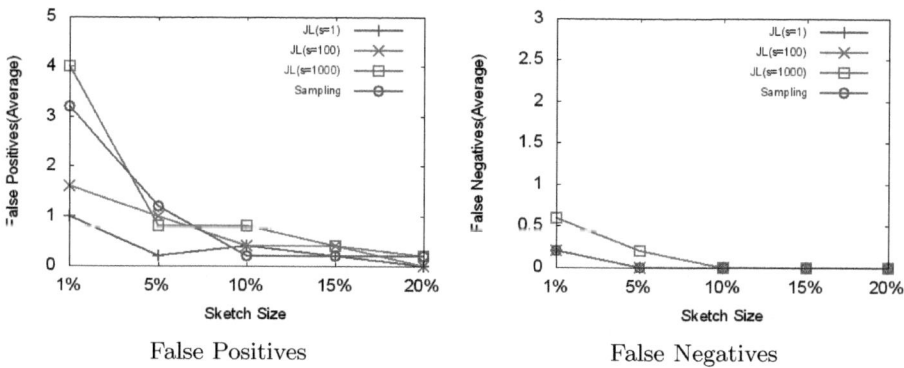

False Positives   False Negatives

Fig. 2. Association Mining Results Varying Sketch Size

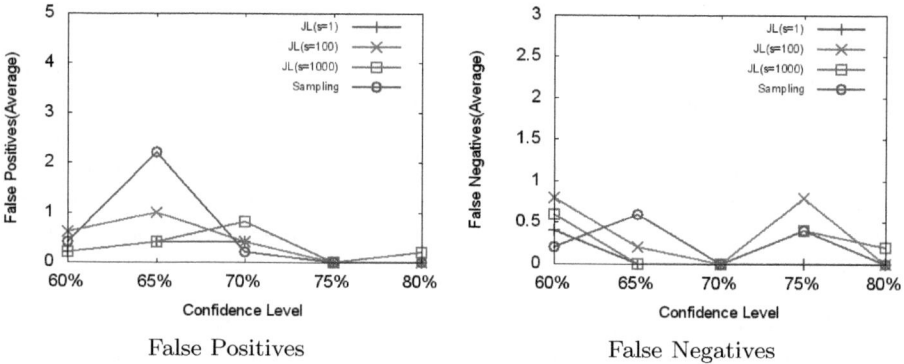

False Positives   False Negatives

Fig. 3. Association Mining Results Varying Confidence

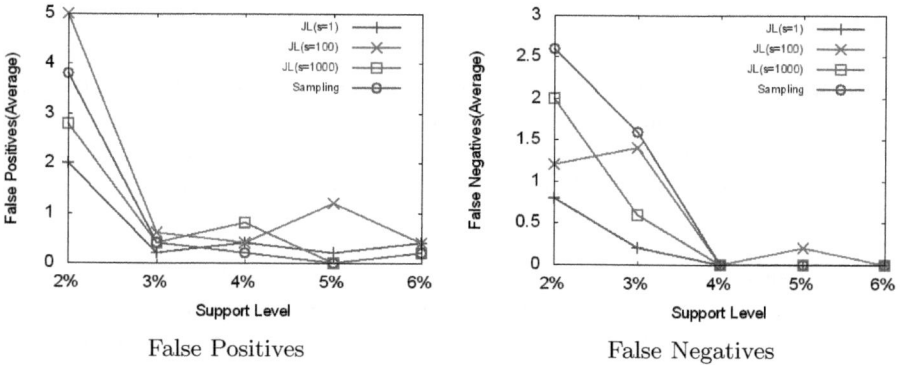

False Positives   False Negatives

Fig. 4. Association Mining Results Varying Support

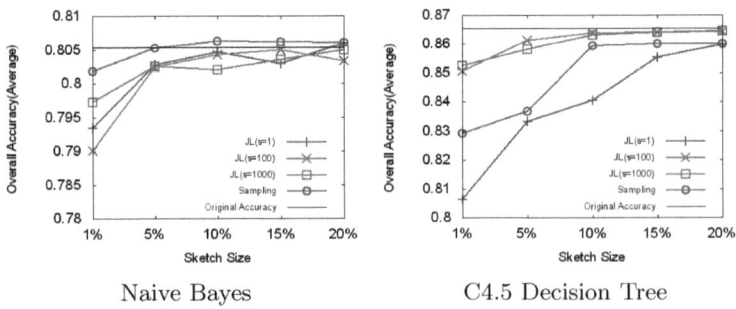

Naive Bayes   C4.5 Decision Tree

Fig. 5. Naive Bayes and C4.5 Results Varying Sketch Size

2 shows the results when we varied the compaction ratio. JL and sampling are very similar in terms of accuracy, with a slight overall edge to JL. Note that by the time we reach a compression ratio of 10%, no more false negatives arise in any JL sketching (regardless of sparsity), or in the sampling protocol.

Figures 3 and 4 show the results varying the required confidence and required support, respectively. As one might expect, there is no discernable correlation between these variables and the accuracy of the approximation for it. A larger error rate generally indicates that there are more itemsets near the exact required value, which means a smaller error in the dot product might result in the incorrect rejection or acceptance of an itemset. This is especially true for a support value of 2%, since below 2%, the number of supported itemsets increases dramatically.

Naive Bayes Classification. Figure 5 (left side) shows the results for naive Bayes classification. JL and sampling, again, perform quite similiarly. The accuracy, as expected, increases with the sketch size. The thin black line on the graph represents the accuracy of the naive Bayes classification on the original, uncompacted data. The accuracy of the approximation for both JL and sampling hovers right around the original accuracy, and in some cases performs better. This is understandable due to the machine learning phenomenon of *overfitting*. When a model is built on some data, it performs quite well on the data it was trained with, but the model will not perform as well on test data. When this happens, the model is said to overfit the training data. Often some noise is added to the model to remove the overfitting problem. The approximation of the dot product can provide such noise. Thus, the approximations can achieve higher accuracy than the exact result.

C4.5 Decision Tree. Figure 5 (right side) shows the results for C4.5 decision tree classification. The results are consistent with our findings in other tasks. Interestingly enough, the more sparse versions of JL outperformed the unabridged ($s = 1$) version. This is likely due to the fact that the sparse vectors provided slightly less distortion in the multiplication, resulting in a closer approximation for the dot product. In this case, as opposed to the naive Bayes case, the original tree provides a higher degree of accuracy, mainly because the C4.5 algorithm implements noise introduction by pruning the tree after building it.

6.2   Efficiency

In order to gauge the efficiency of our sketching protocols, we ran several timing experiments. The machine used was an AMD Athlon(tm) 64X2 dual core processor 4800T at 2.5 GHz with 2GB of RAM, running Windows Vista, and running on the Java 6 Standard runtime environment update 24. As our sub-protocol for exact dot product computation, we use the protocol of Goethals, et al [11], as it is provably secure, and lends itself well to improvement from our sketching protocols.

First, we ran several timing experiments computing the complete dot product of zero-one vectors of size 1000. The average time for the computation was 105 seconds. To ensure that the algorithm scaled linearly, we then ran it on vectors of size 2000, and the average computation time was 211 seconds. So, we determined the time-per-element in the dot product protocol to be .105 seconds. From this point forward, we computed the runtime of the approximate protocol in terms of the run time of the exact protocol by counting the time not involved in the computation of dot products, then adding it to the estimated dot product calculation time based on the previous timing experiments. The actual formula used was:

$$\frac{t_i + .105s \cdot n_d \cdot compactionRatio \cdot n}{.105s \cdot n_d \cdot n}$$

Where t_i is the time involved in the sketching, n_d is the number of dot products performed, n is the length of the vectors involved, and $compactionRatio$ is the fraction of the original vector's size which is retained by the sketching protocol. The results for three different sketching algorithms and five different compaction ratios are be are seen in figure 7.

In all cases, the algorithms are much faster than the exact algorithm. Because it produces a matrix with 1 or -1 for every value, JL with $s = 1$ has a large amount of pre-processing before it can apply the projection to each vector, which again, takes time. This runtime can be improved by using the sparsity factor. We chose, however, to present the worst case, as it is still much better than the original runtime. The association rules mining process involved the fewest number of dot products computed. Therefore, the preprocessing and other portions of the algorithms took up a greater percentage of the time in association rules mining. The Naive Bayes process had orders of magnitude more dot product calculations, so the overall time was dominated by the number of dot product calculations necessary.

In the decision tree case, the number of dot products computed varied with the algorithm involved. This is because we use the dot products to determine if a node is to be split. If a split is found to be not useful, the split will not occur. The compaction introduced enough error into the calculation that splits with very little information gain were not even attempted, resulting in much fewer dot products being calculated. The different algorithms all calculated far fewer dot products at every compaction level, resulting in a much greater efficiency increase.

Mining Task	Sketching Protocol	Compaction Ratio				
		1%	5%	10%	15%	20%
Association Mining	JL(s=1)	1.23301%	5.97532%	12.06352%	17.93215%	23.90036%
	JL(s=100)	1.10189%	5.50682%	11.01542%	16.62495%	22.19586%
	JL(s=1000)	1.09683%	5.43911%	10.97853%	16.49222%	22.01157%
	Sampling	1.07975%	5.09715%	10.08799%	15.07924%	20.08823%
Naive Bayes	JL(s=1)	1.10388%	5.50989%	11.01977%	16.52684%	22.03317%
	JL(s=100)	1.09024%	5.36809%	10.86241%	16.24925%	21.25196%
	JL(s=1000)	1.08882%	5.33216%	10.71943%	15.98638%	21.05157%
	Sampling	1.01317%	5.01338%	10.01391%	15.01437%	20.01472%
C4.5 Decision Tree	JL(s=1)	0.20356%	0.22841%	0.65546%	1.89563%	2.72094%
	JL(s=100)	0.18926%	0.19452%	0.59234%	1.71828%	2.64378%
	JL(s=1000)	0.17586%	0.19623%	0.58419%	1.65025%	2.61224%
	Sampling	0.02198%	0.19146%	0.80031%	1.44452%	2.53887%

Fig. 6. Efficiency Results: Percent of the Exact Algorithm Runtime

7 Conclusions

We have presented several interesting approximation techniques for the secure compuation of the dot product of two vectors. These protocols can be applied to many different data mining tasks, and can provide an efficiency increase to any protocol that uses a secure dot product as a sub-protocol.

7.1 Future Work

In the future, we plan to explore the use of these dot product protocols in other data mining tasks, such as support vector machines, neural networks, and clustering. We also plan to consider carefully the notion of a secure approximation, and determine to what extent the restrictions posed by our security model can be relaxed.

Acknowledgements. This work was partially supported by Air Force Office of Scientific Research MURI Grant FA9550-08-1-0265, National Institutes of Health Grant 1R01LM009989, National Science Foundation (NSF) Grant Career-CNS-0845803, and NSF Grants CNS-0964350, CNS-1016343. It is also based upon work supported by the AFOSR in-house project No. 11RI01COR, the AFRL in-house Job Order Number GGIHZORR, with AFRL/RI Information Information Institute VFRP No. 57739-1095380-1.

References

1. Achlioptas, D.: Database-friendly random projections: Johnson-lindenstrauss with binary coins. Journal of Computer and System Sciences 66(4), 671–687 (2003)
2. Aggarwal, C., Yu, P.: A general survey of privacy-preserving data mining models and algorithms. In: Privacy-Preserving Data Mining, pp. 11–52 (2008)

3. Agrawal, R., Srikant, R.: Privacy-preserving data mining. ACM Sigmod Record 29, 439–450 (2000)
4. Asuncion, A., Newman, D.: UCI machine learning repository (2007)
5. Clifton, C., Kantarcioglu, M., Vaidya, J., Lin, X., Zhu, M.: Tools for privacy preserving distributed data mining. ACM SIGKDD Explorations Newsletter 4(2), 28–34 (2002)
6. Du, W., Atallah, M.: Privacy-preserving cooperative statistical analysis. In: Proceedings of the 17th Annual Computer Security Applications Conference, p. 102. IEEE Computer Society (2001)
7. Dwork, C.: Differential Privacy: A Survey of Results. In: Agrawal, M., Du, D.-Z., Duan, Z., Li, A. (eds.) TAMC 2008. LNCS, vol. 4978, pp. 1–19. Springer, Heidelberg (2008)
8. Feigenbaum, J., Ishai, Y., Malkin, T., Nissim, K., Strausse, M., Wright, R.: Secure multiparty computation of approximations. ACM Transactions on Algorithms (TALG) 2(3), 435–472 (2006)
9. Fradkin, D., Madigan, D.: Experiments with random projections for machine learning. In: Proceedings of the Ninth ACM SIGKDD International Conference on Knowledge Discovery and Data Mining, pp. 517–522. ACM (2003)
10. Goethals, B.: Frequent itemset mining implementations repository (2005)
11. Goethals, B., Laur, S., Lipmaa, H., Mielikäinen, T.: On Private Scalar Product Computation for Privacy-Preserving Data Mining. In: Park, C.-S., Chee, S. (eds.) ICISC 2004. LNCS, vol. 3506, pp. 104–120. Springer, Heidelberg (2005)
12. Hoeffding, W.: Probability inequalities for sums of bounded random variables. Journal of the American Statistical Association 58(301), 13–30 (1965)
13. Huang, Z., Du, W., Chen, B.: Deriving private information from randomized data (2005)
14. Ioannidis, I., Grama, A., Attallah, M.: A secure protocol for computing the dot-products in clustered and distributed environments. In: International Conference on Parallel Processing, 2002, pp. 379–384. IEEE (2002)
15. Johnson, W., Lindenstrauss, J.: Extensions of lipschitz mappings into a hilbert space. Contemporary Mathematics 26(189-206), 1 (1984)
16. Kantarcioglu, M., Clifton, C.: Privacy-preserving distributed mining of association rules on horizontally partitioned data. IEEE Transactions on Knowledge and Data Engineering 16(9), 1026–1037 (2004)
17. Kantarcioglu, M., Nix, R., Vaidya, J.: An Efficient Approximate Protocol for Privacy-Preserving Association Rule Mining. In: Theeramunkong, T., Kijsirikul, B., Cercone, N., Ho, T.-B. (eds.) PAKDD 2009. LNCS, vol. 5476, pp. 515–524. Springer, Heidelberg (2009)
18. Kargupta, H., Datta, S., Wang, Q., Sivakumar, K.: On the privacy preserving properties of random data perturbation techniques. In: Third IEEE International Conference on Data Mining, ICDM 2003, pp. 99–106. IEEE (2003)
19. Li, P., Hastie, T., Church, K.: Very sparse random projections. In: Proceedings of the 12th ACM SIGKDD International Conference on Knowledge Discovery and Data Mining, pp. 287–296. ACM (2006)
20. Lindell, Y., Pinkas, B.: Privacy Preserving Data Mining. In: Bellare, M. (ed.) CRYPTO 2000. LNCS, vol. 1880, pp. 36–54. Springer, Heidelberg (2000)
21. Liu, K., Giannella, C., Kargupta, H.: An Attacker's View of Distance Preserving Maps for Privacy Preserving Data Mining. In: Fürnkranz, J., Scheffer, T., Spiliopoulou, M. (eds.) PKDD 2006. LNCS (LNAI), vol. 4213, pp. 297–308. Springer, Heidelberg (2006)

22. Liu, K., Kargupta, H., Ryan, J.: Random projection-based multiplicative data perturbation for privacy preserving distributed data mining. IEEE Transactions on Knowledge and Data Engineering, 92–106 (2006)
23. Menezes, A., Van Oorschot, P., Vanstone, S.: Handbook of applied cryptography. CRC (1997)
24. Pinkas, B.: Cryptographic techniques for privacy-preserving data mining. ACM SIGKDD Explorations Newsletter 4(2), 12–19 (2002)
25. Qiu, L., Li, Y., Wu, X.: Preserving privacy in association rule mining with bloom filters. Journal of Intelligent Information Systems 29(3), 253–278 (2007)
26. Ravikumar, P., Cohen, W., Feinberg, S.: A secure protocol for computing string distance metrics. In: Proceedings of the Workshop on Privacy and Security Aspects of Data Mining at the International Conference on Data Mining, pp. 40–46. IEEE (2004)
27. Vaidya, J., Clifton, C.: Privacy preserving association rule mining in vertically partitioned data. In: Proceedings of the Eighth ACM SIGKDD International Conference on Knowledge Discovery and Data Mining, pp. 639–644. ACM (2002)
28. Vaidya, J., Clifton, C.: Privacy preserving naive bayes classifier for vertically partitioned data. In: 2004 SIAM International Conference on Data Mining, Lake Buena Vista, Florida, pp. 522–526 (2004)
29. Vaidya, J., Clifton, C.: Privacy-Preserving Decision Trees over Vertically Partitioned Data. In: Jajodia, S., Wijesekera, D. (eds.) Data and Applications Security 2005. LNCS, vol. 3654, pp. 139–152. Springer, Heidelberg (2005)
30. Vaidya, J., Clifton, C.: Secure set intersection cardinality with application to association rule mining. Journal of Computer Security 13(4), 593–622 (2005)
31. Wang, W., Garofalakis, M., Ramchandran, K.: Distributed sparse random projections for refinable approximation. In: Proceedings of the 6th International Conference on Information Processing in Sensor Networks, pp. 331–339. ACM (2007)

Security Limitations of Using Secret Sharing for Data Outsourcing

Jonathan L. Dautrich and Chinya V. Ravishankar

University of California, Riverside
{dautricj,ravi}@cs.ucr.edu

Abstract. Three recently proposed schemes use secret sharing to support privacy-preserving data outsourcing. Each secret in the database is split into n shares, which are distributed to independent data servers. A trusted client can use any k shares to reconstruct the secret. These schemes claim to offer security even when k or more servers collude, as long as certain information such as the finite field prime is known only to the client. We present a concrete attack that refutes this claim by demonstrating that security is lost in all three schemes when k or more servers collude. Our attack runs on commodity hardware and recovers a 8192-bit prime and all secret values in less than an hour for $k = 8$.

1 Introduction

As cloud computing grows in popularity, huge amounts of data are being outsourced to cloud-based database service providers for storage and query management. However, some customers are unwilling or unable to entrust their raw sensitive data to cloud providers. As a result, privacy-preserving data outsourcing solutions have been developed around an *honest-but-curious* database server model. In this model, the server is trusted to correctly process queries and manage data, but may try to use the data it manages for its own nefarious purposes.

Private outsourcing schemes keep raw data hidden while allowing the server to correctly process queries. Queries can be issued only by a trusted client, who has insufficient resources to manage the database locally. Most such schemes use specialized encryption or complex mechanisms, ranging from order-preserving encryption [1], which has limited security and high efficiency, to oblivious RAM [2], which has provable access pattern indistinguishability but poor performance.

Three recent works [3–5] propose outsourcing schemes based on Shamir's secret sharing algorithm [6] instead of encryption. We refer to these works by their authors' initials, HJ [3], AAEMW [4], and TSWZ [5], and in aggregate as the HAT schemes. The AAEMW scheme was also published in [7]. In secret sharing, each sensitive data element, called a *secret*, is split into n *shares*, which are distributed to n data servers. To recover the secret, the client must combine shares from at least k servers. Secret sharing has perfect information-theoretic security when at most $k - 1$ of the n servers *collude* (exchange shares) [6].

Since secret sharing requires only k multiplications to reconstruct a secret, proponents argue that HAT schemes are faster than encryption based schemes.

N. Cuppens-Boulahia et al. (Eds.): DBSec 2012, LNCS 7371, pp. 145–160, 2012.

Other benefits of the HAT schemes include built-in redundancy, as only k of the n servers are needed, and additive homomorphism, which allows SUM queries to be securely processed by the server and returned to the client as a single value. Each of the HAT schemes makes two security claims:

Claim 1. *The scheme achieves perfect information-theoretic security when at most $k - 1$ servers collude.*

Claim 2. *When k or more servers collude, the scheme still achieves adequate security as long as certain information used by the secret sharing algorithm, namely a prime p and a vector $\boldsymbol{X}$, are kept private, known only to the client.*

It is doubtful that the HAT schemes truly fulfill Claim 1, as they sort shares by secret value, which certainly reveals some information about the data [8]. Further, the AAEMW scheme [4] uses correlated coefficients in the secret sharing algorithm instead of random ones, which also contradicts this claim. Nevertheless, for the purposes of this work, we assume that Claim 1 holds.

We are primarily concerned with evaluating Claim 2, which asserts that even k or more colluding servers cannot easily recover secrets. Claim 2 is stated in Sect. 4.5 of [3], Sect. 3 of [4], Sects. 2.2 and 6.1 of [5], and Sect. 3 of [7].

1.1   Our Contribution

Our contribution is to demonstrate that all three HAT schemes [3–5] fail to fulfill Claim 2. We give a practical attack that can reconstruct all secrets in the database when k servers collude, even when p and $\boldsymbol{X}$ are kept private. Our attack assumes that the servers know, or can discover, at least $k + 2$ secrets. To limit data storage costs, k is kept small ($k \approx 10$), so discovering $k + 2$ secrets is feasible (see Sect. 4.2). All three HAT schemes argue that they fulfill Claim 2, so our result provides a much-needed clarification of their security limitations.

The TSWZ scheme [5] argues that if p is known, secrets could be recovered. However, it provides no attack description, and argues that large primes are prohibitively expensive to recover. Our attack recovers 8192-bit primes in less time than [5] needed to recover 32-bit primes. In fact, we can generally recover large primes in less time than the client takes to generate them (Sect. 5.1).

In Sect. 2 we review Shamir's secret sharing algorithm and how it is used for private outsourcing. In Sect. 3 we give assumptions and details of our attack, and we show how to align shares and discover secrets in Sect. 4. We give experimental runtime results in Sect. 5, and discuss possible attack mitigations in Sect. 6. We discuss related work in Sect. 7 and conclude with Sect. 8.

2   Data Outsourcing Using Secret Sharing

We now review Shamir's secret sharing scheme and show how it is used for private data outsourcing in the HAT schemes [3–5]. We use an employee table with m records as our driving example, where queries are issued over the salary attribute. Each salary $s_1, \ldots, s_m$ is a secret that is shared among the n data servers (see Fig. 1).

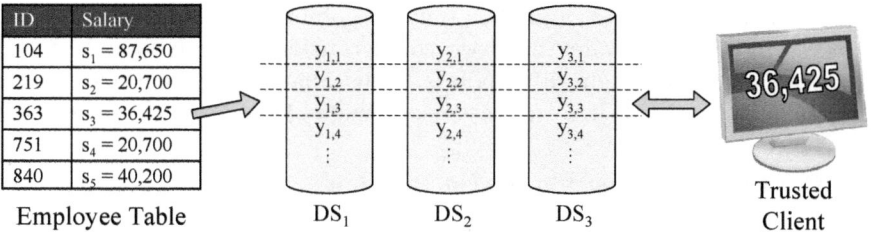

Fig. 1. Secret (salary) data from an employee table is split into shares and distributed to multiple data servers. A trusted client queries shares from the data servers and combines them to recover the secrets.

2.1 Shamir's Secret Sharing

Shamir's secret sharing scheme [6] is designed to share a single secret value s_j among n servers such that shares must be obtained from any k servers in order to reconstruct s_j. The scheme's security rests on the fact that at least k points are needed to uniquely reconstruct a polynomial of degree $k-1$. Theoretically, the points and coefficients used in Shamir's scheme can be taken from any field $\mathbb{F}$. However, to use the scheme on finite-precision machines, we require $\mathbb{F}$ to be the finite field $\mathbb{F}_p$, where p is a prime.

To share s_j, we choose a prime $p > s_j$, and $k-1$ coefficients $a_{1,j}, \ldots, a_{k-1,j}$ selected randomly from $\mathbb{F}_p$. We then construct the following polynomial:

$$q_j(x) = s_j + \sum_{h=1}^{k-1} a_{h,j} x^h \quad \bmod p \tag{1}$$

We then generate a vector $\boldsymbol{X} = (x_1, \ldots, x_n)$ of distinct elements in $\mathbb{F}_p$, and for each data server DS_i, we compute the *share* $y_{i,j} = q_j(x_i)$. Together, x_i and $y_{i,j}$ form a point $(x_i, y_{i,j})$ through which polynomial $q_j(x)$ passes.

Given any k such points $(x_1, y_{1,j}), \ldots, (x_k, y_{k,j})$, we can reconstruct the polynomial $q_j(x)$ using Lagrange interpolation:

$$q_j(x) = \sum_{i=1}^{k} y_{i,j} \ell_i(x) \quad \bmod p \tag{2}$$

where $\ell_i(x)$ is the Lagrange basis polynomial:

$$\ell_i(x) = \prod_{1 \le j \le k, j \ne i} (x - x_j)(x_i - x_j)^{-1} \quad \bmod p \tag{3}$$

and $(x_i - x_j)^{-1}$ is the multiplicative inverse of $(x_i - x_j)$ modulo p.

The secret s_j is the polynomial q_j evaluated at $x = 0$, so we get:

$$s_j = \sum_{i=1}^{k} y_{i,j} \ell_i(0) \quad \bmod p \tag{4}$$

Given only $k' < k$ shares, and thus only k' points, we cannot learn anything about s, since for any value of s, we could construct a polynomial of degree $k - 1$ that passes through all k' points. Thus Shamir's scheme offers perfect, information-theoretic security against recovering s_j from fewer than k shares [6].

2.2 Data Outsourcing via Secret Sharing

We now describe the mechanism used by all three HAT schemes to support private outsourcing via secret sharing. We first choose a single prime p and a vector $\boldsymbol{X}$, which are the same for all secrets and will be stored locally by the client. For each secret s_j, we generate coefficients $a_{1,j}, \ldots, a_{k-1,j}$, and produce a polynomial $q_j(x)$ as in (1). We then use q_j to split s_j into n shares $y_{1,j}, \ldots, y_{n,j}$, where $y_{i,j} = q_j(x_i)$, and distribute each share $y_{i,j}$ to server DS_i, as in Fig. 1.

An important distinction is that the AAEMW scheme performs secret sharing over the real number field $\mathbb{R}$, so there is no p to choose. However, since the scheme must run on finite precision hardware, any implementation will suffer from roundoff error. Our attack works over $\mathbb{R}$, and is efficient because the field is already known. However, we expect that in practice, the AAEMW scheme will switch to a finite field $\mathbb{F}_p$, so we do not treat it as a special case.

When the client issues a point query for the salary s_j of a particular employee, he receives a share from each of the n servers. Using any k of these shares, he can recover s_j using the interpolation equation (4). Other query types, including range and aggregation queries, are supported by the HAT schemes. We give some relevant details in Sect. 2.4, but the rest can be found in [3–5].

$\boldsymbol{X}$ and p are re-used across secrets for two reasons. First, storing distinct $\boldsymbol{X}$ or p on the client for each secret would require at least as much space as storing the secret itself. Second, when the same $\boldsymbol{X}$ and p are used, the secret sharing scheme has additive homomorphism. That is, if each server DS_i adds shares $y_{i,1} + y_{i,2}$, and we interpolate using those sums, the recovered value is the sum $s_1 + s_2$. With additive homomorphism, when the client issues a SUM query, the server can sum the relevant shares, and return a single value to the client, instead of returning shares separately and having the client perform the addition. Using a different $\boldsymbol{X}$ or p for each secret breaks additive homomorphism.

2.3 Security

If $\boldsymbol{X}$ and p are public, and k or more servers collude, then the HAT schemes are clearly insecure, as the servers could easily perform the interpolation themselves. On the other hand, if at most $k - 1$ servers collude, and coefficients are chosen independently at random from $\mathbb{F}_p$, then the servers learn nothing about a secret by examining its shares, and Claim 1 is fulfilled.

The HAT schemes state that by keeping $\boldsymbol{X}$ [3, 4] and p [5] private, they achieve security even when k or more servers collude (Claim 2). Our attack shows that any k colluding servers can recover all secrets in the database, even when $\boldsymbol{X}$ and p are unknown (Sect. 3), contradicting Claim 2.

2.4 Supporting Range and Aggregation Queries

We can use the mechanisms that support range and SUM queries in the HAT schemes to reveal the order of the shares on each server according to their corresponding secret values. We then use these orders to align corresponding shares across colluding servers, and to discover key secret values (see Sect. 4).

The AAEMW scheme [4] crafts coefficients such that the shares preserve the order of their secrets. The HJ and TSWZ schemes [3, 5] both use a single B$^+$ tree to order each server's shares and facilitate range queries. TSWZ assumes the tree is accessible to all servers, while HJ assumes it is on a separate, non-colluding index server. HJ obscures share order from the servers, but we can reconstruct it by observing range queries over time (see Sect. 4.3).

3   Attack Description

We now show that the HAT schemes are insecure when k or more servers collude, even if X and p are kept private. Our attack efficiently recovers all secret values (salaries in Fig. 1) stored in the database, and relies on the following assumptions:

1. At least k servers collude, exchanging shares or other information.
2. The number of servers k and the number of bits b in prime p are modest: $k \approx 13$, $b \approx 2^{13}$. None of the HAT schemes give recommended values for k or b, with the exception of a brief comment in [4] alluding to 16-bit primes originally suggested by Shamir. In practice, primes with more than 2^{13} bits take longer for the client to generate than for our attack to recover, and the cost of replicating data to every server keeps k small.
3. X and p are unknown, and are the same for each secret (see Sect. 2.2).
4. Each set of k corresponding shares can be *aligned*. That is, the colluding servers know which shares correspond to the same secret, without knowing the secret itself. We can align shares if we know share orders (see Sect. 4.1).
5. At least $k + 2$ secrets, and which shares they correspond to, are known or can be discovered. Since k is modest, knowing $k + 2$ secrets is reasonable, especially when the number of secrets m is large (see Sect. 4.2).

In Sect. 6, we show that modifying the HAT schemes to violate these assumptions sacrifices performance, functionality, or generality, eroding the schemes' slight advantages over encryption based techniques.

3.1 Recovering Secrets When p Is Known and X Is Private

As a stepping stone to our full attack, we show how to recover secrets if p is already known. Without loss of generality, let $s_1, \ldots, s_k$ be known secrets, and let $DS_1, \ldots, DS_k$ be the colluding servers. For each secret s_j, we have shares $y_{1,j}, \ldots y_{k,j}$, generated by evaluating $q_j(x)$ at $x_1, \ldots, x_k$, respectively. We therefore have a system of k^2 equations of the form $y_{i,j} = s_j + \sum_{h=1}^{k-1} a_{h,j} x_i^h \mod p$, as in (1). The system has $k(k-1)$ unknown coefficients $a_{h,j}$, and k unknown

x_i, giving k^2 equations in k^2 unknowns. Thus, it would seem we can solve for the relevant values of $\boldsymbol{X}$, which would allow us to recover the remaining secrets. Unfortunately, the system is non-linear, so naively solving it directly requires expensive techniques such as Groebner basis computation [9].

Instead, we can recover the remaining secrets without solving for $\boldsymbol{X}$. Consider the following system of equations obtained by applying the interpolation equation (4) to each of the k secrets:

$$
\begin{aligned}
y_{1,1}\ell_1(0) + y_{2,1}\ell_2(0) + \cdots + y_{k,1}\ell_k(0) - s_1 &\equiv 0 \pmod{p} \\
y_{1,2}\ell_1(0) + y_{2,2}\ell_2(0) + \cdots + y_{k,2}\ell_k(0) - s_2 &\equiv 0 \pmod{p} \\
&\vdots \\
y_{1,k}\ell_1(0) + y_{2,k}\ell_2(0) + \cdots + y_{k,k}\ell_k(0) - s_k &\equiv 0 \pmod{p}
\end{aligned}
\tag{5}
$$

If we treat each basis polynomial value $\ell_i(0)$ as an unknown, we get k unknowns $\ell_1(0), \ldots, \ell_k(0)$, which we call *bases*, in k linear equations. Since we know p, we can easily solve (5) using Gaussian elimination and back-substitution. We can then use the bases to recover the remaining secrets in the database via (4).

We can construct (5) since we know that all shares from a given server DS_i were obtained from the same x_i, and thus should be multiplied by the same base $\ell_i(0)$. The client could obscure the correspondence between shares by mixing shares among servers, but would be forced to store i with each share in order to properly reconstruct the secret. To completely hide the correspondence, i itself would need to be padded and encrypted, which is precisely what secret sharing tries to avoid. Further, mixing the shares would break additive homomorphism.

3.2   Recovering p When X and p Are Private

Let b be the number of bits used to represent p. We can easily have $b > 2^6$, so enumerating possible values for p is not practical. However, we can recover p by exploiting known shares and the $k + 2$ known secrets. Our attack identifies two composites δ_1 and δ_2 both divisible by p ($p|\delta_1, p|\delta_2$), such that the remaining factors of δ_1, δ_2 are largely independent. We then take δ' to be the greatest common divisor of δ_1 and δ_2, and factor out small primes from δ', leaving us with $\delta' = p$ with high probability. Once p is known, we can use the attack from Sect. 3.1 to recover the bases and the remaining, unknown secrets.

Computing δ_1, δ_2. Without loss of generality, we let $s_1, \ldots, s_{k+2}$ be the known secrets. To compute δ_γ, $\gamma \in \{1, 2\}$, we consider the system of interpolation equations for secrets $s_\gamma, \ldots, s_{\gamma+k}$ as in (5), represented by the following $(k + 1) \times (k + 1)$ matrix:

$$
\begin{bmatrix}
y_{1,\gamma} & y_{2,\gamma} & \cdots & y_{k,\gamma} & -s_\gamma \\
y_{1,\gamma+1} & y_{2,\gamma+1} & \cdots & y_{k,\gamma+1} & -s_{\gamma+1} \\
\vdots & & \ddots & & \vdots \\
y_{1,\gamma+k-1} & y_{2,\gamma+k-1} & \cdots & y_{k,\gamma+k-1} & -s_{\gamma+k-1} \\
y_{1,\gamma+k} & y_{2,\gamma+k} & \cdots & y_{k,\gamma+k} & -s_{\gamma+k}
\end{bmatrix}
\tag{6}
$$

Since p is unknown, we cannot compute inverses modulo p and thus cannot divide as in standard Gaussian elimination. However, we can still convert (6) to upper triangular (row echelon) form using only multiplications and subtractions.

We start by eliminating coefficients for $\ell_1(0)$ from all but the first row ($j = \gamma$). To eliminate $\ell_1(0)$ from row $j > \gamma$, we multiply the contents of row γ through by $y_{1,j}$, and of row j by $y_{1,\gamma}$, producing a common coefficient for $\ell_1(0)$ in both rows. We then subtract the multiplied row γ from the multiplied row j, canceling the coefficient for $\ell_1(0)$. Row 1 is left unchanged, but row j now has coefficient 0 for $\ell_1(0)$, and coefficient $(y_{i,j})(y_{1,\gamma}) - (y_{i,\gamma})(y_{1,j})$ for $\ell_i(0), i \geq 2$:

$$
\begin{bmatrix}
y_{1,\gamma} & y_{2,\gamma} & \cdots & -s_\gamma \\
0 & (y_{2,\gamma+1})(y_{1,\gamma}) - (y_{2,\gamma})(y_{1,\gamma+1}) & \cdots & (-s_{\gamma+1})(y_{1,\gamma}) - (-s_\gamma)(y_{1,\gamma+1}) \\
\vdots & & \ddots & \vdots \\
0 & (y_{2,\gamma+k})(y_{1,\gamma}) - (y_{2,\gamma})(y_{1,\gamma+k}) & \cdots & (-s_{\gamma+k})(y_{1,\gamma}) - (-s_\gamma)(y_{1,\gamma+k})
\end{bmatrix}
$$

We then repeat the process, eliminating successive coefficients from lower rows, until the matrix is in upper triangular form:

$$
\begin{bmatrix}
y_{1,\gamma} & y_{2,\gamma} & \cdots & y_{k,\gamma} & -s_\gamma \\
0 & c_{2,\gamma+1} & \cdots & c_{k,\gamma+1} & c_{k+1,\gamma+1} \\
\vdots & & \ddots & & \vdots \\
0 & 0 & \cdots & c_{k,\gamma+k} & c_{k+1,\gamma+k} \\
0 & 0 & \cdots & 0 & \delta_\gamma
\end{bmatrix}
\tag{7}
$$

We use $c_{i,j}$ values to denote constants. In the last row of (7), the coefficient for every $\ell_i(0)$ is 0, so the row represents the equation $\delta_\gamma \equiv 0 \pmod{p}$. Thus, $p|\delta_\gamma$.

Size of δ_1, δ_2. As coefficients for successive $\ell_i(0)$ are eliminated, each non-zero cell below the ith row is set to the difference of products of two prior cell values, doubling the number of bits required by the cell. Thus, the number of bits per cell in (7) is given by:

$$
\begin{bmatrix}
b & b & \cdots & b & b \\
0 & 2^1 b & \cdots & 2^1 b & 2^1 b \\
\vdots & & \ddots & & \vdots \\
0 & 0 & \cdots & 2^{k-1} b & 2^{k-1} b \\
0 & 0 & \cdots & 0 & 2^k b
\end{bmatrix}
$$

As a result, each of δ_1, δ_2 has at most $2^k b$ bits. This is closely related to the result that in the worst case, simple integer Gaussian elimination leads to entries that are exponential in the matrix size [10].

Recovering p from δ_1, δ_2. Since δ_1, δ_2 have $2^k b$ bits, and p has only b bits, it is likely that δ_1, δ_2 both have some prime factors larger than p, so factoring them directly is not feasible. Instead, we take $\delta' = gcd(\delta_1, \delta_2)$, where gcd is the greatest common divisor function, which can be computed using the traditional Euclidean algorithm, or more quickly using Stein's algorithm [11].

Since δ_1 and δ_2 were obtained using different elimination orders and sets of secrets, they rarely share large prime factors besides p, so all other prime factors of δ' should be small. Thus, we can factor δ' by explicitly dividing out all prime factors with at most β bits, leaving behind only p, with high probability. We know that p is larger than all shares, so to avoid dividing out p itself, we never divide out primes that are larger than the largest known share. We have found empirically that the probability that δ_1, δ_2, as computed above, share a factor with more than β bits can be approximated by $\frac{2^{(k-2)/4}}{2^{\beta+1}}k$ for the values of β, k we are interested in (Sect. 5.2). Our attack fails if δ_1, δ_2 share such a factor, but we can make the failure rate arbitrarily low by increasing β.

3.3 Attack Complexity

Since δ_1 and δ_2 are both $(2^k b)$-bit integers, the time required to find $gcd(\delta_1, \delta_2)$ is in $O(2^{2k}b^2)$ [11]. As k grows, storing δ_1, δ_2 and computing their gcd quickly become the dominant space and time concerns, respectively. Thus, recovering p has space complexity $O(2^k b)$ and time complexity $O(2^{2k}b^2)$.

Recovering the bases, once p is known, has space complexity $O(k^2 b)$ for storing the matrix, and time complexity dominated either by computing $O(k^3)$ b-bit integer multiplications during elimination, or $O(k)$ modular inverses during back-substitution. Clearly, these costs are dominated by the costs of recovering p. Once p and the bases have been recovered, the time spent recovering a secret is the same for the colluding servers as it is for the trusted client.

3.4 Example Attack for $k = 2$

We now demonstrate our attack on a simple dataset with $m = 6$ records shared over $n = k = 2$ servers. We choose the 6-bit prime $p = 59$ and select $x_1 = 17, x_2 = 39$. We then generate secrets, coefficients, and shares as follows:

$s_1 = 18$	$a_{1,1} = 18$	$q(x_1, s_1) = 29$	$q(x_2, s_1) = 12$
$s_2 = 36$	$a_{1,2} = 5$	$q(x_1, s_2) = 3$	$q(x_2, s_2) = 54$
$s_3 = 22$	$a_{1,3} = 17$	$q(x_1, s_3) = 16$	$q(x_2, s_3) = 36$
$s_4 = 10$	$a_{1,4} = 28$	$q(x_1, s_4) = 14$	$q(x_2, s_4) = 40$
$s_5 = 39$	$a_{1,5} = 31$	$q(x_1, s_5) = 35$	$q(x_2, s_5) = 9$
$s_6 = 57$	$a_{1,6} = 51$	$q(x_1, s_6) = 39$	$q(x_2, s_6) = 40$

We assume s_1, s_2, s_3, s_4 are known. We first generate the matrix in (6) using s_1, s_2, s_3 ($\gamma = 1$), and do the following elimination to get $\delta_1 = 307980$:

$$\begin{bmatrix} 29 & 12 & -18 \\ 3 & 54 & -36 \\ 16 & 36 & -22 \end{bmatrix} \rightarrow \begin{bmatrix} 29 & 12 & -18 \\ 0 & 1530 & -990 \\ 0 & 852 & -350 \end{bmatrix} \rightarrow \begin{bmatrix} 29 & 12 & -18 \\ 0 & 1530 & -990 \\ 0 & 0 & 307980 \end{bmatrix}$$

We do the same with s_2, s_3, s_4 ($\gamma = 2$) to get $\delta_2 = -33984$:

$$\begin{bmatrix} 3 & 54 & -36 \\ 16 & 36 & -22 \\ 14 & 40 & -10 \end{bmatrix} \rightarrow \begin{bmatrix} 3 & 54 & -36 \\ 0 & -756 & 510 \\ 0 & -636 & 474 \end{bmatrix} \rightarrow \begin{bmatrix} 3 & 54 & -36 \\ 0 & -756 & 510 \\ 0 & 0 & -33984 \end{bmatrix}$$

We then compute $\delta' = gcd(\delta_1, \delta_2) = 2124$, and factor δ' by dividing out the small prime factors $2 \cdot 2 \cdot 3 \cdot 3$, to get $p = 59$, as expected. Now we can recover the bases using the following system of equations as in (5):

$$29\ell_1(0) + 12\ell_2(0) - 18 \equiv 0 \pmod{59}$$
$$3\ell_1(0) + 54\ell_2(0) - 36 \equiv 0 \pmod{59}$$

We eliminate $\ell_1(0)$ from the second equation, giving $55\ell_2(0) \equiv 46 \pmod{59}$. We then compute the inverse $(55)^{-1} \pmod{59} = 44$, giving $\ell_2(0) = 46 \cdot 44 \mod 59 = 18$. We then back-substitute to get $\ell_1(0) = 42$. To verify, we compute s_5 and s_6 using (4), giving $s_5 = 35 \cdot 42 + 9 \cdot 18 \pmod{59} = 39$, and $s_6 = 39 \cdot 42 + 40 \cdot 18 \pmod{59} = 57$, as expected.

4   Aligning Shares and Discovering Secrets

In order to mount our attack, we must be able to *align* shares across colluding servers. That is, given the set of shares from each of k servers, we must be able to identify which subsets of k shares, one from each server, were obtained from the same polynomial q_j (1), even if we do not know the secret value s_j itself. Further, we must know, or be able to discover, at least $k + 2$ secret values and the subset of k shares to which they correspond. We now show how we can satisfy these assumptions for the HAT schemes [3–5] using knowledge of *share order*.

In the AAEMW [4] and TSWZ [5] schemes, the shares on each server are explicitly, totally ordered (Sect. 2.4). The share order sorts the shares in non-decreasing order of their corresponding secrets. If two shares are obtained from distinct polynomials, but the same secret, they have the same relative order on each server. In the HJ scheme [3], shares are totally ordered, but the order is hidden from the data servers. In this case, we can infer a partial share order by observing queries over time.

4.1   Aligning Shares

When the total share order on each data server is known, we simply align the jth share from each server. If only a partial order is known, as in the HJ scheme, the alignment of some shares will be ambiguous. To recover secrets for such shares, we must either try multiple alignments, or wait for more queries to arrive, and use them to refine the partial order and eliminate the ambiguity (see Section 4.3).

4.2   Discovering $k + 2$ Secrets

We have shown that given $k + 2$ secrets and their corresponding shares, our attack can recover all remaining secrets. This weakness is a severe limitation of the HAT schemes, and contradicts Claim 2 (Sect. 1). In practice, $k \ll m$, where m is the number of secrets, so assuming $k + 2$ known secrets is reasonable. Our attack is independent of the mechanism used to discover these secrets.

Simple methods for learning $k + 2$ secrets include a known plaintext attack, where we convince the trusted client to insert $k + 2$ known secrets, and a known ciphertext attack, where the client reveals at least $k+2$ secrets retrieved by some small range query. Since shares are ordered according to their secret values, we can easily identify which subsets of shares from the query go with each secret.

We can also infer secret values using share order. Consider an employee table with secret salaries, as in Fig. 1. If at least $k + 2$ employees earn a well-known minimum-wage salary, then the share order reveals that the first $k+2$ shares have this known salary. Alternatively, there may be $k+2$ employees who anonymously post their salaries. If we can estimate the distribution of salaries in the database, we can guess roughly where the known salaries fall in the order, and run the attack for nearby guesses until we get a solution with a recoverable prime and recovered secrets that fit the expected order.

4.3   Inferring Order in the HJ Scheme

If a scheme hides the share order from the data server, share alignment and secret discovery become harder. The HJ scheme [3] stores the share order for each data server on a single index server that ostensibly does not collude with any data servers. The client sends each query to the index server, and the response tells the client which shares to request from each data server.

In the simplest case, we can align shares by observing point queries, which return only one share from each server. If the colluding servers all observe an isolated request for a single share at the same time, they can assume the shares satisfy a point query, and thus that they all correspond to the same secret. Given enough point queries, we can align enough shares to mount our attack. However, if point queries are rare, this technique will take too long to be useful.

More generally, we can order shares on each server by observing overlapping range queries. In the HJ scheme, a range query appears to the data server as set of unordered share requests. Since range queries request shares that have secrets inside a given range, we know that secrets of requested shares are contiguous. We use this information to order shares according to their secret values.

Consider an example where a client issues two range queries to the same data server. The first query returns shares $\{y_1, y_2\}$, and the second, shares $\{y_3, y_2\}$. Each query is a range query, so the server knows that no secret can fall between the secrets of y_1 and y_2 or of y_3 and y_2. Since y_2 appears in both queries, the server knows that the secret of y_2 comes between the secrets of y_1 and y_3, but is not sure whether the secret of y_1 or of y_3 is smaller. Thus, the true share order contains either subsequence $y_1y_2y_3$ or $y_3y_2y_1$, and we say that the server knows the share order of $\{y_1, y_2, y_3\}$ up to symmetry (see Fig. 2).

Query {y₁, y₂}   Query {y₃, y₂}

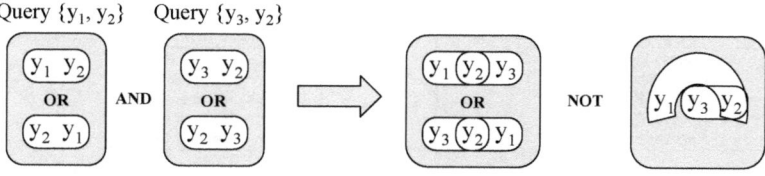

Fig. 2. Range queries indicate that the secrets of shares y_1, y_2 are contiguous, as are those of y_3, y_2. Thus, the secret of y_2 falls between the secrets of y_1 and y_3, though either y_1 or y_3 may have the smallest secret.

We can extend this technique to additional range queries of varying sizes. Given enough queries, we can reconstruct the entire share order on each server up to symmetry. The full reconstruction algorithm uses PQ-trees [12], but is out of scope for this paper. We can link reconstructed share orders across servers, and thereby align shares, by observing that if a query issued to one data server requests the jth share, then the same query must also request the jth share from every other server. If we use the share order to discover secrets, we must make twice as many guesses, since we still only know the order up to symmetry.

5   Attack Implementation and Experiments

We implemented our attack in Java, and ran each of our attack trials using a single thread on a 2.4GHz Intel® Core™2 Quad CPU. All trials used less than 2GB RAM. We used two datasets. The first consists of $m = 1739$ maximum salaries of Riverside County (California, USA) government employees as of February, 2012 [13]. The second is a set of $m = 10^5$ salaries generated uniformly at random from the integer range $[0, 10^7)$.

5.1   Time Measurements

Our first set of experiments measures the time required to run the full attack as described in Sect. 3. Each experiment varies the number of servers k or the number of bits b in the hidden prime p. The total number of servers n has no effect on the attack runtime, so we let $n = k$. All times are averaged over 10 independent trials, and averages are rounded up to a 1ms minimum. In each trial, we divide out primes with at most $\beta = 16$ bits (Sect. 3.2), and we successfully recover p, all k bases ($\ell_i(0)$ values), and all m secrets.

Each plot gives the times spent by the client finding a random b-bit prime p and creating k shares for each of the m secrets. We then plot the times spent by the colluding servers recovering p and the k bases. We also give the time spent recovering all m secrets, which is the same for the colluding servers as it is for the client. From Sect. 3.3, we know that the time needed to recover p is in $O(2^{2k}b^2)$. Thus, incrementing k or $\log_2 b$ increases prime recovery time by a factor of 4. Since k and $\log_2 b$ have similar effects on prime recovery time, we plot against $\log_2 b$ instead of b on the x axis.

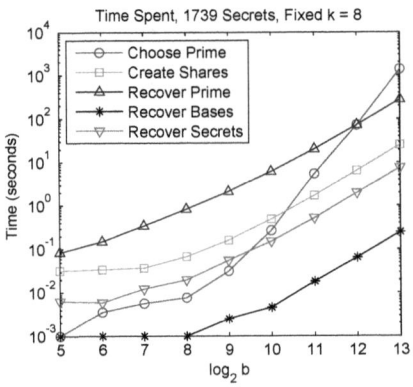 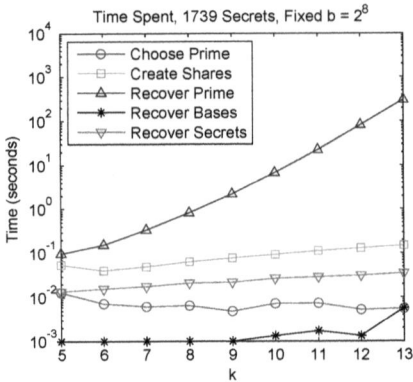

Fig. 3. Riverside dataset times, varied b **Fig. 4.** Riverside dataset times, varied k

Figures 3 and 4 plot times using the Riverside dataset with fixed $b = 2^8$ and $k = 8$, respectively. Figures 5 and 6 give corresponding times for the random dataset. Times to create shares and recover secrets are proportional to m, and so are higher for the larger, random dataset. Times to generate p, recover p, and recover bases depend only on b and k, and so are dataset-independent.

Figures 3 and 5 show that when k is held constant, increasing b costs the client more than it costs the colluding servers. Both prime recovery and modular multiplication take time proportional to b^2, so prime recovery time is a constant factor of share generation time. Further, the time to choose a random b-bit prime using the Miller-Rabin primality test is in $O(b^3)$ [14], so as b grows past 2^{12}, the cost to generate p quickly outstrips the cost to recover it. Thus it is entirely impractical to thwart our attack by increasing b.

In the TSWZ scheme [5], the measured time to recover a prime with less than 2^5 bits was over 1500 seconds. In contrast, our method recovers primes with 2^{13} bits in under 500 seconds on comparable hardware, for $k = 8$. As long as $k \ll b$, as is likely in practice, our method is far faster.

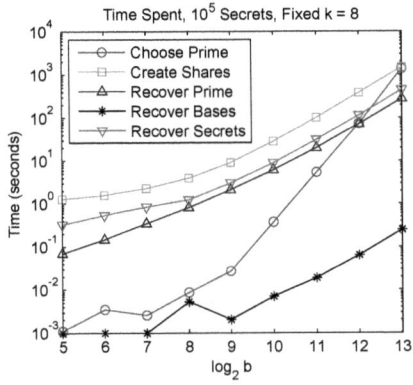 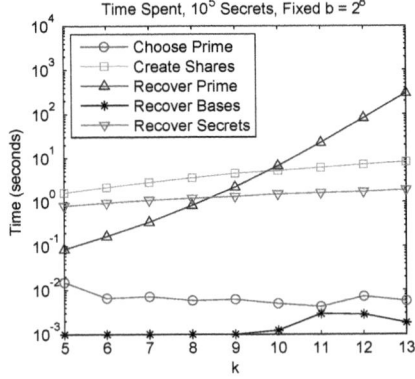

Fig. 5. Synthetic dataset times, varied b **Fig. 6.** Synthetic dataset times, varied k

Figures 4 and 6 show that when b is fixed, most times are in $O(k)$, with the exception of prime recovery time, which is in $O(2^{2k})$. Thus, by increasing k, the attack can be made arbitrarily expensive at a relatively small cost to the client. However, as we discuss in Sect. 6, even $k = 10$ may be impractical.

5.2 Failure Rate Measurements

Since we only factor out small primes with at most β bits (Sect. 3.2), our attack fails if δ_1, δ_2 share any prime factor, other than p, that has more than β bits. Thus, our attack's failure rate r_f is the probability that $\delta_1/p, \delta_2/p$ share a prime with more than β bits. Since δ_1, δ_2 are not independent random numbers, it is difficult to compute r_f analytically, so we measure it empirically. The results of our experiment are shown in Fig. 7.

We found that r_f is largely independent of b, but depends heavily on k and β. To measure r_f, we conducted several trials in which we generated a prime p of $b = 32$ bits, and ran our attack using $k + 2$ randomly generated secrets. For $k = 2$, we ran 10^6 trials, and were able to get meaningful failure rates up through $\beta \approx 16$. Trials with larger k were much more expensive, so we only ran 10^3 trials for $k = 6$ and $k = 10$, and the results are accurate only through $\beta \approx 10$.

From our results, we derived the approximate expression $r_f \approx \frac{2^{(k-2)/4}}{2^{\beta+1}} k$. We then plotted this estimated r_f in Fig. 7, denoted by *est*. The approximation is adequate for the range of β we're considering. The dependence of r_f on $2^{-(\beta+1)}$ is expected, as the probability that a factor of $\beta+1$ bits found in one random d-bit number is found in another random d-bit number is roughly $\frac{2^{d-(\beta+1)}}{2^d} = 2^{-(\beta+1)}$. The nature of the dependence on k is unclear, but it may be related to the fact that k of the $k+1$ equations used to compute δ_1 are also used to compute δ_2.

Using our approximation for r_f, we estimate the worst-case failure rate for our timing experiments, where $\beta = 16$ and $k = 13$, to be $r_f \approx \frac{2^{(13-2)/4}}{2^{16+1}} 13 = 2^{-14.25} 13 \approx 6.67 \times 10^{-4}$. If necessary, we can lower r_f further by increasing β.

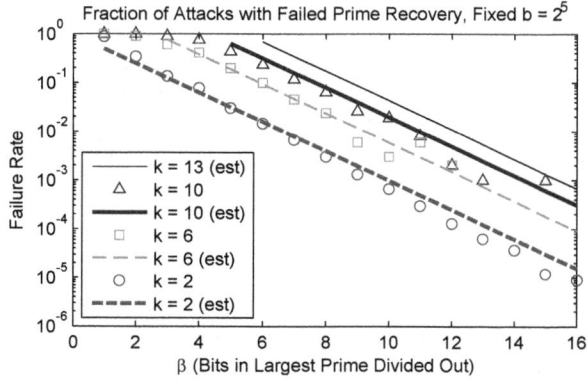

Fig. 7. Attack failure rates for varied k and β

6   Attack Mitigations

We now discuss possible modifications a client can make to the HAT schemes that may improve security. In order to mitigate our attack, a modification must cause at least one of the attack assumptions listed in Sect. 3 to be violated.

Assumption: At Least k Servers Collude. The simplest way to thwart our attack is to ensure that no more than $k - 1$ servers are able to collude. Only in such cases can secret sharing schemes hope to achieve perfect, information-theoretic security. However, if the number of colluding servers must be limited, secret sharing schemes cannot be applied to the honest-but-curious server threat model commonly used for data outsourcing [1, 2, 8, 15, 16].

Assumption: b, k Modest. In Sect. 5, we showed that increasing b costs the client more than it costs the colluding servers, so a large b is impractical. With limited resources, we successfully mounted attacks for $k = 13$ in under 500 seconds, so k must be substantially larger ($k > 20$) to achieve security in practice. For each server, the client pays a storage cost equal to that of storing his data in plaintext. If $k \geq 10$, the combined storage cost exceeds that of many encryption-based private query techniques [1, 2, 15], so increasing k is also impractical.

Assumption: Same X, p for Each Secret. Storing a distinct X or prime p on the client for each secret is at least as expensive as storing the secret itself. An alternative is to use a strong, keyed hash h_j to generate a distinct vector $X' = h_j(X)$ for each secret s_j. Using this method, each secret requires different basis polynomials for interpolation, so mounting an attack would be much harder. Unfortunately, it also eliminates additive homomorphism, removing support for server-side aggregation, which is cited as a reason for adopting secret sharing.

Assumption: Corresponding Shares can be Aligned. Hiding share order from data servers as in [3] can hinder share alignment, but if the scheme supports range or point queries, share alignment can eventually be inferred (Sect. 4.3). Schemes could use re-encryption or shuffling to obscure order as in [2], but the cost of such techniques outweighs the performance advantages of secret sharing.

Assumption: $k + 2$ Known Secrets. It is difficult to keep all secrets hidden from an attacker. Known plaintext/ciphertext attacks for small amounts of data are always a threat, and if we known the real-world distribution of the secrets, we can guess them efficiently (Sect. 4.2). The client could encrypt secrets before sharing, but doing so adds substantial cost and eliminates additive homomorphism.

7   Related Work

Privacy-preserving data outsourcing was first formalized in [16] with the introduction of the *Database As a Service* model. Since then, many techniques have been proposed to support private querying [1, 2, 15, 17, 18], most based on specialized encryption techniques. For example, order-preserving encryption [1] supports efficient range queries, while [15] supports server-side aggregation through additively homomorphic encryption. Other schemes are based on fragmentation, where only links between sensitive and identifying data are encrypted [17, 18].

As far as we know, the schemes we discuss in this paper [3–5, 7] are the first to use secret sharing to support private data outsourcing, though secret sharing has been used for related problems, such as cooperative query processing [19]. Prior works, such as [8], have addressed various security issues surrounding data outsourcing schemes, but as far as we know, ours is the first to reveal the specific limitations of schemes based on secret sharing.

8 Conclusion

Private data outsourcing schemes based on secret sharing have been advocated because of their slight advantages over existing encryption-based schemes. Such advantages include security, speed, and support for server-side aggregation. All three outsourcing schemes based on secret sharing [3–5] claim that security is maintained even when k or more servers collude. To the contrary, we have shown that all three schemes are highly insecure when k or more servers collude, regardless of whether X and p are kept secret.

We described and implemented an attack that reconstructs all secret data when only $k + 2$ secrets are known initially. In less than 500 seconds, our attack recovers a hidden 256-bit prime for $k \leq 13$ servers, or an 8192-bit prime for $k \leq 8$. We discussed possible modifications to mitigate our attack and improve security, but any such modifications sacrifice generality, performance, or functionality.

We conclude that secret sharing outsourcing schemes are not simultaneously secure and practical in the honest-but-curious server model, where servers are not trusted to keep data private. Such schemes should only be used when the client is absolutely confident that at most $k - 1$ servers can collude.

Acknowledgements. This work was supported in part by contract number N00014-07-C-0311 with the Office of Naval Research.

References

1. Agrawal, R., Kiernan, J., Srikant, R., Xu, Y.: Order preserving encryption for numeric data. In: Proc. ACM SIGMOD, pp. 563–574 (2004)
2. Stefanov, E., Shi, E., Song, D.: Towards practical oblivious RAM. In: Proc. NDSS (2012)
3. Hadavi, M., Jalili, R.: Secure Data Outsourcing Based on Threshold Secret Sharing; Towards a More Practical Solution. In: Proc. VLDB PhD Workshop, pp. 54–59 (2010)
4. Agrawal, D., El Abbadi, A., Emekci, F., Metwally, A., Wang, S.: Secure Data Management Service on Cloud Computing Infrastructures. In: Agrawal, D., Candan, K.S., Li, W.-S. (eds.) Information and Software as Services. LNBIP, vol. 74, pp. 57–80. Springer, Heidelberg (2011)
5. Tian, X., Sha, C., Wang, X., Zhou, A.: Privacy Preserving Query Processing on Secret Share Based Data Storage. In: Yu, J.X., Kim, M.H., Unland, R. (eds.) DASFAA 2011, Part I. LNCS, vol. 6587, pp. 108–122. Springer, Heidelberg (2011)

 6. Shamir, A.: How to share a secret. Communications of the ACM, 612–613 (1979)
 7. Agrawal, D., El Abbadi, A., Emekci, F., Metwally, A.: Database Management as a Service: Challenges and Opportunities. In: Proc. ICDE Workshop on Information and Software as Services, pp. 1709–1716 (2009)
 8. Kantarcıoğlu, M., Clifton, C.: Security Issues in Querying Encrypted Data. In: Jajodia, S., Wijesekera, D. (eds.) Data and Applications Security 2005. LNCS, vol. 3654, pp. 325–337. Springer, Heidelberg (2005)
 9. Buchberger, B., Winkler, F.: Gröbner bases and applications. Cambridge University Press (1998)
10. Fang, X., Havas, G.: On the worst-case complexity of integer gaussian elimination. In: Proceedings of the 1997 International Symposium on Symbolic and Algebraic Computation, pp. 28–31. ACM (1997)
11. Stein, J.: Computational problems associated with racah algebra. Journal of Computational Physics 1(3), 397–405 (1967)
12. Booth, K.S., Lueker, G.S.: Testing for the consecutive ones property, interval graphs, and graph planarity using PQ-tree algorithms. J. Comput. System Sci. 13(3), 335–379 (1976)
13. County of riverside class and salary listing (February 2012), http://www.rc-hr.com/HRDivisions/Classification/tabid/200/ItemId/2628/Default.aspx
14. Rabin, M.: Probabilistic algorithm for testing primality. Journal of Number Theory 12(1), 128–138 (1980)
15. Mykletun, E., Tsudik, G.: Aggregation Queries in the Database-As-a-Service Model. In: Damiani, E., Liu, P. (eds.) Data and Applications Security 2006. LNCS, vol. 4127, pp. 89–103. Springer, Heidelberg (2006)
16. Hacigümüş, H., Iyer, B., Li, C., Mehrotra, S.: Executing SQL over encrypted data in the database-service-provider model. In: Proc. ACM SIGMOD, pp. 216–227 (2002)
17. Ciriani, V., De Capitani di Vimercati, S., Foresti, S., Jajodia, S., Paraboschi, S., Samarati, P.: Keep a Few: Outsourcing Data While Maintaining Confidentiality. In: Backes, M., Ning, P. (eds.) ESORICS 2009. LNCS, vol. 5789, pp. 440–455. Springer, Heidelberg (2009)
18. Nergiz, A.E., Clifton, C.: Query Processing in Private Data Outsourcing Using Anonymization. In: Li, Y. (ed.) DBSec 2011. LNCS, vol. 6818, pp. 138–153. Springer, Heidelberg (2011)
19. Emekci, F., Agrawal, D., Abbadi, A., Gulbeden, A.: Privacy preserving query processing using third parties. In: Proc. ICDE, p. 27. IEEE (2006)

Privacy-Preserving Subgraph Discovery

Danish Mehmood[1], Basit Shafiq[1,2], Jaideep Vaidya[2], Yuan Hong[2], Nabil Adam[2], and Vijayalakshmi Atluri[2]

[1] Lahore University of Management Sciences, Pakistan
[2] CIMIC, Rutgers University, USA
{danish.mehmood,basit}@lums.edu.pk,
{jsvaidya,yhong,adam,atluri}@cimic.rutgers.edu

Abstract. Graph structured data can be found in many domains and applications. Analysis of such data can give valuable insights. Frequent subgraph discovery, the problem of finding the set of subgraphs that is frequent among the underlying database of graphs, has attracted a lot of recent attention. Many algorithms have been proposed to solve this problem. However, all assume that the entire set of graphs is centralized at a single site, which is not true in a lot of cases. Furthermore, in a lot of interesting applications, the data is sensitive (for example, drug discovery, clique detection, etc). In this paper, we address the problem of privacy-preserving subgraph discovery. We propose a flexible approach that can utilize any underlying frequent subgraph discovery algorithm and uses cryptographic primitives to preserve privacy. The comprehensive experimental evaluation validates the feasibility of our approach.

1 Introduction

Today, graph structured data is ubiquitous. All types of relationships, including, spatial, topological, and other characteristics, can be modeled through the graph abstraction, thus capturing different data semantics. Graph-based analysis gives valuable insights into the data and has been successfully applied to various application domains including protein analysis [17],chemical compound analysis [3], link analysis[12] and web searching[15]. One of the most useful forms of analysis is to find frequent subgraphs from a large database of graphs. This has application in many different domains including cheminformatics (drug disovery), transportation networks, etc. Unlike standard frequent itemset discovery, used for association rule mining in transactional databases, frequent subgraph discovery is a much tougher problem due to underlying fundamental problems, such as canonical labeling of graphs and subgraph isomorphism.

In recent years, this has attracted a lot of attention with many efficient algorithms being developed to solve this problem. However, all of these algorithms assume that the set of graphs is either public or available at a single site. In reality, in many valuable cases, the set of graphs may be distributed between multiple parties, with each party owning a subset of the graphs. Chemical compound databases are one such example. Many pharmaceutical companies have local databases of pharmaceutical drugs which can be represented as graphs. Furthermore, in a lot of interesting applications, the data is sensitive (for example, drug discovery, clique detection, etc). Therefore, due to privacy

N. Cuppens-Boulahia et al. (Eds.): DBSec 2012, LNCS 7371, pp. 161–176, 2012.

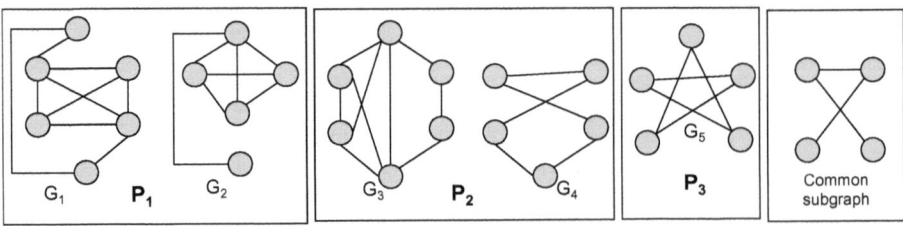

Fig. 1. Graphs owned by three parties (P_1, P_2, and P_3) and the subgraph present in all the graphs of each party

and security concerns, the parties may not wish to reveal their individual graphs to each other or to a central site. In this case, a distributed privacy-preserving algorithm must be developed to enable mining in such cases. In this paper, we develop such an algorithm using cryptographic primitives to preserve privacy. Our algorithm uses any known subgraph discovery method as a subroutine, and therefore can be enhanced in the future as well. We have implemented our approach and the comprehensive experimental evaluation validates the feasibility of our approach.

Illustrative Example: Consider the case depicted in Figure 1, where 3 parties together own 5 graphs (Party 1 owns 2, Party 2 owns 2, and Party 3 owns 1). Figure 1 also depicts the common subgraph that is common to all of the five graphs. Therefore, even with a support threshold of 5, this graph will be detected when subgraph mining is done on the global set of graphs. Note that in this case, we assume that the graph is unlabeled (i.e., neither nodes nor edges are labeled). However, our approach is agnostic to the labeling - either the nodes, the edges, both, or neither could be labeled, based on the data semantics. As long as the underlying subgraph discovery algorithm can handle these cases, our approach will be able to take all of these requirements correctly into account.

1.1 Problem Statement

The basic problem is to discover frequent subgraphs in a privacy-preserving way from a set of graphs owned by different parties. This can be formalized as follows:

Definition 1. *Given k parties $P_1, \ldots, P_k$, each of which own a set of graphs S_i (let $S = \bigcup S_i$), and a global threshold δ ($0 \leq \delta \leq 1$), find the set of frequent subgraphs FS in a privacy-preserving fashion, wherein the global support of each subgraph in FS is over δ. Thus, for each subgraph $fs_j \in FS$, $\sum_i support(S_i, fs_j) \geq \delta |S|$, where $support(S_i, fs_j) = \#$ graphs in S_i that include fs_j as a subgraph.*

Note that in the definition above, we simply require that the set of frequent subgraphs is found in a privacy-preserving fashion. Under the framework of secure multiparty computation[24,5], this equates to not leaking any additional information to any party beyond what can be inferred (in polynomial time) through the local input and output.

Instead of strictly following Definition 1, our protocol satisfies a relaxed form of this definition that allows efficient computation at the expense of leaking additional

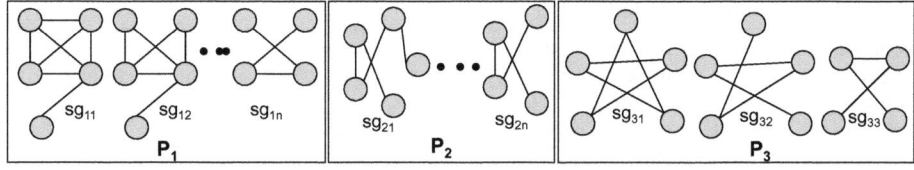

Fig. 2. Local candidate sets with respect to the example in Fig. 1

information. Below is the revised formulation with relaxed privacy requirement that the protocol satisfies.

Definition 2. *Given k parties $P_1, \ldots, P_k$, each of which own a set of graphs S_i (let $S = \bigcup S_i$), and a global threshold δ ($0 \leq \delta \leq 1$), find the set of frequent subgraphs FS (wherein the global support of each subgraph in FS is over δ) without leaking any additional information beyond the set of all local candidates, their support counts, and the number of parties owning them.*

2 Proposed Approach

In this section, we present our proposed approach for privacy-preserving discovery of frequent subgraphs in the set of graphs distributed among multiple parties. The proposed approach essentially involves three key steps:

1. Generation of local candidates – each party computes a candidate set of frequent subgraphs from it local graph set.
2. Generation of a global candidate set of subgraphs – secure union of local candidates to form a global candidate set – the global candidate set is generated by performing secure union of the local candidate sets.
3. Removal of non-frequent subgraphs from the global candidate set – the frequency count of each subgraph in the global candidate set is compared against a global threshold to check if this candidate is a real result.

The overall algorithm encompassing the above steps for subgraph discovery is given in Algorithm 1.This algorithm implements the distributed protocol involving k parties and a coordinator site. Each party owns a local set of graphs, and N is the total number of graphs, which is the sum of the number of graphs in the local set of all parties. The algorithms also uses a commutative cryptography system [14] for computing secure union and a homomorphic cryptography system [13] for computing secure sum. Correspondingly, each party and coordinator site has a pair of commutative encryption and decryption keys; a public homomorphic encryption key which is shared among all the parties and coordinator; a private homomorphic decryption key derived from the public key. In addition, the algorithm requires the user to specify the support threshold (s_T), which is the percentage of total graphs in which the computed subgraphs are present. The user also specifies the minimum size of these subgraphs that need to be computed. This minimum size is specified in terms of the number of nodes ($Node_{min}$) and number of edges ($Edge_{min}$).

Algorithm 1. SubgraphDiscovery

Require: k parties $(P_1, \ldots, P_k)$ each owning a set of graphs and a coordinator site $Coord$
Require: P_i owns the graph set, $GS^{(i)} = G_1^{(i)}, \ldots, G_m^{(i)}$
Require: $N = |GS^{(1)}| + \ldots + |GS^{(k)}|$
Require: $E_C^{(i)}, D_C^{(i)}$ Commutative encryption and decryption keys of party P_i
Require: E_H A public Homomorphic encryption key
Require: $D_H^{(i)}$ Homomorphic decryption key of party P_i
Require: s_T, support threshold, percentage of total graphs in which the resulting subgraph(s) is/are present
Require: $Node_{min}$, minimum number of nodes in each of the subgraphs
Require: $Edge_{min}$, minimum number of edges in each of the subgraphs
Ensure: FSG, frequent subgraph set
1: {**STEP 1:** Generation of Local Candidates}
2: **At each party** P_i:
3: {select an appropriate support threshold, $s_T^{(i)}$. This threshold value is used to find a set of candidate frequent subgraphs from the set of local graphs $GS^{(i)}$}
4: {select an appropriate support threshold, $s_T^{(i)}$. This threshold value is used to find a set of candidate frequent subgraphs from the set of local graphs $GS^{(i)}$}
5: $LocalCand^{(i)} \leftarrow filter(gSpan(GS^{(i)}, s_T^{(i)}, Node_{min}, Edge_{min}))$ {run the gSpan algorithm locally to find the candidate frequent subgraphs from the local graph set. The $filter$ function ensures that only those subgraphs that a party is comfortable with are included in the local candidate set.}
6: {**STEP 2:** Generation of Global Candidate Set}
7: $m^{(i)} \leftarrow Encrypt(LocalCand^{(i)}, E_C^{(i)})$. {The encryption method treats $LocalCand^{(i)}$ as a string so that the resulting cipher text $m^{(i)}$ does not reveal the structure of any of the local graphs.}
8: send $m^{(i)}$ to $Coord$ for secure union
9: **At** $Coord$:
10: $FSG \leftarrow \{\}$
11: receive $m^{(i)}$ from each party P_i
12: $M \leftarrow \bigcup_{i=1}^{k} m^{(i)}$
13: $GlobalCand \leftarrow SecureUnion(M)$
14: {**STEP 3:** Removal of Non-frequent Subgraphs}
15: send $GlobalCand$ to each party P_i
16: **At each party** P_i:
17: create an array $Ecount^{(i)}$ with length equal to the length of $GlobalCand$ and initialize all the elements of $Ecount^{(i)}$ to random encryption of 0 using the homomorphic encryption key E_H
18: **for** each subgraph $sg_j \in GlobalCand$ **do**
19:     **for** each local graph $G_x \in GS^{(i)}$ **do**
20:         **if** sg_j is present in G_x **then**
21:             $h_{enc}^1 \leftarrow HomomorphicEncrypt(1, r, E_H)$ {Random encryption of 1 using homomorphic encryption key E_H)}
22:             $Ecount^{(i)}[j] \leftarrow Ecount^{(i)}[j] * h_{enc}^1$
23:         **end if**
24:     **end for**
25: **end for**
26: send $Ecount^{(i)}$ to $Coord$ for secure sum
27: **At** $Coord$:
28: receive $Ecount^{(i)}$ from each party P_i
29: $Count \leftarrow SecureSum(Ecount^{(1)}, \ldots, Ecount^{(k)})$
30: **for** each subgraph $sg_j \in GlobalCand$ **do**
31:     **if** $Count[j] \geq s_T * N$ **then**
32:         $FSG \leftarrow FSG \bigcup sg_j$
33:     **end if**
34: **end for**
35: **return** FSG

Algorithm 2. SecureUnion(M)

Require: k parties ($P_1, \ldots, P_k$) and a coordinator site $Coord$
Require: $E_C^{(i)}, D_C^{(i)}$ Commutative encryption and decryption keys of party P_i
  1: **At** $Coord$:
  2: $EM \leftarrow \{\}$
  3: $GlobalCand \leftarrow \{\}$
  4: {Commutaive encryption of M by all parties}
  5: **for** each $m^{(i)} \in M$ **do**
  6:     $q \leftarrow m^{(i)}$ {$m^{(i)}$ is received from P_i}
  7:     **for** $j = 1 \ldots k$ **do**
  8:         **if** $j \neq i$ **then**
  9:             shuffle and send q to party P_j for commutative encryption with its key $E_C^{(j)}$
 10:             receive eq from P_j
 11:             $q \leftarrow eq$
 12:         **end if**
 13:     **end for**
 14:     $EM = EM \bigcup \{q\}$
 15: **end for**
 16: send commutative encryption complete signal
 17: **At each party** P_i:
 18: **while** commutative encryption complete signal is not received from $Coord$ **do**
 19:     receive q from $Coord$
 20:     send $eq \leftarrow Encrypt(q, E_C^{(i)})$ to $Coord$ {eq is encryption of q with the ommutative encryption key $E_C^{(i)}$}
 21: **end while**
 22: {Decryption of EM by all parties}
 23: **for** each $em \in EM$ **do**
 24:     $q \leftarrow em$
 25:     **for** $j = 1 \ldots k$ **do**
 26:         send dq to party P_j for decryption with its key $D_C^{(j)}$
 27:         receive dq from P_j
 28:         $q \leftarrow dq$
 29:     **end for**
 30:     $GlobalCand = GlobalCand \bigcup \{q\}$
 31: **end for**
 32: send decryption complete signal
 33: remove duplicate elements (subgraphs) from $GlobalCand$
 34: **return** $GlobalCand$
 35: **At each party** P_i:
 36: **while** decryption complete signal is not received from $Coord$ **do**
 37:     receive q from $Coord$
 38:     send $dq \leftarrow Decrypt(q, D_C^{(i)})$ to $Coord$ {dq is decryption of q with the Commutative decryption key $D_C^{(i)}$}
 39: **end while**

Algorithm 3. SecureSUM($Ecount^{(1)}, \ldots, Ecount^{(k)}$)

Require: Threshold-based homomorphic crypto system
Require: E_H A public Homomorphic encryption key
Require: $D_H^{(i)}$ Homomorphic decryption key of party P_i and D_H^{Coord} Homomorphic decryption key of Coord
Require: T, Threshold for homomorphic decryption (No. of parties needed for decryption)
1: **At** $Coord$:
2: Create an array $Ecnt$ with length equal to the length of $GlobalCand$ and initialize all the elements of $Ecnt$ to random encryption of 0 using the homomorphic encryption key E_H
3: **for** $i = 1 \ldots length(Ecnt)$ **do**
4:   **for** $j = 1 \ldots k$ **do**
5:     $Ecnt[i] \leftarrow Ecnt[i] * Ecount^{(j)}[i]$
6:   **end for**
7: **end for**
8: Collaboratively decrypt $Ecnt$ with T parties to get actual frequency count of each subgraph
9: **return** Decrypted $Ecnt$

Below we discuss each of the above three steps and how these steps are implemented in the Subgraph discovery algorithm.

2.1   Generation of Local Candidates

This step of generation of local candidates is implemented in lines 1 - 4 of Algorithm 1. For generation of local candidates, each party runs the frequent subgraph mining algorithm. For frequent subgraph mining, we use the gSpan algorithm [23]. Our approach is agnostic to the underlying frequent subgraph mining algorithm. We chose gSpan since it was easily available and reasonably efficient. gSpan takes a collection of graphs and a minimum support threshold as input and computes all the subgraphs whose frequency is greater than or equal to the given threshold. In addition, we constrain the minimum size of subgraphs to avoid retrieving trivial subgraphs. For this, we use the ($Node_{min}$) and ($edge_{min}$) parameters which are defined globally by the user.

Note that a $filter$ function is applied to the output of $gSpan$ to ensure that only those subgraphs that a party is comfortable with are included in its local candidate set. This improves the privacy protection. For computing the local candidate set, the support threshold needs to be closer to the global support threshold (s_T) or smaller to reduce the number of frequent sub-graphs that are missed from the final solution. Clearly, if a local support threshold corresponding to one graph only (i.e., a subgraph is present in only one of the local graph) is used, there will not be any miss. However, this significantly increases the computational overhead as there will be too many subgraphs in the local candidate set. We analyze this trade-off between performance and accuracy in our experiments discussed in Section 4.2.

Fig. 2 depicts the set of local candidates computed locally by each of the three parties P_1, P_2, and P_3 using their local set of graphs discussed in the Introduction and illustrated in Figure 1. In this Figure, the minimum size of the subgraphs are restricted to 4 nodes and 3 edges.

2.2   Generation of a Global Candidate Set

This step of generation of a global candidate set of frequent subgraphs is implemented in lines 5 - 12 of Algorithm 1. Essentially, the global candidate set is the union of the local candidate sets computed by each party in Step 1. However, due to privacy requirements this union needs to be computed in a secure manner without revealing which candidate subgraph comes from which party. We employ a commutative encryption-based approach for computing the secure union of the local candidate sets.

An encryption algorithm is commutative if plain text data item enciphered with multiple encryption keys in any arbitrary order will have the same enciphered text.Formally, an encryption algorithm is commutative if the following two equations hold for any given encryption keys $K_1, \ldots, K_n \in K$, any data item to be encrypted $m \in M$, and any permutations of $i, j : \forall m_1, m_2 \in M$ such that $m_1 \neq m_2$:

$$E_{K_{i_1}}(...E_{K_{i_n}}(m)...) = E_{K_{j_1}}(...E_{K_{j_n}}(m)...) \tag{1}$$

and for any given $k, \epsilon < 1/2^k$

$$Pr(E_{K_{i_1}}(...E_{K_{i_n}}(m_1)...) = E_{K_{j_1}}(...E_{K_{j_n}}(m_2)...)) < \epsilon \tag{2}$$

Pohlig-Hellman [14] is one example of a commutative encryption scheme (based on the discrete logarithm problem). This or any other commutative encryption scheme would work well for our purposes.

The basic idea of computing the secure union using the commutative encryption protocol is that each subgraph in every local candidate set is encrypted by all the parties using their commutative encryption keys. Then all these encrypted subgraphs are put into a global candidate set with their order shuffled. However, the elements in the encrypted global candidate set would have duplicates that need to be removed for the global candidate set to the union of all the local candidate sets. The encryption method used in the proposed approach treats each element in the local candidate set as a string so that the resulting cipher text mask the structural information of each local subgraph. Without knowing the structural information, duplicate subgraphs in the encrypted global candidate set cannot be removed. The following substeps elaborate on the proposed commutative encryption-based strategy for computing the secure union of local candidate set to form the global candidate set.

Substep 1. Each party P_i represents its local candidate set into a string and applies its commutative encryption key $E_C^{(i)}$ on the resulting string. This encrypted local candidate set is then sent to $Coord$ for secure union. (Lines 5 and 6 of Algorithm 1)

Substep 2. The $Coord$ receives the encrypted local candidate set from each party and routes each candidate set to all other parties for commutative encryption. When all the local candidate sets are encrypted by all parties, the $Coord$ combines them into one global encrypted set (Lines 7 - 11 of Algorithm 1 and lines 1 - 16 of Algorithm 2).

Substep 3. The encrypted global candidate set is sent by the coordinator to each party for decryption. Each party upon receiving the encrypted global candidate set shuffles its order and then applies its commutative decryption key. After all parties have applied their decryption keys, the structural information in the global candidate set is restored. After this the duplicate subgraphs in the global candidate set are removed (lines 17 - 39 of Algorithm 2)). For duplicate removal, we perform a pairwise comparison between the subgraphs in the global candidate set using gSpan.

This strategy computes the global candidate set without revealing private information about the local subgraphs of any party to other parties – specifically, which subgraphs belong to which party. This is because during the commutative encryption substeps (substeps 1 and 2), the local candidate set of each party is encrypted by the encryption key of the party owning the graph. During the decryption phase all the local subgraphs are merged into one set with their order shuffled. Therefore, inferring which subgraph belongs to which party based on its position in the global candidate set is also not possible. The drawback of removing duplicates after decryption is that the coordinator would know how many parties have a given subgraph present in their graphs. We discuss this issue further in Section 3.

2.3   Removal of Non-frequent Subgraphs

The global candidate set is the union of the local candidate set. As discussed in Section 2.1 the support threshold for generation of subgraphs in the local candidate set may be smaller than the global support threshold. Therefore, all those subgraphs that do not satisfy the global threshold need to be removed from the global candidate set. This requires computing the frequency count (number of graphs in which a given subgraph is present) for each subgraph. This frequency count also needs to be computed in a secure and distributed manner as there is no global set of graphs and all the graphs are with their owner parties.

The step of removal of non-frequent subgraphs from the global candidate set is implemented in lines 13 - 33 of Algorithm 1. intuitively in this step, each party P_i computes the frequency count of each subgraph in the global set with respect to its local graph set $GS_i^{(i)}$ (lines 17 - 24 of Algorithm 1). This frequency count is stored in a vector which is sent to the coordinator. The coordinator after receiving the frequency count vector from all parties computes the sum for each subgraph in the global candidate set. If this sum is less than the given global support threshold, the corresponding subgraph is removed from the global candidate set (lines 22 - 33 of Algorithm 1).

For secure computation and summation of the frequency counts, we employ an additive homomorphic cryptosystem such as the Paillier cryptosystem [13]. An additive homomorphic encryption is semantically-secure public-key encryption which has the additional property that given any two encryptions $E(A)$ and E(B), there exists an encryption $E(A+B)$ such that $E(A)*E(B) = E(A+B)$, where * denotes multiplication operator.

Following this homomorphic encryption property, each party initializes its frequency count vector by putting a random homomorphic encryption of '0' in each element of its frequency count vector (line 16 of Algorithm 1). When computing the frequency count of each subgraph in the nested for loop of lines 17 - 24 of Algorithm 1, if a match is found the party increments the value of the corresponding element of the frequency count vector by '1'. This is done by multiplying the value of such element with a random homomorphic encryption of '1' (line 21 of Algorithm 1). Similarly, the coordinator employs a secure sum protocol to compute the sum of the frequency count vectors received from each party and employs threshold based decryption to decrypt the values of the global count for each subgraph. The reason for using threshold-based decryption strategy is to prevent a single party (coordinator) to decrypt the values in the frequency count vector received from each party. Threshold-based decryption with threshold of T requires T parties to collaboratively decrypt the encrypted values.

3   Complexity and Security Analysis

We now analyze the computation and communication complexity of our algorithms as well as the security provided through our approach.

3.1   Computation Cost

The computation cost of the distributed algorithm is actually comparable to the cost incurred in the centralized case. The main cost is incurred due to three steps, the local calls to the gSpan algorithm to find the local candidates, the commutative encryption based protocol to find the global candidate set, and the second round of local calls to the gSpan algorithm to find the frequency count of each candidate subgraph. Compared to these steps, the cost of the secure sum to find the global frequencies is negligible and can be ignored. Let us now consider the cost of each step.

Essentially, in the first step, even though each party invokes gSpan independently, it does so, only over the local set of graphs. Therefore the total computation cost is comparable to the cost of running gSpan over the entire set of graphs in a centralized case. In the second step, the secure union protocol is used to create a global candidate set. Essentially, each candidate subgraph is represented by a string, which is encrypted by all the parties, and then decrypted after merging into a combined set. Thus, assuming a total of l candidate graphs in the global set, the total cost is that of $O(kl)$ encryptions and decryptions. In the third step, gSpan is run over each pair of candidate subgraph, and local graph. Thus, gSpan is invoked $l|S|$ times, where $|S|$ is the total number of graphs. Note that each invocation of gSpan in step 3 takes much less time than in step 1 as only two graphs are being compared.

3.2   Communication Cost

Communication between the parties only occurs when the local candidates are merged into a global set and sent to all of the parties, and in the final frequency determination phase. For the secure union, there are a total of $O(kl)$ messages, for encryption and decryption. For the secure sum, it is the same with $O(kl)$ messages being exchanged.

3.3   Security Analysis

Consider Algorithm 1. Step 1 is completely local, so no private information is disclosed. In step 2, the global candidate set is generated from the set of local candidates. Encryption is used to obscure the link between the candidates and the party generating them. The Secure Union protocol is used to securely combine the candidate sets. Assuming that this protocol is secure, nothing is leaked through the combination process, though all parties will learn the global candidate set (and can therefore identify candidates that do not belong to them, though they cannot identify which party the candidates come from). Note that, in reality, the secure union does leak some additional information. In the secure union (Algorithm 2), after merging local sub-graphs, duplicate sub-graphs are removed. However, the duplicates are removed after decryption. Therefore, the co-ordinator would know the number of parties that support each candidate sub-graph. In the extreme case where all parties support a particular subgraph, the coordinator would now know that the particular subgraph is frequent for all parties (though it still would not know which local subgraphs does it belong to for each party, as long as there are at least 3 parties).

In step 3, the non-frequent subgraphs are removed. For this, the support count of each of the candidate subgraphs is computed using the secure sum algorithm. Assuming this is secure, nothing is leaked, except for the support count (though, again, it is not clear which graphs contributed to this support count). In total, the overall process simply leaks the set of global candidates to all parties (along with the number of parties supporting each candidate, though only to the coordinator), as well as the frequency count (again, only to the coordinator).

Assuming that this additional information is given to the simulator, we can prove that the algorithm does not leak anything further. The question is whether this constitutes too much information. Let us consider each independently. Our algorithm leaks the set of global candidates, from which the final set of frequent subgraphs is picked. However, since each party locally mines its frequent subgraphs in step 1, it can easily refrain from including any of the subgraphs that it is uncomfortable with. This makes it difficult for any party to learn the entire graph or any unique / identifying subgraph of other parties. In the case of the frequency counts, if these are considered sensitive, it can be easily handled through the use of a simple add and compare[18] protocol that can check whether the global count of the candidate is above the threshold or not. Such a protocol can easily be implemented through the VIFF system[1].

Given a large set of graphs, this extra information can be considered to be acceptable and worth allowing, given the gain in efficiency that is obtained, as compared to generic secure multiparty computation techniques which would leak no information but be extremely slow.

4   Experimental Evaluation

4.1   Implementation Details

In this section we cover the details of our implementation of the privacy-preserving subgraph discovery algorithm on modern hardware and present experimental results demonstrating the usability of this algorithm.

The general model of privacy preserving subgraph is as follows. The coordinator initiates a request for subgraph discovery. The coordinator could be a separate entity or one of the graph owner parties. There is one global coordinating class/interface that

---

[1] http://viff.dk/

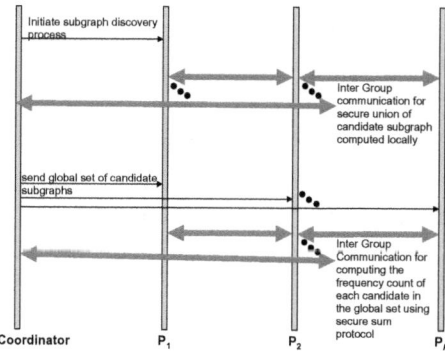

Fig. 3. Basic Interaction Diagram

provides access to the subgraph discovery functionality. We call this the RmtMain interface. However, since this interface is present at a user site, the class implementing it should have access to no private information about the graphs of other parties. The class implementing this is the SiteCoordinator class which is initialized with the appropriate site information. There is another interface called, RmtSlave interfaces. All other parties that are involved in the sub-graph discovery process are treated as slave sites. The coordinator site coordinates with the slave sites to perform the required action.

Figure 3 demonstrates the basic interaction diagram. Java RMI is used to implement the distributed protocol between all of the sites.

We used Pohlig-Hellman encryption scheme [14] for implementing commutative encryption and Paillier Crypto system [13] for implementing Homomorphic encryption.

4.2   Experimental Evaluation

We now present experimental results demonstrating the usability of the proposed algorithms. We ran our experiments on two real graph datasets [23]: i) Chemical 340; ii) Compound 422. The Chemical 340 dataset 340 chemical compounds, 24 different atoms, 66 atom types, and 4 types of bonds. The average graph size in this dataset is 27 nodes and 28 edges. The largest graph in this dataset has 214 nodes and 214 edges. The Compound 422 dataset has 422 graphs with average graph size of 40 nodes and 42 edges. The largest graph in this dataset has 189 nodes and 196 edges.

Figure 4 shows the computation time and accuracy results of the *SubgraphDiscovery* algorithm for the Chemical 340 and Compound 422 datasets. The global threshold was set to 12% for both data sets. For the Chemical 340 dataset, the minimum size of the frequent subgraph was set to 5 nodes and 5 edges. The total number of frequent subgraphs in the Chemical 340 dataset satisfying the global threshold and minimum graph size requirements was 550. For the Compound 422 dataset, the minimum size of the frequent subgraph was also set to 5 nodes and 5 edges. In addition, we also constrain the maximum size to be 7 nodes and 7 edges. The total number of frequent subgraphs in the Compound 422 dataset satisfying the global threshold and minimum graph size requirements was 562. The graphs in both datasets were randomly distributed among three sites with each site having almost equal number of graphs.

Chemical 340							
Dataset	Total graphs = 340; Average graph size = (27 nodes, 28 edges); Largest graph size = (214 nodes, 214 edges)						
Frequent Subgraphs	Size($Node_{min}$=5, $Edge_{min}$=5); Global Threshold = 12%; Number of frequent subgraphs = 550						
Sites	Randomly distributed among 3 sites with each site having equal number of graphs						
No.	**Local Threshold (%age)**	**Average subgraphs per site**	**Step 1 Time (sec)**	**Step 2 Time (sec)**	**Step 3 Time (sec)**	**Total Time (sec)**	**Accuracy (%age)**
1	9	1308	23	8,871	1,537	10,423	100
2	13	513	14	2,761	458	3,233	99.82
3	15	393	12	2,087	359	2,458	95.82
4	18	264	11	1336	249	1,596	71.09
5	19	203	10	1008	205	1,223	58.55
6	22	120	11	578	151	740	33.82

Compound 422							
Dataset	Total graphs = 422; Average graph size = (40 nodes, 42 edges); Largest graph size = (189 nodes, 196 edges)						
Frequent Subgraphs	Size ($Node_{min}$=5, $Edge_{min}$ edges=5; $Node_{max}$=7, $Edge_{max}$ edges=7); Global Threshold = 12%; Number of frequent subgraphs = 562						
Sites	Randomly distributed among 3 sites with each site having equal number of graphs						
No.	**Local Threshold (%age)**	**Average subgraphs per site**	**Step 1 Time (sec)**	**Step 2 Time (sec)**	**Step 3 Time (sec)**	**Total Time (sec)**	**Accuracy (%age)**
1	9	728	14	3768	661	4433	100
2	12	571	13	2739	513	3265	100
3	15	442	12	2094	437	2543	95.20
4	18	351	14	1631	341	1986	78.29

Fig. 4. Computation time and accuracy vs. local threshold for Chemical 340 and Compound 422 datasets

Figure 4 shows the computation time and accuracy results against different local threshold values. Note that the running time depends much more on the local threshold level rather than the global threshold level, since the local threshold determines the number of candidates which in turn determines the time taken by steps 2 and 3. For both datasets, the computation time decreases as the local threshold value increases. This is because increasing the local threshold results in smaller number of local candidate subgraphs and consequently the size of the global candidate set decreases. Also, it is obvious from the results that the computation overhead of step 2 (Generation of global candidate set) dominates all other steps. This step involves encryption of the local candidate set, computing secure union, and removing duplicates.

As expected the accuracy is much higher for local threshold values that are closer to the global threshold or smaller. For example in both datasets, local threshold value of 9% yields 100% accuracy.

The appropriate local threshold is set by the parties in order to generate a reasonable set of candidates. From the security perspective, higher thresholds are better than lower. Therefore one possibility is to start from high threshold and progressively lower it to get an interesting set of results. This incremental computation does not incur any additional

Chemical 340 - multiple sites						
Dataset	Total graphs = 340; Average graph size = (27 nodes, 28 edges); Largest graph size = (214 nodes, 214 edges); Randomly distributed among sites.					
Frequent Subgraphs	Size($Node_{min}$=5, $Edge_{min}$=5); Global Threshold = 20%; Local Threshold = 20%; Number of frequent subgraphs = 134					
Sites	**Average subgraphs per site**	**Step 1 Time (sec)**	**Step 2 Time (sec)**	**Step 3 Time (sec)**	**Total Time (sec)**	**Accuracy (%age)**
3	171	10	830	188	1028	100
4	263	13	2455	336	2804	100
5	284	16	4188	605	4809	100

Fig. 5. Computation time and accuracy vs. number of sites

privacy loss since the results obtained at a higher threshold level are a subset of the results obtained at a lower threshold.

Figure 5 compares the computation time results for the Chemical 340 dataset distributed among 3, 4, and 5 sites. For this experiment both local and global threshold was set to 20%. The computation time increases with the number of sites. This is mainly due to the increase in number of messages for commutative encryption and decryption in step 2. Moreover, as the number of sites increases the coordinator has to interact with more sites for receiving the frequency count vector and summing them up for removal of non-frequent subgraphs (step 3).

5   Related Work

Privacy-Preserving Data Mining (PPDM). The Work in PPDM has followed two major directions: i) randomization/perturbation; and ii) secure multiparty computation.

In the perturbation approach data is locally perturbed by adding "noise" before mining is done. For example, if we add a random number chosen from a Gaussian distribution to the real data value, the data miner no longer knows the exact value. However, important statistics on the collection (e.g., average) will be preserved. Agrawal and Srikant [2] introduced this notion as PPDM to the data mining community. Zhu and Lei [25] study the problem of optimal randomization for privacy-preserving data mining and demonstrate the construction of optimal randomization schemes for density estimation.

The alternative approach of using cryptographic techniques to protect privacy was first utilized for the construction of decision trees by Lindell and Pinkas[11]. Later, these techniques were utilized in many subfields of data mining, e.g. association rule mining [21], clustering[8], classification [4,19,22], outlier detection [20] and regression [9,16]. Our work presents a secure method for frequent subgraph mining, which also follows the same line of research.

All of the cryptographic work falls under the theoretical framework of Secure Multiparty Computation [24,5].

Frequent Subgraph Discovery. The graph mining techniques, in general, can be categorized into two categories:i) apriori-based approaches and pattern-growth based approaches.

In the first category, the apriori-based approaches follow the idea of apriori algorithm in frequent pattern mining for itemsets [1] – all the subgraphs of a discovered frequent subgraph are also frequent. AGM (apriori-based graph mining) [7], FSG (frequent subgraph discovery) [10] and PM (path mining) [6] enumerate candidate subgraphs using vertices, edges and edge-disjoint paths respectively. Specifically, AGM [7] discovers frequent subgraphs that occur above the percentage threshold of all graphs and uses a canonical representation of subgraphs for improving the efficiency of checking the subgraph isomorphism. FSG [10] generates candidates with the edges which is shown in the adjacency matrix. The class of subgraphs discovered to connected subgraphs has been limited, and several heuristics have been proposed in [10] to improve the efficiency of computing the subgraph support. Meanwhile, the efficiency of generating pattern candidates is also guaranteed. Similar to AGM and FSG, PM [6] also generates candidate subgraph patterns using breadth-first enumeration. Nevertheless, this approach utilizes edge-disjoint paths to generate candidate patterns which reduces the number of iterations while still maintaining the completeness of the search space.

In the second category, the algorithm of gSpan (graph-based Substructure pattern mining) [23] discovers frequent subgraphs without candidate generation. It encodes the tree representation of graphs rather than the adjacency matrix using depth-first search code which provides a lexicographical order for searching the candidate patterns (subgraphs). gSpan algorithm performs efficiently not only on reducing the runtime cost but also saving memory space.

6   Conclusions and Future Work

In this paper, we have looked at the problem of finding frequent sub-graphs from a large distributed set of graphs in a privacy-preserving fashion. Our algorithm is flexible and can use any underlying subgraph discovery approach as a subroutine. We have implemented our approach and the experimental evaluation shows that our approach is effective and allows a trade-off between utility and computation time. While we conducted the experimental evaluation with pharmaceutical data that have relatively small graph size, we plan to follow on with experiments on social network data. In the future, we also plan to consider the case of a single global graph, which is distributed between multiple parties (this happens in many cases such as transactions shared between financial organizations, call graphs, etc.) Here, you can find local frequent substructures as described in our paper, however, the inter-edges cause a problem. This could perhaps be solved using the graph duality restructuring approach (by building the dual of the graph, with nodes becoming edges, and vice versa). We plan to explore this in the future.

Acknowledgements. The work of Mehmood and Shafiq is supported in part by the LUMS Departmental Research Grant. The work of Vaidya is supported in part by the National Science Foundation under Grant No. CNS-0746943. The work of Atluri is supported through the IR/D by the National Science Foundation.

References

1. Agrawal, R., Srikant, R.: Fast algorithms for mining association rules. In: Proceedings of the 20th International Conference on Very Large Data Bases, September 12-15, pp. 487–499. VLDB, Santiago (1994),
   http://www.vldb.org/dblp/db/conf/vldb/vldb94-487.html
2. Agrawal, R., Srikant, R.: Privacy-preserving data mining. In: Proceedings of the 2000 ACM SIGMOD Conference on Management of Data, May 14-19, pp. 439–450. ACM, Dallas (2000),
   http://doi.acm.org/10.1145/342009.335438
3. Chittimoori, R.N., Holder, L.B., Cook, D.J.: Applying the subdue substructure discovery system to the chemical toxicity domain. In: Proceedings of the Twelfth International Florida Artificial Intelligence Research Society Conference, pp. 90–94. AAAI Press (1999),
   http://dl.acm.org/citation.cfm?id=646812.707494
4. Du, W., Zhan, Z.: Building decision tree classifier on private data. In: Clifton, C., Estivill-Castro, V. (eds.) IEEE International Conference on Data Mining Workshop on Privacy, Security, and Data Mining, December 9, vol. 14, pp. 1–8. Australian Computer Society, Maebashi City (2002), http://crpit.com/Vol14.html
5. Goldreich, O., Micali, S., Wigderson, A.: How to play any mental game - a completeness theorem for protocols with honest majority. In: Proceedings of the 19th ACM Symposium on the Theory of Computing, pp. 218–229. ACM, New York (1987),
   http://doi.acm.org/10.1145/28395.28420
6. Gudes, E., Shimony, S.E., Vanetik, N.: Discovering frequent graph patterns using disjoint paths. IEEE Trans. on Knowl. and Data Eng. 18, 1441–1456 (2006),
   http://dx.doi.org/10.1109/TKDE.2006.173
7. Inokuchi, A., Washio, T., Motoda, H.: Complete mining of frequent patterns from graphs: Mining graph data. Mach. Learn. 50, 321–354 (2003),
   http://dl.acm.org/citation.cfm?id=608108.608123
8. Jagannathan, G., Wright, R.N.: Privacy-preserving distributed k-means clustering over arbitrarily partitioned data. In: Proceedings of the 2005 ACM SIGKDD International Conference on Knowledge Discovery and Data Mining, August 21-24, pp. 593–599. ACM, Chicago (2005)
9. Karr, A.F., Lin, X., Sanil, A.P., Reiter, J.P.: Secure regressions on distributed databases. Journal of Computational and Graphical Statistics 14, 263–279 (2005)
10. Kuramochi, M., Karypis, G.: Frequent subgraph discovery. In: Cercone, N., Lin, T.Y., Wu, X. (eds.) ICDM, pp. 313–320. IEEE Computer Society (2001)
11. Lindell, Y., Pinkas, B.: Privacy preserving data mining. Journal of Cryptology 15(3), 177–206 (2002)
12. Mukherjee, M.: Graph-based data mining for social network analysis. In: Proceedings of the ACM KDD Workshop on Link Analysis and Group Detection (2004)
13. Paillier, P.: Public-Key Cryptosystems Based on Composite Degree Residuosity Classes. In: Stern, J. (ed.) EUROCRYPT 1999. LNCS, vol. 1592, pp. 223–238. Springer, Heidelberg (1999)
14. Pohlig, S.C., Hellman, M.E.: An improved algorithm for computing logarithms over GF(p) and its cryptographic significance. IEEE Transactions on Information Theory IT-24, 106–110 (1978)
15. Rakhshan, A., Holder, L.B., Cook, D.J.: Structural web search engine. International Journal on Artificial Intelligence Tools 13(1), 27–44 (2004)

16. Sanil, A.P., Karr, A.F., Lin, X., Reiter, J.P.: Privacy preserving regression modelling via distributed computation. In: KDD 2004: Proceedings of the Tenth ACM SIGKDD International Conference on Knowledge Discovery and Data Mining, pp. 677–682. ACM Press, New York (2004)

17. Su, S., Cook, D.J., Holder, L.B.: Application of knowledge discovery to molecular biology: Identifying structural regularities in proteins. In: Pacific Symposium on Biocomputing, pp. 190–201 (1999)

18. Vaidya, J., Clifton, C.: Privacy-preserving k-means clustering over vertically partitioned data. In: The Ninth ACM SIGKDD International Conference on Knowledge Discovery and Data Mining, August 24-27, pp. 206–215. ACM, Washington, DC (2003), http://doi.acm.org/10.1145/956750.956776

19. Vaidya, J., Clifton, C.: Privacy preserving naïve bayes classifier for vertically partitioned data. In: 2004 SIAM International Conference on Data Mining, April 22-24, pp. 522–526. SIAM, Philadelphia (2004)

20. Vaidya, J., Clifton, C.: Privacy-preserving outlier detection. In: Proceedings of the Fourth IEEE International Conference on Data Mining (ICDM 2004), November 1-4, pp. 233–240. IEEE Computer Society Press, Los Alamitos (2004)

21. Vaidya, J., Clifton, C.: Secure set intersection cardinality with application to association rule mining. Journal of Computer Security 13(4), 593–622 (2005)

22. Vaidya, J., Clifton, C., Kantarcioglu, M., Patterson, A.S.: Privacy-preserving decision trees over vertically partitioned data. ACM Trans. Knowl. Discov. Data 2(3), 1–27 (2008)

23. Yan, X., Han, J.: gspan: Graph-based substructure pattern mining. In: ICDM, pp. 721–724 (2002)

24. Yao, A.C.: How to generate and exchange secrets. In: Proceedings of the 27th IEEE Symposium on Foundations of Computer Science, pp. 162–167. IEEE Computer Society, Los Alamitos (1986)

25. Zhu, Y., Liu, L.: Optimal randomization for privacy preserving data mining. In: KDD 2004: Proceedings of the Tenth ACM SIGKDD International Conference on Knowledge Discovery and Data Mining, pp. 761–766. ACM Press, New York (2004)

Decentralized Semantic Threat Graphs

Simon N. Foley and William M. Fitzgerald

Cork Constraint Computation Centre,
Computer Science Department, University College Cork, Ireland
s.foley@cs.ucc.ie, wfitzgerald@4c.ucc.ie

Abstract. Threat knowledge-bases such as those maintained by MITRE
and NIST provide a basis with which to mitigate known threats to an
enterprise. These centralised knowledge-bases assume a global and uniform level of trust for all threat and countermeasure knowledge. However,
in practice these knowledge-bases are composed of threats and countermeasures that originate from a number of threat providers, for example
Bugtraq. As a consequence, threat knowledge consumers may only wish
to trust knowledge about threats and countermeasures that have been
provided by a particular provider or set of providers. In this paper, a trust
management approach is taken with respect to threat knowledge-bases.
This provides a basis with which to decentralize and delegate trust for
knowledge about threats and their mitigation to one or more providers.
Threat knowledge-bases are encoded as Semantic Threat Graphs. An
ontology-based delegation scheme is proposed to manage trust across a
model of distributed Semantic Threat Graph knowledge-bases.

Keywords: Decentralized Threat Management, Security Configuration.

1 Introduction

Threat management is a process used to help implement a security configuration
that mitigates known enterprise (security) threats. Centralised threat knowledge-
bases, such as NIST's *National Vulnerability Database (NVD)* [1] are an integral
part of the threat management process. However, in practice threat knowledge-
bases are composed of threats, vulnerabilities and countermeasures that originate
from multiple providers, for example US-Cert [2], Bugtraq [3] and/or vendors
(such as Cisco and Microsoft). As a consequence, threat knowledge may only
be trusted if it has been asserted by a particular provider or set of providers.
For example, a consumer of the NVD may only want to trust knowledge about
software buffer-overflow vulnerabilities that have been asserted by Bugtraq and
corresponding countermeasures asserted by Microsoft. However, access to a centralised threat knowledge-base implies a global or uniform level of trust for all
knowledge about threats, vulnerability and countermeasures indiscriminately.

This paper adopts a trust management approach with respect to threat
knowledge-bases. The advantages are twofold. The first is that it provides a basis
with which a consumer may delegate authority to trusted providers for knowledge about threats, vulnerabilities and countermeasures. Secondly, it provides

N. Cuppens-Boulahia et al. (Eds.): DBSec 2012, LNCS 7371, pp. 177–192, 2012.

a basis to decentralize a threat knowledge-base where trusted threat knowledge may reside with the originating provider and/or be distributed across other trusted provider threat knowledge-bases.

Threat Trees/Graphs, such as [20, 28, 29], are used to represent, structure and analyse what is known about threats and their countermeasures. In this paper, threat knowledge-bases are encoded as Semantic Threat Graphs [20]. We argue that using an ontology-based framework provides a natural approach to constructing, reasoning about and managing decentralized Semantic Threat Graphs. An ontology provides a conceptual model of a domain of interest. It provides a framework for distributed and extensible structured knowledge founded on formal logic [7]. In recent years, research in computer security has seen an increase in the use of ontologies. For example, ontologies have been applied to the areas of information security (common security vocabulary) [24], security management (threats, vulnerabilities and countermeasures) [17], access control [15], policy management [25] and trust management [33]. The decentralized Semantic Threat Graphs (ontology fragments) are implemented in OWL-DL, a language subset of OWL which is a W3C standard that includes Description Logic reasoning semantics [30].

Distributed fragments of Semantic Threat Graphs that are naturally composable under the open-world assumption, provide a unified view of the threat-based domain. Ontologies provide for separation of concerns, whereby consumers and providers of threat-based knowledge can be separately developed, with reasoning and deployment over their composition done locally. It also means that information about new threats, vulnerabilities and countermeasures can be incorporated as new facts within existing Semantic Threat Graph knowledge-bases.

The paper is outlined as follows. Section 2 provides an overview of Description Logic. Section 3 outlines the Semantic Threat Graphs model and an encoding of standards of best practice for threat mitigation. Section 4 describes the delegation scheme used to manage trust across a model of distributed threat knowledge-bases. A model for decentralized Semantic Threat Graphs is presented in Section 5. Section 6 provides use case scenarios that demonstrate how the approach may work in practice. Related research is discussed in Section 7.

2   Description Logic and Knowledge-Bases

The Description Logic $\mathcal{SHOIN}^{(\mathcal{D})}$ is a decidable portion of First Order Logic that can be used to represent and reason about application knowledge and is commonly implemented as OWL-DL [7]. Knowledge is described in terms of *concepts* about *individuals* and their relationships. For example, suppose that concept DOS describes denial of service threats and concept TCPcntr describes TCP-stack based countermeasures such as syn-cookie and syn-cache, then the concept

$$(\exists_1 \, mitigates.\text{DOS}) \sqcap \text{TCPcntr}$$

can be considered to characterize TCPcntr countermeasure concepts that mitigate denial of service threats. In this case *mitigates* is a *property* that has been defined between DOS threats and countermeasures. For example, the individual countermeasure syn-cache is related (property *mitigates*) to the individual DOS threat syn-flood. A Description Logic concept corresponds to a unary predicate; intuitively, it represents a set, and concept conjunction ($\sqcap$) and disjunction ($\sqcup$) provide set intersection and union, respectively. A Description Logic property (role) corresponds to a binary predicate. The concept ($\exists_{\geq 1} p.\phi$) defines individuals related to at least one individual of concept ϕ via property p.

A knowledge-base comprises a terminological component, hereafter referred to as its *TBox*, and an assertional component (its *ABox*). In addition to describing concepts, a TBox may define concept and property relationships. These are given as axioms of the form $\phi_1 \sqsubseteq \phi_2$, given concepts ϕ_1 and ϕ_2 and where subsumption $\sqsubseteq$ can be interpreted as subset. For example, the TBox containing the concepts from the previous paragraph could include the axiom DOS $\sqsubseteq$ Threat, indicating that denial of service is a kind of threat. Property subsumption axioms may similarly be defined, which we also represent as $\phi_1 \sqsubseteq \phi_2$ if no ambiguity arises.

3 Semantic Threat Graphs

Semantic Threat Graphs [20], a variation of the traditional Threat/Attack Tree, are encoded within an ontology-based framework. Figure 1 provides an abstract model for Semantic Threat Graphs and identifies the key concepts and relationships.

Enterprise IT assets are represented as individuals of the *Asset* concept. An asset may have one or more *hasWeakness*'s (property relationship) that relate to individuals categorised in the *Vulnerability* concept. Individuals of the *Vulnerability* concept are exploitable (*exploitedBy*) by a threat or set of threats (*Threat* concept). As a consequence, an asset that has a vulnerability is, therefore, also *threatenedBy* a corresponding *Threat*. A countermeasure *mitigates* particular vulnerabilities. Countermeasures are deemed to be kinds-of assets and thus are defined as a *subConceptOf Asset*.

Semantic Threat Graphs can be used to encode standards and best practices for threat mitigation using firewalls [20]. Mis-configured firewall security configurations have the potential to expose both internal servers and the firewalls themselves to threats. For example, consider the following scenario where a webServer is susceptible to a $\text{threat}_{\text{SynFlood}}$ attack via the $\text{vul}_{\text{WebTCPConnOverflow}}$ weakness. An individual $\text{vul}_{\text{WebTCPConnOverflow}}$ is representative of a weakness in the TCP stack where it is possible exceed the maximum number of socket connections permitted by the TCP protocol due to a syn flood attack [32]. An iptables rule, represented as individual $\text{iptr}_{\text{SynThresholdWeb}}$, mitigates the vulnerability $\text{vul}_{\text{WebTCPConnOverflow}}$ on the Web server (webServer). Note that we use human-interpretable names (in a typewriter font) for individuals in the ontology as a way of suggesting their meaning. For example, individual $\text{iptr}_{\text{SynThresholdWeb}}$ represents an iptables rule (iptr) that limits TCP syn packet connections to the Web server (SynThresholdWeb).

The Semantic Threat Graph model presented in Figure 1 can be further refined with sub-concepts. For example, the *Threat* concept can define a

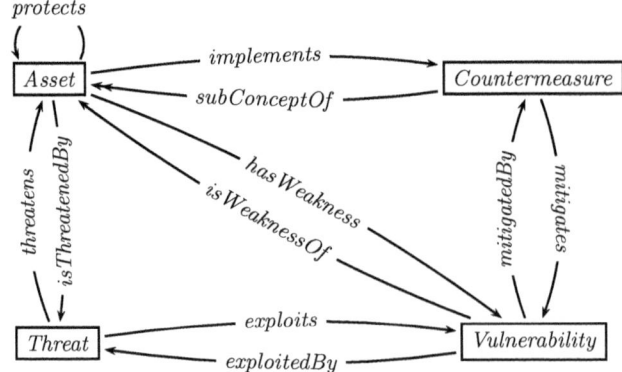

Fig. 1. Abstract Semantic Threat Graph Model

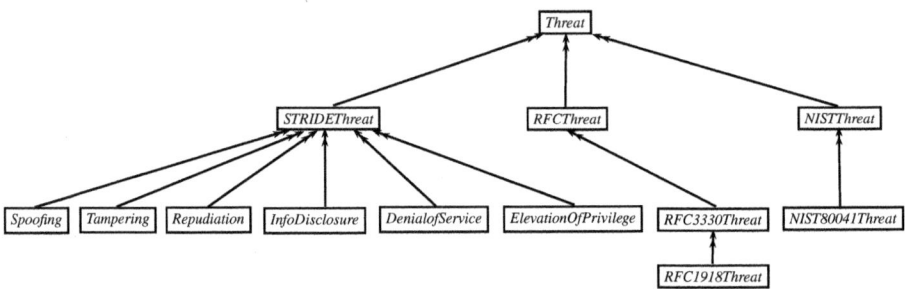

Fig. 2. Fragment of Threat Hierarchy

number of sub-concepts in accordance with best practice, such as the Microsoft STRIDE standard whereby threats are categorised as (Figure 2): *Spoofing identity*, *Tampering with data*, *Repudiation*, *Information disclosure*, *Denial of service* and *Elevation of privilege* [23]. A similar hierarchy is adopted for the corresponding vulnerability and countermeasure concepts.

A best practice standard is a high-level document, written in natural language (typically English text), that defines a set of recommended best practices (countermeasures) to protect sensitive and critical system resources. Best practice standards for network access control, including NIST for secure Web-servers [34] and Internet RFCs for anti-bogon (RFC3330) are encoded as Semantic Threat Graphs (ontologies). How these best practice standards are encoded in terms of Semantic Threat Graphs is described in [20]. For example, Table 1 provides a Semantic Threat Graph interpretation for part of the NIST-800-41 standard [35] for firewall configuration. FBP-1 recommends that (spoofed) packets arriving on an external interface claiming to have originated from either of the three RFC1918 reserved internal IP address ranges should be dropped. Such traffic indicates a denial of service attack typically involving the TCP syn flag. Therefore, *Threat*

Table 1. Ontology Extract for NIST-800-41: Guidelines on Firewalls & Firewall Policy

ID	Recommendation Description		
FBP-1	Deny *"Inbound or Outbound traffic from a system using a source address that falls within the address ranges set aside in RFC1918 as being reserved for private networks"* [35].		
	Threat	**Vulnerability**	**Countermeasure**
	threat$_{\text{Inbound192.168.0.0/16SrcIPPkt}}$	vul$_{\text{UnAuthenInbound192.168.0.0/16PktToFW}}$	iptr$_{\text{DropIn192.168.0.0/16SrcIPPktInputChain}}$
	threat$_{\text{Outbound192.168.0.0/16SrcIPPkt}}$	vul$_{\text{UnAuthenOutbound192.168.0.0/16PktFromFW}}$	iptr$_{\text{DropOut192.168.0.0/16SrcIPPktOutputChain}}$
	threat$_{\text{Inbound192.168.0.0/16SrcIPPkt}}$	vul$_{\text{UnAuthenInbound192.168.0.0/16PktToHost}}$	iptr$_{\text{DropIn192.168.0.0/16SrcIPPktForwardChain}}$
	threat$_{\text{Outbound192.168.0.0/16SrcIPPkt}}$	vul$_{\text{UnAuthenOutbound192.168.0.0/16PktFromHost}}$	iptr$_{\text{DropOut192.168.0.0/16SrcIPPktForwardChain}}$
	threat$_{\text{Inbound10.0.0.0/8SrcIPPkt}}$	vul$_{\text{UnAuthenInbound10.0.0.0/8PktToFW}}$	iptr$_{\text{DropIn10.0.0.0/8SrcIPPktInputChain}}$
	threat$_{\text{Outbound10.0.0.0/8SrcIPPkt}}$	vul$_{\text{UnAuthenOutbound10.0.0.0/8PktFromFW}}$	iptr$_{\text{DropOut10.0.0.0/8SrcIPPktOutputChain}}$
	threat$_{\text{Inbound10.0.0.0/8SrcIPPkt}}$	vul$_{\text{UnAuthenInbound10.0.0.0/8PktToHost}}$	iptr$_{\text{DropIn10.0.0.0/8SrcIPPktForwardChain}}$
	threat$_{\text{Outbound10.0.0.0/8SrcIPPkt}}$	vul$_{\text{UnAuthenOutbound10.0.0.0/8PktFromHost}}$	iptr$_{\text{DropOut10.0.0.0/8SrcIPPktForwardChain}}$
	threat$_{\text{Inbound172.16.0.0/12SrcIPPkt}}$	vul$_{\text{UnAuthenInbound172.16.0.0/12PktToFW}}$	iptr$_{\text{DropIn172.16.0.0/12SrcIPPktInputChain}}$
	threat$_{\text{Outbound172.16.0.0/12SrcIPPkt}}$	vul$_{\text{UnAuthenOutbound172.16.0.0/12PktFromFW}}$	iptr$_{\text{DropOut172.16.0.0/12SrcIPPktOutputChain}}$
	threat$_{\text{Inbound172.16.0.0/12SrcIPPkt}}$	vul$_{\text{UnAuthenInbound172.16.0.0/12PktToHost}}$	iptr$_{\text{DropIn172.16.0.0/12SrcIPPktForwardChain}}$
	threat$_{\text{Outbound172.16.0.0/12SrcIPPkt}}$	vul$_{\text{UnAuthenOutbound172.16.0.0/12PktFromHost}}$	iptr$_{\text{DropOut172.16.0.0/12SrcIPPktForwardChain}}$
ID	**Recommendation Description**		
FBP-2	Deny *"Inbound traffic containing ICMP (Internet Control Message Protocol) traffic"* [35].		
	Threat	**Vulnerability**	**Countermeasure**
	threat$_{\text{ICMPNetworkScan}}$	vul$_{\text{InfoDisclosureICMPReplyPktFromFW}}$	iptr$_{\text{DropInICMPPktInputChain}}$
	threat$_{\text{ICMPNetworkScan}}$	vul$_{\text{InfoDisclosureICMPReplyPktFromHost}}$	iptr$_{\text{DropInICMPPktForwardChain}}$

individual threat$_{\text{Inbound192.168.0.0/16SrcIPPkt}}$, is asserted to be a member of sub-concepts *Spoofing*, *DenialOfService* and *RFC*1918*Threat*. Figure 2 illustrates a partial hierarchy of threats.

4   Knowledge Delegation as Subsumption

A (distributed) system may comprise of a number of separately managed knowledge bases. Each knowledge-base is assumed to have a unique name that indicates its controlling/owning principal. We assume that each atomic concept (or property) ϕ syntactically embeds the name $(\phi)^{\mathcal{N}}$ of the principal that describes the concept (or property). For example, a TBox owned by principal A includes a concept A:DOS where $(\text{A:DOS})^{\mathcal{N}} = A$. A principal P has *jurisdiction* over any concept (or property) ϕ if $(\phi)^{\mathcal{N}} = P$; this means that P is considered to be the original authority on ϕ.

Note that while a TBox may contain concepts originating from different principals, all concepts referenced in a concept expression are required to have the same name. For example, $(\text{A:Threat} \sqcap \text{A:DOS})^{\mathcal{N}} = A$. This ensures a consistent interpretation for our *syntatic* approach to referencing (the originator of) concepts. Future research will consider how a permission-naming logic such as [21] can be used to provide a consistent treatment for the originators of a concept such as $(\text{A:Threat} \sqcap \text{B:DOS})$.

Principals may make public assertions about terminological knowledge. A public assertion $P \approx (\phi_1 \sqsubseteq \phi_2)$ is a statement by principal P about concept (or property) subsumption. For example, $A \approx (\text{B:DOS} \sqsubseteq \text{A:Threat})$ is an assertion by A that its concept of threat includes B's concept of denial of service. These ontology mappings can be used to implement delegation of jurisdiction. Given $(\phi_1)^{\mathcal{N}} = Q$ and $(\phi_2)^{\mathcal{N}} = P$ then $P \approx (\phi_1 \sqsubseteq \phi_2)$ is an assertion by P that it trusts Q when it describes ϕ_1 to the extent that Q's concept of ϕ_1 can be considered as a kind of ϕ_2 concept over which P's has jurisdiction. For example, suppose that principal B provides a vulnerability reporting service, then assertions $A \approx\!\!\!\sim (\texttt{B:DOS} \sqsubseteq \texttt{A:DOS})$ and $A \approx\!\!\!\sim (\texttt{B:mitigates} \sqsubseteq \texttt{A:mitigates})$ mean that A trusts B's mitigation knowledge for denial of service attacks.

Transitive subsumption in $\mathcal{SHOIN}^{(\mathcal{D})}$ can be used to reason over chains of delegation statements. For example, public assertion $B \approx\!\!\!\sim (\texttt{C:mitigates} \sqsubseteq \texttt{B:mitigates})$ indicates that the vulnerability reporting service B trusts mitigation recommendations provided by a software developer C. Continuing the example, principal A can use these public assertions to deduce that $\texttt{C:mitigates} \sqsubseteq \texttt{A:mitigates}$ and thus be happy to trust mitigation recommendations from the software developer.

The following rule defines the conditions under which an arbitrary principal may import, into its TBox, a public assertion (delegation) from P.

$$\frac{P \approx\!\!\!\sim (\phi_1 \sqsubseteq \phi_2); (\phi_2)^{\mathcal{N}} = P}{\texttt{import } \phi_1 \sqsubseteq \phi_2 \texttt{ into TBox}}$$

This does not extend the semantics of $\mathcal{SHOIN}^{(\mathcal{D})}$, rather, it is a syntactic treatment whereby delegation statements translate to concept axioms that can in turn be reasoned over within $\mathcal{SHOIN}^{(\mathcal{D})}$. This treatment is easily modeled within OWL-DL. The URI of an OWL-DL document provides its principal/namespace. A public assertion $P \approx\!\!\!\sim (\phi_1 \sqsubseteq \phi_2)$ is an ontology document that is trusted to originate from the namespace of P: this trust can be achieved by P signing the document. The ontology-import constructor $\texttt{owl:imports}$ is used to import a public assertion that is confirmed to come from a from another namespace.

The assertional component of a knowledge base, hereafter referred to as its *ABox*, contains assertions about named concept individuals. A concept assertion $\phi(i)$, indicates that named individual i is a member of concept ϕ; a role (property) assertion $p(i, j)$ indicates that named individual i is related to named individual j under property p. For example, ABox assertion $\texttt{DOS(syn-flood)}$ describes that individual $\texttt{syn-flood}$ is a $\texttt{DOS}$ threat and ABox role assertion $\texttt{mitigates(syn-cache,syn-flood)}$ describes that the $\texttt{syn-cache}$ countermeasure mitigates a $\texttt{syn-flood}$ threat.

We use a similar naming scheme for individuals whereby $(i)^{\mathcal{N}}$ indicates a principal/namespace syntactically embedded in the identifier of the individual i. Principals may make public assertions about individuals. A public assertion $P \sim\!\!\!\vdash \phi(i)$ is a statement by P that named individual i is a member of concept ϕ. A principal may not make public assertions about ABox knowledge (concept and individual) that is not under its jurisdiction. However, a principal may use subsumption to infer ABox knowledge that is effectively under the jurisdiction of others. The following rule defines the conditions under which an arbitrary principal may import, into its ABox, a public assertion from P about a named individual i and concept ϕ.

$$\frac{P \sim\!\!\!\vdash \phi(i); (\phi)^{\mathcal{N}} = (i)^{\mathcal{N}} = P}{\texttt{import } \phi(i) \texttt{ into ABox}}$$

A similar rule can be defined for public ABox property assertions.

Continuing the above example, $\text{B} \hspace{1pt}\vdash\hspace{-3pt}\sim\hspace{-2pt}(\text{B:DOS(B:syn-flood)})$ is a statement by the vulnerability reporting service B that B:syn-flood is a B:DOS threat. On importing this ABox assertion and the TBox assertion $\text{A} \hspace{1pt}\vDash\hspace{-3pt}\approx\hspace{-2pt}(\text{B:DOS} \sqsubseteq \text{A:DOS})$, it is possible for A to deduce A:DOS(B:syn-flood), that is, B:syn-flood is an A:DOS threat.

These public ABox assertions are a syntactic treatment that do not extend $\mathcal{SHOIN}^{(\mathcal{D})}$. In practice, OWL-DL individuals include reference to their namespace (principal) and a public ABox assertion $P \hspace{1pt}\vdash\hspace{-3pt}\sim\hspace{-2pt} \phi(i)$ is an ontology document that has been signed by its originator P and a valid $\phi(i)$ can be imported into another ontology using the owl:imports constructor.

In general, a public TBox assertion $P \hspace{1pt}\vDash\hspace{-3pt}\approx\hspace{-2pt}(\phi_1 \sqsubseteq \phi_2)$ is effectively a delegation certificate that can be understood as a statement $P \hspace{1pt}\vDash\hspace{-3pt}\approx\hspace{-2pt}(\phi_2 \supset \phi_1)$ in a delegation logic such as [4], where ϕ_1 and ϕ_2 are unary (or binary) predicates that refer to static principals $(\phi_1)^{\mathcal{N}}$ and $(\phi_2)^{\mathcal{N}}$, respectively. A public ABox assertion $P \hspace{1pt}\vdash\hspace{-3pt}\sim\hspace{-2pt} \phi(i)$ is a signed value of type ϕ that can be effectively considered to be a form of a signed permission that cannot be forged by another principal that does not hold jurisdiction.

Note that for ease of presentation, delegation of trust is assumed transitive. Non-transitive trust can be incorporated into the model, for example, by adding a SPKI-style [16] delegation-bit to delegation certificates. Alternatively, a KeyNote-style [12] Action_Authorizers concept could be added to the ontology to constrain the delegation.

5   Delegation in Semantic Threat Graphs

A decentralized Semantic Threat Graph (STG) uses subsumption to model the delegation of jurisdiction over (potentially distributed) fragments of a Semantic Threat Graph. These fragments can include concepts such as *Vulnerability*, *Threat* and *Countermeasure*, assertions about membership of these concepts and assertions about properties between the concepts. In this section, we outline how the delegation involving these fragments is encoded using subsumption; Section 6 will provide complete examples of decentralized Semantic Threat Graphs.

STG Concept Delegation. Subsumption is used to define delegation of *Threat*, *Vulnerability* and *Countermeasure* concepts between principals. For example,

$$Y \vDash\hspace{-3pt}\approx X : DenialOfService \sqsubseteq Y : Threat$$

defines that principal Y trusts principal X, concerning the identification (naming) of denial of service attacks. Suppose that X in turn asserts

$$X \vdash\hspace{-3pt}\sim X : DenialOfService(X : \text{threat}_{\text{SynFlood}})$$

that is, $X : \text{threat}_{\text{SynFlood}}$ is a *DenialOfService* individual, as identified by X. In this case, as a result of the delegation by subsumption above, principal Y can safely deduce that

$$Y : Threat(X : \text{threat}_{\text{SynFlood}})$$

that is, $X : \mathtt{threat}_{\mathtt{SynFlood}}$ can be identified as an individual of Y's *Threat* concept. Similar assertions can be made about other Semantic Threat concepts including *Vulnerability* and *Countermeasure*.

STG Property Delegation. Like concepts, properties can be hierarchical, and over which, principals may make jurisdiction assertions. For example, the property delegation

$$Y \mathrel{\mspace{1mu}\not\approx\mspace{1mu}} X : exploits \sqsubseteq Y : exploits$$

is a statement by Y that it is willing to trust properties from the knowledge-base of X that relate how vulnerabilities are exploited by threats. For example, suppose that principal X asserts

$$X \mathrel{\mspace{1mu}\vdash\mspace{-6mu}\sim\mspace{1mu}} X : exploits(X : \mathtt{threat}_{\mathtt{SynFlood}}, X : \mathtt{vul}_{\mathtt{WebTCPConnMax}})$$

then principal Y, trusting X's assertions on exploits ($X : exploits \sqsubseteq Y : exploits$) can infer the following statements in its knowledge base.

$$Y : Threat(X : \mathtt{threat}_{\mathtt{SynFlood}}), Y : Vulnerability(X : \mathtt{vul}_{\mathtt{WebTCPConnMax}}),$$
$$Y : exploits(X : \mathtt{threat}_{\mathtt{SynFlood}}, X : \mathtt{vul}_{\mathtt{WebTCPConnMax}})$$

STG Property Restriction Delegation. Restrictions ('quantifier' and 'hasValue') can be applied to properties and are used to constrain an individual's membership to a specific concept. A property restriction effectively describes an anonymous or unnamed concept that contains all the individuals that satisfy the restriction. For example, principal Y asserts

$$Y \mathrel{\mspace{1mu}\not\approx\mspace{1mu}}(\exists_{\geq 1} X : exploits.X : Vulnerability \sqsubseteq Y : Threat)$$

meaning that threat individuals that are members of an unnamed threat concept $\exists_{\geq 1} X : exploits.X : Vulnerability$ defined within principal X's knowledge-base are considered as trusted individuals of concept $Y : Threat$ in principal Y's knowledge-base.

A 'hasValue' restriction, denoted by $\ni$, describes a set of individuals that are members of an anonymous concept (domain) that are related to a specific individual along the range of a given property. For example, in

$$Y \mathrel{\mspace{1mu}\not\approx\mspace{1mu}} X : exploits \ni X : \mathtt{vul}_{\mathtt{WebTCPConnMax}} \sqsubseteq Y : Threat$$

Y asserts that it trusts principal X's knowledge about threats that exploit the $X : \mathtt{vul}_{\mathtt{WebTCPConnMax}}$. Therefore, threat is $X : \mathtt{threat}_{\mathtt{SynFlood}}$ is considered a threat (a member of concept $Y : Threat$) within principal Y's knowledge-base. Note, concept $X : exploits \ni X : \mathtt{vul}_{\mathtt{WebTCPConnMax}}$ is a sub-concept of the $\exists_{\geq 1} X : exploits.X : Vulnerability$.

6   Decentralized Semantic Threat Graphs: Use Cases

In this section, a series of examples are presented in order to demonstrate how decentralized Semantic Threat Graphs may work in practice and in which we

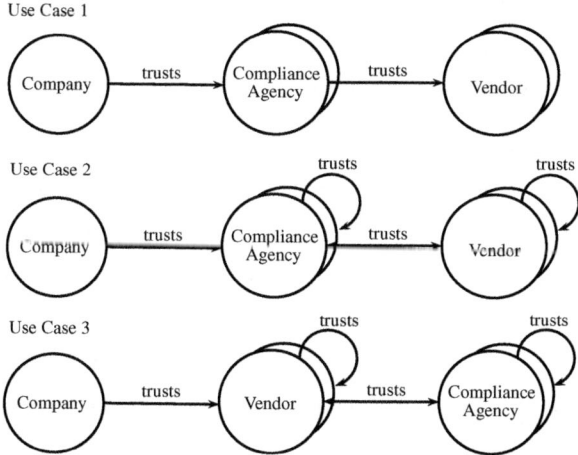

Fig. 3. Example Delegation Usage Patterns

identify some potential business usage patterns (Figure 3). We believe this to be an improvement over the current centralised approach where trust between different providers is not discriminated.

Knowledge about threats, vulnerabilities and countermeasures while originating from a number of providers, is typically managed within centralised threat knowledge-bases such as NIST's *National Vulnerability Database (NVD)* [1]. The advantage of decentralising a threat knowledge-base means that threat knowledge may reside with the originating provider and/or be distributed across other trusted third party providers. A disadvantage of a centralised approach is that it implies on the part of a consumer a global level of trust for all threat knowledge indiscriminately. However, a consumer may only wish to trust threat knowledge that has been provided by a particular provider or set of providers. For example, a consumer of the NVD may only want to trust knowledge about network-based Denial of Service threats that have been provided by US-Cert and corresponding (firewall) countermeasures provided by Redhat. The advantage of a trust management approach is that it provides a basis with which a consumer may delegate authority to trusted providers for threat knowledge. In practice, providers/producers construct fragments of Semantic Threat Graphs and consumers make assertions about the conditions under which they trust the fragments provided by providers. These are defined as a combination of STG Concept, Property and Property Restriction delegation assertions described in the previous section.

Use Case 1. A company (for example, *ACME* Inc.) having a consumer role, delegates jurisdiction for knowledge about threats, vulnerabilities and countermeasures to one or more trusted compliance agencies (for example, *NIST*) having a role of a provider. A compliance agency, having a consumer role in this instance, in turn delegates jurisdiction for knowledge about threats, vulnerabilities and countermeasures to one or more trusted vendors (for example Redhat). A vendor (provider) may then notify the company with respect to Semantic Threat Graph knowledge it has jurisdiction over. The company can import these STG assertions into its knowledge-base and use the knowledge during reasoning. For example, principal *ACME* asserts:

$$NIST : NIST80041\,Threat \sqsubseteq ACME : Threat$$

$$NIST : NIST80041\,Vulnerability \sqsubseteq ACME : Vulnerability$$

$$NIST : NIST80041\,Countermeasure \sqsubseteq ACME : Countermeasure$$

$$NIST : exploits \sqsubseteq ACME : exploits$$

$$NIST : mitigates \sqsubseteq ACME : mitigates$$

indicating that it trusts countermeasure recommendations made by *NIST* regarding threats and vulnerabilities that are to be mitigated in order to be compliant with NIST-800-41 [35] firewall best practice. Note that, when no ambiguity can arise, we drop the turnstile notation "$\approx$" and "$\sim$" and infer the principal from the statement.

Principal *NIST*, in turn, asserts:

$$NIST : Spoofing \sqsubseteq NIST : NIST80041\,Threat$$

$$NIST : UnAuthPkt \sqsubseteq NIST : NIST80041\,Vulnerability$$

$$RH : Spoofing \sqsubseteq NIST : Spoofing$$

$$RH : UnAuthPkt \sqsubseteq NIST : UnAuthPkt$$

$$RH : IptablesRule \sqsubseteq NIST : NIST80041\,Countermeasure$$

$$RH : exploits \sqsubseteq NIST : exploits$$

$$RH : mitigates \sqsubseteq NIST : mitigates$$

indicating that principal *RH* (Redhat vendor) is trusted to specify Linux iptables firewall countermeasures used to mitigate spoofing threats and associated vulnerabilities. Intuitively, the NIST compliance agency has outsourced the instantiation of spoof-based threats and recommended firewall countermeasures to Redhat. Note, the delegation statement made by *NIST* also includes additional knowledge about its threat and vulnerability hierarchy. For example concept *NIST : Spoofing* is a subsumed by concept *NIST : NIST80041 Threat*.

Delegation chains (transitive subsumption) constructed in terms of concept and property subsumption can be reasoned over within OWL-DL to establish if received ABox statements are to be trusted and imported. For example, Redhat (principal *RH*) asserts the following anti-bogon IP spoofing threat information:

$RH : Spoofing(RH : \mathtt{threat}_{\mathtt{Inbound192.168.0.0/16SrcIPPkt}})$,

$RH : UnAuthPkt(RH : \mathtt{vul}_{\mathtt{UnAuthenInbound192.168.0.0/16PktToHost}})$,

$RH : IptablesRule(RH : \mathtt{iptr}_{\mathtt{DropIn192.168.0.0/16SrcIPPktForwardChain}})$,

$RH : exploits(RH : \mathtt{threat}_{\mathtt{Inbound192.168.0.0/16SrcIPPkt}},$

$\qquad\qquad\qquad RH : \mathtt{vul}_{\mathtt{UnAuthenInbound192.168.0.0/16PktToHost}})$,

$RH : mitigates(RH : \mathtt{iptr}_{\mathtt{DropIn192.168.0.0/16SrcIPPktForwardChain}},$

$\qquad\qquad\qquad RH : \mathtt{vul}_{\mathtt{UnAuthenInbound192.168.0.0/16PktToHost}})$

On receipt of this ABox statement principal *ACME* has no prior trust relationship with *RH*. However, given the set of known delegation statements, *ACME* can form the following delegation (trust) chains.

RH: *Spoofing* $\sqsubseteq$ *NIST*: *Spoofing* $\sqsubseteq$ *NIST*: *NIST80041Threat* $\sqsubseteq$ *ACME*: *Threat*
RH: *UnAuthPkt* $\sqsubseteq$ *NIST*: *UnAuthPkt* $\sqsubseteq$ *NIST*: *NIST80041Vul* $\sqsubseteq$ *ACME*: *Vulnerability*
RH: *IptablesRule* $\sqsubseteq$ *NIST*: *NIST80041Countermeasure* $\sqsubseteq$ *ACME*: *Countermeasure*
RH: *exploits* $\sqsubseteq$ *NIST*: *exploits* $\sqsubseteq$ *ACME*: *exploits*
RH: *mitigates* $\sqsubseteq$ *NIST*: *mitigates* $\sqsubseteq$ *ACME*: *exploits*

As a consequence, *ACME* can deduce that the ABox mitigation knowledge received from *RH* is trusted. It then becomes possible for *ACME* to deduce a new concept hierarchy within its local Semantic Threat Graph knowledge-base, for example:

RH: *Spoofing* $\sqsubseteq$ *NIST*: *Spoofing* $\sqsubseteq$ *NIST*: *NIST80041Threat* $\sqsubseteq$ *ACME*: *Threat*

in addition to the following inferred concept membership and property relations:

$ACME$: *Threat*(RH: $\texttt{threat}_{\texttt{Inbound192.168.0.0/16SrcIPPkt}}$),
$ACME$: *Vulnerability*(RH: $\texttt{vul}_{\texttt{UnAuthenInbound192.168.0.0/16PktToHost}}$),
$ACME$: *Countermeasure*(RH: $\texttt{iptr}_{\texttt{DropIn192.168.0.0/16SrcIPPktForwardChain}}$),
$ACME$: *exploits*(RH: $\texttt{threat}_{\texttt{Inbound192.168.0.0/16SrcIPPkt}}$,
$\qquad\qquad\qquad\qquad RH$: $\texttt{vul}_{\texttt{UnAuthenInbound192.168.0.0/16PktToHost}}$),
$ACME$: *mitigates*(RH: $\texttt{iptr}_{\texttt{DropIn192.168.0.0/16SrcIPPktForwardChain}}$,
$\qquad\qquad\qquad\qquad RH$: $\texttt{vul}_{\texttt{UnAuthenInbound192.168.0.0/16PktToHost}}$)

Use Case 2. As in the previous use case, a company may delegate jurisdiction for knowledge about threats, vulnerabilities and countermeasures to one or more trusted compliance agencies. However, rather than a compliance agency delegating jurisdiction over threats, vulnerabilities and countermeasures as a collection to one or more vendors such as Redhat or Cisco, it may instead decide to delegate certain fragments (for example threats) to one or more additional compliance agencies and other fragments (for example countermeasures) to one or more vendors. Vendors in turn may also trust one or more compliance agencies.

In this example, *ACME* makes the same TBox statement for delegation of jurisdiction to *NIST* defined in the previous scenario. Principal *NIST* delegates jurisdiction to *CVE* (compliance agency) for knowledge about NIST-800-41 spoofing threats and vulnerabilities only.

$NIST$: *Spoofing* $\sqsubseteq$ *NIST*: *NIST80041Threat*
$NIST$: *UnAuthPkt* $\sqsubseteq$ *NIST*: *NIST80041Vulnerability*
CVE: *Spoofing* $\sqsubseteq$ *NIST*: *Spoofing*
CVE: *UnAuthPkt* $\sqsubseteq$ *NIST*: *UnAuthPkt*
CVE: *exploits* $\sqsubseteq$ *NIST*: *exploits*

Principal *NIST* also asserts the following delegation statement stating that principal *RH* is trusted for NIST-800-41 based iptables firewall countermeasures.

$$RH\!:\!iptablesRule \sqsubseteq NIST\!:\!NIST80041Countermeasure$$
$$RH\!:\!mitigates \sqsubseteq NIST\!:\!mitigates$$

Note, trust is not bidirectional. Given that *RH* has not been given jurisdiction over relevant threats and vulnerabilities with which to make iptables recommendations, it must also delegate jurisdiction to *NIST* for this knowledge.

$$NIST\!:\!Spoofing \sqsubseteq RH\!:\!Spoofing$$
$$NIST\!:\!UnAuthPkt \sqsubseteq RH\!:\!UnAuthPkt$$
$$NIST\!:\!exploits \sqsubseteq RH\!:\!exploits$$

Principal *RH* receives the following ABox statements from *CVE* for which it has no prior trust relationship.

$CVE\!:\!Spoofing(CVE\!:\!\texttt{threat}_{\texttt{Inbound192.168.0.0/16SrcIPPkt}})$,
$CVE\!:\!UnAuthPkt(CVE\!:\!\texttt{vul}_{\texttt{UnAuthenInbound192.168.0.0/16PktToHost}})$,
$CVE\!:\!exploits(CVE\!:\!\texttt{threat}_{\texttt{Inbound192.168.0.0/16SrcIPPkt}}$,
$\qquad\qquad\qquad\qquad\qquad CVE\!:\!\texttt{vul}_{\texttt{UnAuthenInbound192.168.0.0/16PktToHost}})$

Principal *RH* can form a chain of trust based on its trust for *NIST* and *NIST*'s trust for *CVE*. As a consequence, *RH* can define a suitable iptables rule (countermeasure) that mitigates the vulnerability of unauthenticated 192.168.0.0/16 subnet packets exploited by spoofed packets of the same source IP range.

$RH\!:\!IptablesRule(RH\!:\!\texttt{iptr}_{\texttt{DropIn192.168.0.0/16SrcIPPktForwardChain}})$,
$RH\!:\!mitigates(RH\!:\!\texttt{iptr}_{\texttt{DropIn192.168.0.0/16SrcIPPktForwardChain}}$,
$\qquad\qquad\qquad\qquad\qquad CVE\!:\!\texttt{vul}_{\texttt{UnAuthenInbound192.168.0.0/16PktToHost}})$

Principal *ACME* in turn receives the following ABox statements from *RH* for which it has no prior trust relationship.

$RH\!:\!Spoofing(CVE\!:\!\texttt{threat}_{\texttt{Inbound192.168.0.0/16SrcIPPkt}})$,
$RH\!:\!UnAuthPkt(CVE\!:\!\texttt{vul}_{\texttt{UnAuthenInbound192.168.0.0/16PktToHost}})$,
$RH\!:\!IptablesRule(RH\!:\!\texttt{iptr}_{\texttt{DropIn192.168.0.0/16SrcIPPktForwardChain}})$,
$RH\!:\!exploits(CVE\!:\!\texttt{threat}_{\texttt{Inbound192.168.0.0/16SrcIPPkt}}$,
$\qquad\qquad\qquad\qquad\qquad CVE\!:\!\texttt{vul}_{\texttt{UnAuthenInbound192.168.0.0/16PktToHost}})$,
$RH\!:\!mitigates(RH\!:\!\texttt{iptr}_{\texttt{DropIn192.168.0.0/16SrcIPPktForwardChain}}$,
$\qquad\qquad\qquad\qquad\qquad CVE\!:\!\texttt{vul}_{\texttt{UnAuthenInbound192.168.0.0/16PktToHost}})$

Principal *ACME* can form a chain of trust based on its trust for *NIST* and *NIST*'s trust for *CVE* and *RH*. For example:

$CVE\!:\!Spoofing \sqsubseteq NIST\!:\!Spoofing \sqsubseteq NIST\!:\!NIST80041Threat \sqsubseteq ACME\!:\!Threat$
$CVE\!:\!UnAuthPkt \sqsubseteq NIST\!:\!UnAuthPkt \sqsubseteq NIST\!:\!NIST80041Vul \sqsubseteq ACME\!:\!Vulnerability$
$RH\!:\!IptablesRule \sqsubseteq NIST\!:\!NIST80041Countermeasure \sqsubseteq ACME\!:\!Countermeasure$

Use Case 3. This scenario is a variation of use case 2. A company may trust one or more vendors for Semantic Threat Graph ABox statements where each vendor may in turn trust other vendors and/or compliance agencies for ABox Semantic Threat Graph statements. For reasons of space, we do not provide example TBox delegation statements and Abox statements.

7 Related Research

The delegation scheme proposed in this paper is based on managing trust across a model of distributed knowledge-bases. While the model is simple, it closely resembles the OWL-DL approach to modular ontologies [22] using the `owl:imports` constructor with a URI based namespace. Future research will explore representing and reasoning about distributed trust in other modular Description Logic languages such as [14, 36]. The TBox intensional reasoning provided by existing OWL-DL reasoners is relatively scalable, however, extensional reasoning is poor for medium to large-sized ABoxes.

A large body of research results exist on Trust Management and decentralized authorization systems, for example, [9, 13, 16, 27]. However, there has been little consideration regarding how it might be applied to managing trust relationships across knowledge-bases, which is considered in this paper. In [31], a centralised reference ontology is developed to represent trust requirements. Agarwal and Rudolph [5] present an ontology for SPKI/SDSI certificates. In [5] SPKI names are represented as concept names while public keys are represented as individuals. However, once the ontology is constructed any reasoning over delegation chains for the purpose of compliance checking is performed outside of the ontology framework.

Trust Management systems typically describe policy and authorization in terms of discrete permissions and/or assertions. In this paper, authority (about STGs) is defined in terms of Description Logic concepts. Description Logic has been used to describe and reason about RBAC [18] and XACML policies [26] with subsumption providing role/authorization hierarchies, but do not consider the jurisdiction that principals may have over the ontologies in their local knowledge-bases. Semantic SPKI [6] uses subsumption to define SDSI local name bindings, however an external certificate discovery algorithm implements name reduction. In our paper, public keys are used to uniquely identify principals and their name spaces. Future work will extend this to support SDSI naming, based on the logic described in [21].

The requirements of Distributed Semantic Threat Graphs determined our particular use of Trust Management and effectively corresponds to a conventional compliance check [13]: *for a given delegation network, is a principal trusted for some action?* This check returns a binary answer and we believe that the model could be extended to support forms of quantitative trust, by incorporating KeyNote-style [12] compliance values or weights [11, 19] in the delegation statements. We also assume that trust is monotonic, for example, it is safe to rely on a Semantic Threat Graph delegation chain provided by a vendor since the model does not permit other principals to make conflicting assertions about concepts that originated from the vendor's namespace. Supporting non-monotonic trust, including inter-policy-conflicts such as [10], is non-trivial and effectively requires synchronization of the distributed ABox/TBoxes. Providing support for distributed ontologies is an active research topic [8]. The extent to which these other forms of reasoning over the distributed ontology are applicable, and could be supported by extending our current model, is a topic for future research.

8 Conclusion

In this paper, a trust management approach is proposed to decentralize and delegate knowledge for threats and their mitigation (encoded as Semantic Threat Graphs) to one or more trusted providers. That is, the ability to trust-manage the (delegation) relationships that may exist between the providers.

The ontology-based delegation scheme used subsumption to model the delegation of jurisdiction over (potentially distributed) fragments of a Semantic Threat Graph. This paper extends the model from [20] — which did not consider the possibility that threat catalogues may originate from different trusted providers — to a decentralized trust model.

In this paper, the Semantic Threat Graphs knowledge-bases comprised of knowledge about standards and best practices for threat mitigation using firewalls. The applicability of the (centralised) approach of encoding numerous best-practices is demonstrated in [20]. We argue that the effort of decentralizing this cataloging exercise is comparable. Future work will consider constructing Semantic Threat Graphs from additional threat knowledge-bases such as NVD.

Acknowledgements. The authors would like to thank the anonymous reviewers for their valuable feedback. This research has been supported by Science Foundation Ireland grant 08/SRC/11403.

References

1. http://www.nist.gov/
2. http://www.us-cert.gov/
3. http://www.securityfocus.com
4. Abadi, M., Burrows, M., Lampson, B., Plotkin, G.: A calculus for access control in distributed systems. ACM Trans. Program. Lang. Syst. 15, 706–734 (1993), http://doi.acm.org/10.1145/155183.155225
5. Agarwal, S., Rudolph, S.: Semantic Description of Behavior and Trustworthy Credentials of Web Services. In: 6th International Semantic Web Conference, Busan, Korea (November 2007)
6. Agudo, I., Lopez, J., Montenegro, J.A.: Enabling attribute delegation in ubiquitous environments. Mobile Netw. Appl., 1–13 (July 2008), http://www.springerlink.com/content/q845pp64672m3586/
7. Baader, F., Calvanese, D., McGuinness, D.L., Nardi, D., Patel-Schneider, P.: The Description Logic Handbook: Theory, Implementation and Applications. Cambridge University Press (March 2003)

8. Bao, J., Voutsadakis, G., Slutzki, G., Honavar, V.: Package-Based Description Logics. In: Stuckenschmidt, H., Parent, C., Spaccapietra, S. (eds.) Modular Ontologies. LNCS, vol. 5445, pp. 349–371. Springer, Heidelberg (2009)
9. Becker, M., Fournet, C., Gordon, A.: Design and semantics of a decentralized authorization language. In: 20th IEEE Computer Security Foundations Symposium (January 2007)
10. Bertino, E., Jajodia, S., Samarati, P.: Supporting multiple access control policies in database systems. In: Proceedings of the 1996 IEEE Conference on Security and Privacy, SP 1996, pp. 94–107. IEEE Computer Society, Washington, DC (1996), http://dl.acm.org/citation.cfm?id=1947337.1947353
11. Bistarelli, S., Martinelli, F., Santini, F.: A Semantic Foundation for Trust Management Languages with Weights: An Application to the RT Family. In: Rong, C., Jaatun, M.G., Sandnes, F.E., Yang, L.T., Ma, J. (eds.) ATC 2008. LNCS, vol. 5060, pp. 481–495. Springer, Heidelberg (2008)
12. Blaze, M., Feigenbaum, J., Ioannidis, J., Keromytis, A.D.: The keynote trust-management system, version 2 (September 1999)
13. Blaze, M., Feigenbaum, J., Lacy, J.: Decentralized trust management. In: Proceedings of the IEEE Symposium on Research in Security and Privacy, pp. 164–173. IEEE Computer Society Press, Oakland (1996)
14. Borgida, A., Serafini, L.: Distributed Description Logics: Directed Domain Correspondences in Federated Information Sources. In: Meersman, R., et al. (eds.) CoopIS 2002, DOA 2002, and ODBASE 2002. LNCS, vol. 2519, pp. 36–53. Springer, Heidelberg (2002)
15. Cuppens-Boulahia, N., Cuppens, F., de Vergara, J.E.L., Guerra, J., Debar, H., Vazquez, E.: An Ontology-Based Approach to React to Network Attacks. In: 3rd International Conference on Risk and Security of Internet and Systems (CRiSIS), Tozeur, Tunisia (October 2008)
16. Ellison, C., Frantz, B., Lampson, B., Rivest, R.L., Thomas, B., Ylonen, T.: SPKI certificate theory (September 1999)
17. Fenz, S., Goluch, G., Ekelhart, A., Riedl, B., Weippl, E.R.: Information Security Fortification by Ontological Mapping of the ISOIEC 27001 Standard. In: 13th Pacific Rim International Symposium on Dependable Computing (PRDC), Australia (December 2007)
18. Finin, T., Joshi, A., Kagal, L., Niu, J., Sandhu, R., Winsborough, W.H., Thuraisingham, B.: ROWLBAC - Representing Role Based Access Control in OWL. In: 13th Symposium on Access Control Models and Technologies, Colorado, USA (June 2008)
19. Foley, S.N., Mac Adams, W., O'Sullivan, B.: Aggregating Trust Using Triangular Norms in the KeyNote Trust Management System. In: Cuellar, J., Lopez, J., Barthe, G., Pretschner, A. (eds.) STM 2010. LNCS, vol. 6710, pp. 100–115. Springer, Heidelberg (2011)
20. Foley, S.N., Fitzgerald, W.M.: Management of Security Policy Configuration using a Semantic Threat Graph Approach. Journal of Computer Security (JCS) 19(3) (2011)
21. Foley, S.N., Abdi, S.: Avoiding Delegation Subterfuge Using Linked Local Permission Names. In: Barthe, G., Datta, A., Etalle, S. (eds.) FAST 2011. LNCS, vol. 7140, pp. 100–114. Springer, Heidelberg (2012)
22. Grau, B.C., Horrocks, I., Kazakov, Y., Sattler, U.: Modular Reuse of Ontologies: Theory and Practice. Journal of Artificial Intelligence Research 31 (February 2008)
23. Hernan, S., Lambert, S., Ostwald, T., Shostack, A.: Uncover Security Design Flaws Using The STRIDE Approach, http://microsoft.com/

24. Herzog, A., Shahmehri, N., Duma, C.: An Ontology of Information Security. International Journal of Information Security and Privacy (IJISP) 1(4) (2007)
25. Kodeswaran, P.A., Kodeswaran, S.B., Joshi, A., Finin, T.: Enforcing Security in Semantics Driven Policy Based Networks. In: 24th International Conference on Data Engineering Workshops, Secure Semantic Web, Cancun, Mexico (April 2008)
26. Kolovski, V., Hendler, J., Parsia, B.: Analyzing web access control policies. In: Proceedings of the 16th International Conference on World Wide Web, WWW 2007, pp. 677–686. ACM, New York (2007),
http://doi.acm.org/10.1145/1242572.1242664
27. Li, N., Winsborough, W., Mitchell, J.: Distributed credential chain discovery in trustmanagement. Journal of Computer Security 11(3), 35–86 (2003)
28. Ray, I., Poolsapassit, N.: Using Attack Trees to Identify Malicious Attacks from Authorized Insiders. In: De Capitani di Vimercati, S., Syverson, P.F., Gollmann, D. (eds.) ESORICS 2005. LNCS, vol. 3679, pp. 231–246. Springer, Heidelberg (2005)
29. Schneier, B.: Secrets and Lies Digital Security in Networked World. Wiley Publishing (2004)
30. Smith, M.K., Welty, C., McGuinness, D.L.: OWL Web Ontology Language Guide. W3C Recommendation, Technical Report (2004)
31. Squicciarini, A.C., Bertino, E., Ferrari, E., Ray, I.: Achieving Privacy in Trust Negotiations with an Ontology-Based Approach. IEEE Transactions on Dependable and Secure Computing 3(1) (2006)
32. Stevens, R.: Unix Network Programming, Networking API's: Sockets and XTI, 2nd edn., vol. 1. Prentice Hall (1998)
33. Thuraisingham, B.: Building Trustworthy Semantic Webs. AUERBACH (2007)
34. Tracy, M., Jansen, W., Scarfone, K., Winograd, T.: Guidelines on Securing Public Web Servers: Recommendations of the National Institute of Standards and Technology. NIST Special Publication 800-44, Version 2 (September 2009)
35. Wack, J., Cutler, K., Pole, J.: Guidelines on Firewalls and Firewall Policy: Recommendations of the National Institute of Standards and Technology. NIST-800-41 (2002)
36. Wang, Y., Haase, P., Bao, J.: A survey of formalisms for modular ontologies. In: International Joint Conference on Artificial Intelligence (IJCAI 2007) Workshop (2007)

Code Type Revealing Using Experiments Framework

Rami Sharon[1] and Ehud Gudes[2]

[1] The Open University, Ra'anana, Israel
Sharon.Rami@gmail.com
[2] Ben-Gurion University, Beer-Sheva, Israel
ehud@cs.bgu.ac.il

Abstract. Identifying the type of a code, whether in a file or byte stream, is a challenge that many software companies are facing. Many applications, security and others, base their behavior on the type of code they receive as an input.

Today's traditional identification methods rely on file extensions, magic numbers, propriety headers and trailers or specific type identifying rules. All these are vulnerable to content tampering and discovering it requires investing long and tedious working hours of professionals. This study is aimed to find a method of identifying the best settings to automatically create type signatures that will effectively overcome the content manipulation problem.

In this paper we lay out a framework for creating type signatures based on byte N-Grams. The framework allows setting various parameters such as N-Gram sizes and windows, selecting statistical tests and defining rules for score calculations. The framework serves as a test lab that allows finding the right parameters to satisfy a predefined threshold of type identification accuracy. We demonstrate the framework using basic settings that achieved an F-Measure success rate of 0.996 on 1400 test files.

Keywords: File Type, Content type revealing framework, Code type, Byte N-Gram statistical analysis.

1 Introduction

In today's connected environment, most businesses increasingly rely on the Internet as a source of information and a platform for communication and Electronic commerce. One of the main motivating factors driving the increased use of the net is the ability to use technologies based on active content such as Active-x, Java applets, Java Script and Executable files, in order to implement Web Based Applications. The flexibility of these technologies convey great benefits, but at the same time, allow using Web based applications as Malware carriers, capable of harming the organization by damaging its infrastructures, stealing information or performing other illegal activities.

One of the most difficult parts of the attack is to penetrate and infuse the code into the system. Attackers develop new approaches to disguise the true nature of the penetrating file, if by using naive methods such as changing File Extensions or by

N. Cuppens-Boulahia et al. (Eds.): DBSec 2012, LNCS 7371, pp. 193–206, 2012.

manipulating file content such as the file header. Recent research in this area was undertaken in order to find efficient ways to identify the true nature of a code in a file, without relying on external characteristics. One of the most common methods applied for this purpose is the analysis of the N-Grams, which are variable sequences of bytes (usually consecutive but not necessarily), present in the file [1, 2, 8].

The main contribution of this paper is a new Framework, called CTR – Code Type Revealing, that enable in a convenient manner the finding of the most efficient parameters (such as N-Gram size, Statistical classifier and other qualifiers), for the creation of characteristic signatures for different file types. The CTR Framework serves as an infrastructure for automatic creation of a signature for every file type, based on training data constructed from files of that type. Moreover, the CTR Framework can scan unknown files and determine their type based on that signature. Experiments that were made using the CTR Framework, presented very good results with 1394 files out of 1400 that were correctly identified. This is when taking into consideration closely related files, such as EXE and DLL, as one type. The F-Measure value, based on these tests, was 0.996.

The rest of this paper is structured as follows. In Section 2 we discuss related work, in section 3 we describe the CTR framework and in Section 4 the results of the evaluation. Section 5 is the summary.

2   Related Work

Using statistical measures on the content of a file has been investigated as a text classification technique, and later on was used to explore new methods for type classification and malware detection. Some of these techniques are based on statistical measurements and analysis of N-Gram distribution. McDaniel and Heydari [1] focus their early paper on automatic methods of creating file signatures, called file type "fingerprints". They suggest three algorithms based on 1-gram frequency distribution. The first algorithm created a fingerprint using training data of files from the same type. The second algorithm used similar methods but based the fingerprint on cross correlation between byte pairs. The third one simply tested the file header and trailer for repeating patterns and their correlation strength. The first algorithm achieved a success rate of 27.5%, which is not far from a random guess. The second one did a better job (45.83%) but performed much slower. The third algorithm did much better (95.83%) but is much more vulnerable to content manipulation. Wei-Jen Li et al. [2] continued this approach and claimed reaching better results by refining the fingerprints using a set of centroids, which were carefully selected by using clustering methods to find the minimum set that provides good enough performance. Overall, the results were improved compared to [1]. Karresand and Shahmehri [3] continued with developing a similar method using Centroids with the goal of identifying file type based on binary data fragments. The Centroids were based on the Mean and Standard Deviation of byte frequency distribution. Later they extended their work, introducing the interesting concept of 'Rate of Change' between consecutive bytes [4].The calculation of the distance to the Centroid were made using two methods, 1-norm (also known as Manhattan Distance) and the frequency distribution of the Rate of Change of the fragment to be identified. The results were not conclusive. JPEG files achieved the best results, gaining more than 86.8%, but with 20% of False Positives. For other types the results were much worse. For instance, executable file gained 45% - 85% with up to 35% of FP.

Kolter et al. [5] suggested the use of fixed length N-Grams to identify malicious content in executable files. They translated binary content to textual representation and extracted 4-Grams of known malicious and benign binaries, resulting in about 255M unique 4-Grams. A number of classification methods have been used, such as Naive Bayes, Support Vector Machines (SVM) and Decision Trees, which yielded the best results. Dash et al. [6] Continued this approach and introduced the use of variable length N-Grams to identify malicious code. They claimed better performance over fixed size N-Grams due to essential data loss of significant longer N-Grams. Indeed, the results indicate less errors of type I (FP). Irfan et al. [7] suggest two approaches for type identification. First approach uses the cosine similarity for byte frequency comparison; the second is divide-and-conquer approach to group similar files, based on repeating byte patterns, regardless of their type. The results were not conclusive and the later approach improved the classification accuracy of some types, while getting worse results on others, comparing to methods not using divide-and-conquer approach. Moskovitch et al. [8] deal with methods of identifying malicious code based on concepts taken from text categorization. They introduce the class imbalance problem, which basically arises when the classes (in this case, benign and malicious code classes) are not balanced so that most inspected objects belong to one larger class, while other classes remain much smaller. This may result, in some extreme cases, in mistakenly labeling all data as member of the larger class. The results showed that when selecting about 10% of malicious code in the test data, which closely represent real life according to the authors, they achieved over 95% accuracy when using 16.7% of malicious code in the training data. The overall conclusion was that 10% - 40% of malicious code in the training data will provide optimal results in a true life distribution.

As was shown in the cited papers, different file types may require different parameters of analyzing the file content, and the framework discussed next is based on this approach.

3 Identifying File Type by Its Content

The challenge that we are facing, is to find an automatic procedure to create type identifiers, called signatures, which will be accurate enough and resistant to content manipulation up to some degree.

We can deduce, based on previous work, that there is some correlation between the type of a file and its content. To be more precise, we can find a direct link, very strong in some cases and weaker in others, between repeating N-Grams and the type. The problem is that the general N-grams method has several parameters (e.g. the size N) and they are not equivalent for the different file types.

The aim of this paper is to describe the CRT framework, acting as an experiments lab, which will allow performing experiments on automatic type signatures creation, while changing settings such as N-Gram size, Window size, Statistical measures, Score calculation rules and others. In this section we describe the idea and the architecture of the framework. In the following section we present an implementation of the idea, which will serve as a proof of concept, and the results of our tests.

3.1   Implementation

The Framework makes use of the N-Gram Statistics Package (NSP) [9]. The package, which was developed by Pedersen et al., is a suite of Perl utilities, aimed to analyze N-Grams in text files. It collects information on the appearance of N-Grams and allows running association tests such as Fisher's exact test and log likelihood ratio on the collected data. Since the Framework goal is to analyze files of any type, binary or text, a preliminary step was added to translate files content to their hexadecimal textual representation.

3.2   The CTR Framework Architecture

The Framework consist of 2 paths, Signature creating and File test paths, which are schematically presented in Figure 1. The right side flow describes the Signature creation path based on training data files, while the left one describes the test flow. Both paths use the same preparation procedure, which is the "Count N-grams and Calculate statistics" step (see 3.3). We next describe each step in the two paths.

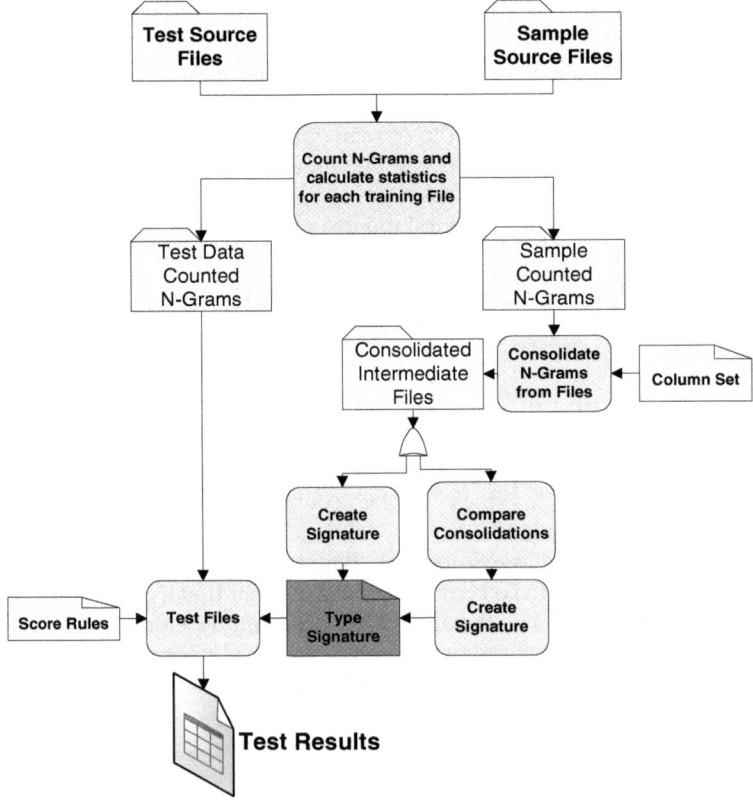

Fig. 1. Framework Workflow

3.3   Count N-Grams and Calculate Statistics

This step is common to both paths. The input for the step is a folder containing files. In case of signature creation path, the files will be training data, representing a type. In the case of the test path, the folder contains files of unknown type. The step accepts, as an input, the following parameters:

1. N-Gram Size. The size of N-Grams, which will be collected.
2. Window Size. Allows extracting non-contiguous N-Grams. For instance, in case of N-Gram with size 2 and Window with size 3 for bytes XYZ, the extracted N-Grams will be XY, XZ, YZ. Window size should be greater or equal to the N-Gram size.
3. N-Gram Threshold. Allows settings the minimum support of the N-Grams that will be counted.
4. Statistical measures, e.g. Fisher's exact test, Log-likelihood, Chi-squared test etc.

```
@count.Ngram=3
@count.WindowSize=3
@count.FrequencyCut=0
@count.RemoveCut=0
@count.InputFilePath=D:\Documents\OpenU\Final_paper_test_data\2_intermediate
Total sample size = 69628
N-gram             Percentage      Frequency Values
------             ----------      ---------------------------
00<>00<>00         12.29821336     8563  17770  17771  17772   11103  10959  11104
ff<>ff<>ff          0.97230999      677  5026   5026   5026    2267   763    2267
ff<>ff<>8d          0.47825587      333  5026   5026   1492    2267   418    343
00<>00<>ff          0.45958522      320  17770  17771  5026    11103  577    371
c6<>45<>fc          0.38921124      271  389    1124   720     276    271    325
01<>00<>00          0.37054059      258  669    17771  17772   328    281    11104
10<>00<>00          0.34612512      241  1880   17771  17772   282    495    11104
00<>10<>00          0.33607170      234  17770  1880   17772   1565   10959  282
ff<>8d<>4d          0.32314586      225  5026   1492   1075    343    282    645
ff<>ff<>8b          0.24271845      169  5026   5026   1576    2267   216    186
00<>10<>e8          0.23410122      163  17770  1880   1828    1565   244    171
00<>65<>00          0.22835641      159  17770  643    17772   160    10959  176
00<>ff<>75          0.22548400      157  17770  5026   1023    371    176    600
00<>00<>8b          0.21543057      150  17770  17771  1576    11103  331    189
c0<>00<>10          0.21543057      150  629    17771  1880    163    153    1565
```

Fig. 2. N-Gram Count file

The result of the step is a set of files, one for each source file, containing a list of extracted N-Grams, together with the collected statistics on each.

A sample output is presented in Figure 2. The first few lines describe the settings, such as N-Gram size (3-Gram, in this case) and required statistics measures. Following are column headers and the data. Each data line presents an N-Gram, followed by the results for the statistic calculation (percentage appearance in the file, in this case) and quantity information on the number of N-Grams that were found in the file. For each N-Gram that was found, the first frequency value states the number of occurrences that the N-Gram appears in the file. The following columns list the number of times that each subset of the N-Gram appeared in its current position. For instance, inspecting the marked line in Figure 2 reveals that '000000' appeared 8563 times in the file, '00' appeared 17770 times on the left hand side, 17771 times in the middle and 17772 times on the right hand side. Also, '0000' appeared 11103 times on the left hand side, and so on.

3.4   Create Signature Path

3.4.1   Consolidate N-Grams from Files
In this step, result files from the Count N-Grams step are collected and consolidated into one summary file per type. The step accepts, as an input, a file describing the columns to be collected from the N-Grams count files, together with some statistical measures that should be made on them. These measures can be Minimum and Maximum value, Standard Deviation etc. The step accepts, the following parameters:

1. Files Threshold. This value set the threshold, in percentage, of files that should contain an N-Gram, in order for it to be counted as part of the signature.
2. Type Name. This value assigns the type to a summary file.

#dll

NGram	Files	Files Percentage	Percentage Mean	Percentage SD	Quantity Sum	Quantity Min	Quantity Max
00<>00<>04	1199	100	0.059667623	0.076436868	91593	1	4323
00<>00<>00	1199	100	18.05606886	16.0723579	23949606	67	1058108
61<>6d<>20	1195	99	0.003094595	0.004334811	2637	1	344
00<>00<>b8	1197	99	0.014303125	0.018074456	32969	1	4228
4c<>cd<>21	1195	99	0.002432365	0.003296389	1215	1	11
ba<>0e<>00	1192	99	0.002454841	0.003269294	1359	1	19
72<>61<>6d	1198	99	0.00675898	0.013321965	17224	1	3011
64<>65<>2e	1191	99	0.002526989	0.003339484	1401	1	16
00<>4c<>01	1193	99	0.0027357	0.003393409	1866	1	21
90<>00<>03	1191	99	0.002445613	0.003252542	1289	1	13
45<>00<>00	1197	99	0.009426805	0.014138348	14945	1	637
cd<>21<>b8	1195	99	0.002431566	0.003276079	1219	1	15
21<>54<>68	1193	99	0.00242683	0.003270916	1206	1	7
74<>00<>00	1197	99	0.020787791	0.027150586	32098	1	758
44<>4f<>53	1190	99	0.002492442	0.003495695	1386	1	67
b8<>00<>00	1195	99	0.008980422	0.011609342	20951	1	586
00<>50<>45	1195	99	0.002541216	0.003792236	2597	1	1107

Fig. 3. Type Consolidation Summary File

A sample output is presented in Figure 3. The type is specified on the first line. Lines, following the column headings, list N-Grams, together with their relevant collected statistics. The second column presents the number of files, the N-Gram was found in and the third column presents this value in percentage of files. Next values present the collected columns for the Count N-Gram step, with statistical measures. For instance, the fourth column presents the Mean of the percentage value, taken from the second column in Figure 2, and the fifth column presents the Standard Deviation of this value.

3.4.2   Create Signature
A Signature can be constructed using two methods. One uses consolidation files as a direct input for the step. The second one allows comparing consolidation files, in order to eliminate N-Grams that repeat in more types then a predefined threshold.

Both methods accept a list of data columns to be collected from consolidation files. In case of choosing the 'Compare Consolidation' method, a threshold value should be added. For instance, if N-Gram 00<>00<>00 appears in 5 different types and the threshold is 4, it will not be included in the Signature. This feature was designed to get more unique type signatures.

pdf	9206		
6e<>74<>65	400	100	1^547
6e<>66<>6f	398	99	1^73
65<>69<>67	395	98	1^160
69<>64<>74	395	98	1^157
65<>61<>72	376	94	1^68
74<>69<>6f	400	100	1^150
73<>74<>72	398	99	1^636
?3<>30<>30	376	94	1^43
74<>73<>20	386	96	1^117
69<>67<>68	395	98	2^160
30<>32<>32	335	83	1^132
dll	10299		
21<>b8<>01	498	99	1^2

Fig. 4. Signature Sample File

A sample of a signature file is shown in Figure 4. It contains the signature information for all types. The first line on each section states the type on the first column and the maximum available score for it on the second column. Following lines list the N-Grams, together with their collected information. This information includes, in the second and third columns, the number of files and percentage that the N-Gram appears in. The fourth column contain the statistics collected from the consolidated files (see 3.4.1), separated by a '^' sign. In Figure 4, the min and max N-Grams appearance was collected. In the first row, for instance, the N-Gram 6e<>74<>65 appeared in each training file between 1 and 547 times.

3.5   Test Files Path

In this path, we use the type signatures to test files and estimate their type. Tested files are processed through the count N-Grams step (see 3.3), which was also applied on training data. It is important to emphasize that, in order to get reliable results, the Count N-Grams step should be set with the same setting as used in the signature path.

The step accepts, as an input, the list of files to be tested, the signature file and a Score rules file.

3.5.1 Score Rules File

The score rules file is assembled from Rules, containing sequential Conditions. Each Rule applies to a specific file type. A Condition, within the rule, determines the score that a found N-Gram will contribute to the final score of a file, in case that the condition will be found to be true. Conditions are ordered and the first Condition that will be found as true will be applied.

Following is a sample set of two conditions in a rule:

```
For a specific N-Gram
  Condition 1:
    If the N-Gram was found X times in the file, where
    X >= 'Min times in the signature' and X <= 'Max times
    in the signature'
      Grant it the score Y.
  Condition 2:
    If the N-Gram was found X times in the file, where
    X > 'Max times in the signature'
      Grant it the score Z.
```

According to these sample conditions, if an N-Gram was found in a file between the min and max times, defined by the signature, it will add Y to the total score, otherwise, if it will be found more than the max, it will add Z to the total score. These Rules and the order between the conditions, which is important, can be created by setting preliminary Rules, creating a signature and using it on test data. Then, repeating this step again and again, using different settings, until the results accuracy is acceptable. Section 3.6 discusses this issue further. Since an N-Gram may be missing from some type training files because of the file threshold, a factor of N-Gram contribution weight is taken into consideration. It is calculated based on the percentage of files, the N-Gram was found in. The following algorithm describes the contribution of a specific N-Gram to the score:

```
For each r ∈ Rules
  If r. Type matches current type
    For each c ∈ r. Conditions
      If c is true
        Add S = (c. Score * NGram weight) to the file
          type total score
        Break
```

S is the score that this N-Gram contributes to the total score of the file for the type. The total score that will be granted to a file for the type is based on the formula:

$$S_{Total} = \frac{\sum S_i W_i}{\sum W_i}$$

Where S_i is the score for N-Gram i and W_i is its weight.

File	Best Match	dll	docx	exe	pdf	rtf	zip
102_381.rtf	rtf	0	1	0	5	60	0
102_694.rtf	rtf	0	1	0	8	69	0
102_517.rtf	rtf	0	1	0	7	71	0
102_607.rtf	rtf	0	1	0	7	71	0
102_383.rtf	rtf	0	2	0	8	61	0
102_137.rtf	rtf	0	1	0	7	69	0
102_268.rtf	rtf	0	1	0	10	70	0
102_290.rtf	rtf	0	1	0	12	78	0

Fig. 5. Test Sample Output file

A test output sample file is shown in Figure 5. The first column list all test files. Second one presents the best guess, meaning, types that received the best score. Following are one column per type, and the score that the file received for it.

3.6   Methodology

In this section we describe the methodology that can be used by the researcher, when creating a new type signature. The methodology can also be adopted by an automatic process. Following are the methodology steps. These steps should be processed for each type separately.

1. Set an accuracy threshold, which serves as an indication for success, when achieved.
2. Choose training and test files from reliable sources. All training set files must be of the same type, which the signature is created for. More files in the training set will result in a smaller and more accurate signature since each additional file in the set may lack of some N-Grams, which otherwise, could be part of the signature. This means that the removed N-Gram could negatively affected the accuracy of the signature. Test set should contain files from different types with existing signatures, in order to identify FP and FN errors.
3. For each set of parameters, run the following steps:
   (a) Choose settings, not tested before, such as N-Gram size, Window size, Statistical measures and Score Rules. For better performance, start from easy settings. e.g., start from smaller N value for the N-Gram size, choose simple statistical measures (min and max for example) before using complex ones etc.
   (b) Create a signature.
   (c) Run Tests.
   (d) If results exceeds success threshold then stop.

The methodology is very simple and should be justified. Generally, if we take all possible options for N-Gram size, window size, possible score rules syntax, parameters of score rules and statistical measures and run an exhaustive test, it will take exponential time. One can use a Genetic algorithm for generating a close to best configuration, but as can be seen from Sections 4.2 and 4.3 below, the above simple methodology gave very good results. It's quite easy to replace it with a genetic algorithm or another learning method.

4   Experiments

We performed the Experiments with very basic tests settings. The goal was to demonstrate the strength of the process, while keeping it simple. All experiments were done on Windows, although the framework does not enforce specific OS. Performance was not taken into consideration, since the framework is not considered to be part of a runtime or production environment, but further development of the test file path, using C# on .NET environment, yielded an average of 70ms scan time for a 0.5MB size file on an i5 core Desktop, which is a quite small overhead.

4.1   Data

Test and training files of 6 different types were collected from a few sources, such as a repository of tested benign files of known types, files resided on a well-protected PC and Google. In order to reduce the chance of random guesses, we collected a large set, containing a total of 3920 files. Collected files distribution is listed in table 1.

Table 1. Files Used for Experiments

File Type	Training Files	Test Files	Total Files
EXE	500	200	**700**
DLL	500	699	**1199**
DOCX	500	182	**682**
PDF	400	19	**419**
RTF	500	255	**755**
ZIP	120	45	**165**
Total	2520	1400	**3920**

4.2   Tests Settings

For the first experiment we used 3-Grams, with the same window size. No statistical measures were collected, except for N-Grams Minimum and Maximum distribution information. Although the Framework allows different settings for different file types, the experiment reported here used the same settings for all types for simplicity. The following simple score rule was applied to all types:

```
If an N-Gram was found then
  If it appeared between Min and Max Times
    Grant the score 100
  Else
    Grant the score 80.
Else
  Grant the score 0.
```

In order to maintain strong signatures, we set the training files percentage threshold, which defines the min percentage of files that an N-Gram should appear in (see 3.4.1, 1), to a value that will assure at least 100 N-Grams in a signature. We succeeded in keeping this rule of thumb for all types except zip files, which seems to have less N-Grams in common due to high entropy of bytes sequences in compressed files. Table 2 lists the threshold for different types.

Table 2. Files Percentage Threshold

Type	Threshold
DLL	99%
PDF	79%
EXE	94%
ZIP	55%
RTF	84%
DOCX	100%

4.3   Results for the First Experiment

Tests were two-folded. The first part was a K-Fold Cross Validation with K = 5. Training files for each type were split into 5 subsets. For each subset, a signature was made from the other 4 and the subset was used as a test set. This step was done on all types except zip, due to the small number of available zip files. DOCX files received the best scores, all in the range between 95 and 100 out of 100, while 92% of the DOCX files received the perfect score 100. RTF files received the worst results. In one subset case, about 33% of the files scored in the range 40 and 49 out of 100 and 17% scored in the range between 50 and 59. But nevertheless, in the actual tests, RTF files detection rate reached 100% success with no FP and FN (see Table 3). This can be explained if we understand that the score does not stand by itself and should be compared to other types score for the tested file, i.e., RTF files received low scores for the type RTF, but much lower for other types, so they were recognized as RTF.

In the second part, file scanning was made on test files, based on signatures that were created using all training data. A total of 1400 test files from all 6 different types were used. 1293 files, which are about 92%, were accurately recognized.

When closely inspecting the remaining 8% files, an interesting picture is raised as shown in Figure 6.

A total of 97 files, which are about 90% of non-matched files, are EXE files that were mistakenly identified as DLL files (one file was also identified as PDF). 5 DLL files (4%) were identified as EXE files. Also, MU_ file, which is a compressed file, was recognized as ZIP. Since DLL and EXE files are executable files with very similar characteristics, we can consider them as executable files. This changes the picture entirely and leaves us with 6 non-recognized files out of 1440, which is about 0.4% error. We also calculated the F-Measure values for the results in order to get a sense of the Precision and Recall. The results are presented in Table 3.

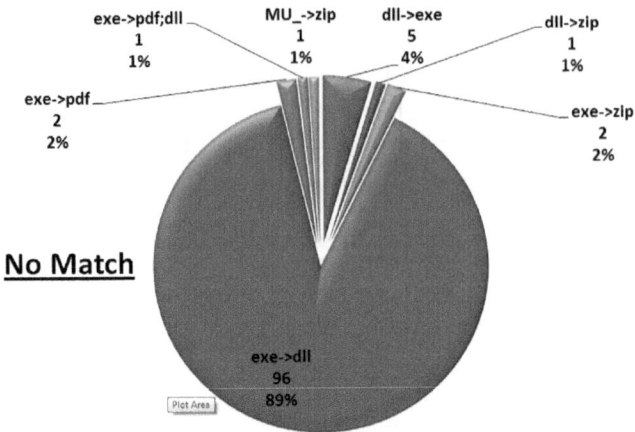

Fig. 6. No Match File and their distribution

Table 3. Results Precision, Recall and F-Measure

Type	Files	TP	FP	FN	Precision	Recall	F-Measure
DOCX	182	182	0	0	1	1	1
RTF	255	255	0	0	1	1	1
PDF	19	19	3	0	0.864	1	0.927
DLL	699	693	97	6	0.877	0.991	0.931
EXE	200	99	5	101	0.952	0.495	0.651
ZIP	45	45	3	0	0.938	1	0.968
Total	1400	1293	108	107	0.923	0.924	0.923
Exec.	899	894	0	6	1	0.993	0.997
Total Exec.	1400	1394	6	6	0.996	0.996	0.996

As can be seen, DOCX and RTF achieved the best results, scoring perfect F-Measure value of 1. PDF, DLL and ZIP files also did well, scoring 0.931-0.968. EXE files gained a poor Recall value of 0.651, but when taking DLL and EXE files as Executable files, the gained Precision is 1 and Recall is 0.993, resulting in an F-Measure value of 0.997.

Overall, in the latter case, the total F-Measure for all 1400 files is 0.996.

4.4   Improving Accuracy

As can be seen in the previous section, when isolating the EXE type, the settings of the first experiment gained poor results, with an F-Measure value of 0.651. Since the result revealed that, in most miss-identifications, EXE files were recognized as DLL and vice versa, we performed two additional experiments, one using 4-Gram and the another using 5-Gram, only for the DLL and EXE types. N-Gram sizes for the rest of the types were left unchanged (3). Also, other settings for the experiment were left unchanged. The precision, recall and F-Measure for the EXE type for the different N-Gram sizes are presented in Table 4.

Table 4. EXE Precision, Recall and F-Measure for different N-Gram sizes

N-Gram Size	Files	TP	FP	FN	Precision	Recall	F-Measure
3	200	99	5	101	0.952	0.495	0.651
4	200	108	7	92	0.939	0.54	0.686
5	200	132	8	68	0.943	0.66	0.777

As can be seen clearly, with the increasing of the N-Gram size, the accuracy of the Framework improved when identifying EXE files. This experiment demonstrates the strength of the Framework in finding the best settings for signature creation.

From these experiments we can identify few factors that contribute to the results accuracy:

- The increasing accuracy when improving the score rules.
- Grouping of file types with a similar structure into categories. As shown, categorizing may contribute for increased accuracy.
- Threshold values, which set the ground rules to determine if an N-Gram will be counted for the signature.

4.5   Identifying Tampered Files

One of the challenges, the Framework is facing, is the ability to correctly identify tampered files. For instance, in many file types, the first few bytes (magic numbers) are used to identify the file type. These can be easily manipulated to obfuscate the type of the file. In order to test the ability of the Framework to overcome this challenge, we performed another experiment.

We randomly picked 24 files, which were correctly identified in the first experiment. The file list contained 4 files from each of the 6 different types. The first 10 bytes of each file were set to zero (0). An attempt to open these files, using the relevant application completed with a failure, as expected. The 4-Gram experiment settings were used for the experiment.

All 24 files were correctly identified by the Framework. This result demonstrates the strength of the method used by the Framework to overcome common cases of file content tampering. More complex forms of content tampering, which affect the whole content of the file, will be dealt with in a future work. Furthermore, the tests were performed on six file types and it is possible that the identification accuracy may be reduced when adding more type signatures. This may be overcome by removing mutual N-Grams of many different type signatures and will be investigated in the future.

5   Conclusions

This paper presents the CTR framework, which is a general framework for revealing the true type of various files. Using very simple settings, the framework demonstrated promising results, successfully identifying 1394 out of 1400 files and achieving

F-Measure value of 0.996, when taking EXE and DLL files into account as executable files. There is no need of prior knowledge or manual process of finding patterns in file structures. The process can be done automatically without any human intervention.

Not all tested types achieved the same signature strength. While DOCX files have many repetitive patterns, strong structure characteristics and similarities, ZIP files, for example, demonstrated very weak signature with small number of descriptive N-Grams. Investigation of additional file types using the CTR framework is planned in future work.

File content tampering is still an option, but it is much harder since the signature usually covers large amount of N-Grams and the scan is made on full content, or at least large portion of the file. This was clearly shown by the results in 4.5.

In future work we intend to use the framework as a basis for identifying anomalies in files, in order to mark them as suspicious or benign. Also, we will explore new directions in text classification, trying to identify content language or identify data leakage.

References

[1] McDaniel, M., Heydari, M.H.: Content Based File Type Detection Algorithms. In: Proceedings for the 36th Hawaii International Conference on System Sciences (2002)
[2] Li, W.-J., Stolfo, S.J., Herzog, B.: Fileprints: Identifying File Types by n-gram Analysis. In: 2005 IEEE Workshop on Information Assurance, West Point, NY (2005)
[3] Karresand, M., Shahmehri, N.: Oscar – File Type Identification of Binary Data in Disk Clusters and RAM Pages. In: Fischer-Hübner, S., Rannenberg, K., Yngström, L., Lindskog, S. (eds.) Security and Privacy in Dynamic Environment. IFIP, vol. 206, pp. 413–424. Springer, Boston (2006)
[4] Karresand, M., Shahmehri, N.: File Type Identification of Data Fragments by Their Binary Structure. In: Proceedings of the 2006 IEEE Workshop on Information Assurance United States Military Academy, West Point, NY (2006)
[5] Kolter, J.Z., Maloof, M.A.: Learning to Detect Malicious Executables in the Wild. In: Tenth ACM SIGKDD International Conference on Knowledge Discovery and Data Mining (2004)
[6] Dash, K.S., Dubba, S.R.K., Pujari, K.A.: New Malicious Code Detection Using Variable Length n-grams. In: Algorithms, Architectures and Information Systems Security, ch. 14, pp. 307–323. World Scientific (2008)
[7] Irfan, A., Kyung, L., Hyunjung, S., ManPyo, H.: Content-Based File-type Identification Using Cosine Similarity and a Divide-and-Conquer Approach. IETE Technical Review 27(4) (July 2010)
[8] Moskovitch, R., et al.: Unknown malcode detection and the imbalance problem. Journal in Computer Virology 5(4), 295–308 (2009)
[9] Pedersen, T., Banerjee, S., Purandare, A., McInnes, B.T., Liu, Y.: NSP - Ngram Statistics Package (2009)

From MDM to DB2:
A Case Study of Security Enforcement Migration

Nikolay Yakovets[1], Jarek Gryz[1], Stephanie Hazlewood[2], and Paul van Run[2]

[1] Department of Computer Science and Engineering, York University, Canada
Centre for Advanced Studies, IBM Canada
{hush,jarek}@cse.yorku.ca
[2] IBM Canada
{stephanie,pvanrun}@ca.ibm.com

Abstract. This work presents a case study of a migration of attribute-based access control enforcement from the application to the database tier. The proposed migration aims to improve the security and simplify the audit of the enterprise system by enforcing information protection principles of the least privileges and the least common mechanism. We explore the challenges of such migration and implement it in an industrial setting in a context of master data management where data security, privacy and audit are subject to regulatory compliance. Based on our implementation, we propose a general, standards-driven migration methodology.

Keywords: Master Data Management, Enterprise Security, Attribute-Based Access Control, Database Security, XACML, DB2.

1 Introduction

Today's enterprise data is complex and heterogeneous, and users who access it are diverse and belong to multiple domains. The access control models have evolved to fit these requirements. Traditional discretionary (DAC, [1]) and mandatory (MAC, [2]) access control models have been replaced by role-based access control (RBAC, [3]) and, more recently, attribute-based access control (ABAC, [4]). Unlike DAC, MAC and RBAC, the ABAC model defines permissions based on just about any security relevant characteristics of requesters, actions, resources, and environment, known as attributes.

The lack of native ABAC support by conventional DBMS has motivated many enterprise developers to implement access control checks at the application tier, bypassing the native database access control. In such architecture, the database connection is established on behalf of the application and is able to access the entire database, while application program logic itself is used to limit the privileges of the end-user. While this approach allows enforcement of more complex and flexible attribute-based policies, it also introduces several problems. First,

N. Cuppens-Boulahia et al. (Eds.): DBSec 2012, LNCS 7371, pp. 207–222, 2012.

because the database connection is at an elevated privilege, the enterprise system is prone to *privilege escalation attacks* [5], such as SQL injection [6]. Second, since the database is not aware of the identity of an end-user issuing the request, it may be difficult for application developers to provide the *end-to-end audit trail* which is essential to comply with regulations such as SOX [7], PCI [8], and HIPAA [9].

In this paper, we describe how to perform a migration of an existing enterprise security enforcement from the application tier to the database tier. In other words, we would like to be able to securely enforce the enterprise security policies at the database, while retaining the flexibility of their management at the application. We see several benefits of such database-level enterprise security enforcement. First, it helps protecting the enterprise system from privilege escalation attacks by complying with information protection principles of the least privileges and the least common mechanism [10]. Further, the placement of security controls on the database allows for full end-to-end audit trail as the identity of an end-user is not hidden behind a common authorization ID of an application layer.

We consider our approach in an industrial setting in a context of master data management (MDM) [11]. A wide variety of MDM solutions are available on the market today from major vendors such as IBM, Oracle, Microsoft and SAP. Such systems aim to provide a centralized environment for maintaining the *master* data about customers, suppliers, products and services. Master data plays an important role in both operational and analytical aspects of the organization operation. However, it is also a tempting target for attackers as it is captured by combining, cleansing and enriching highly confidential data across various data sources in the organization. Further, master data also contains personally identifiable information, making MDM solutions subject to regulatory compliance. Since the proposed migration aims to improve the security and simplify the compliance of the enterprise system, we believe it may be useful, in the first place, in the MDM domain.

In collaboration with IBM Centre for Advanced Studies, we used IBM InfoSphere Master Data Management Server (in short, MDM Server) as an implementation platform. MDM Server is a large scale industrial-grade MDM solution that enforces attribute-based security policies at the application tier while storing the master data in the relational database. This enabled us to implement the proposed migration in MDM Server to prove the feasibility and practical usefulness of our approach.

Lastly, we believe that our MDM Server migration experience may be helpful in similar migration projects in a variety of enterprise applications outside of MDM. To facilitate the implementation of such projects, we propose a general migration methodology that is standards-driven and vendor-independent.

The rest of the paper is organized as follows. Section 2 describes the challenges of the proposed migration, and Section 3 describes our industrial implementation. Section 4 discusses the lessons we learned, and Section 5 presents a general migration methodology. Section 6 discusses some of the related works, and Section 7 concludes.

2   Challenges

The proposed migration aims to place the security enforcement on the database, while keeping the security administration at the application. This comes with several challenges as both end-user identities and security policies need to be timely and efficiently *propagated* from the application tier down to the database. The following subsections provide more detail.

2.1   Identity Propagation

Request processing often involves the propagation of end-user's identity from one enforcement point to another within the enterprise environment. For example, consider a case where end-user is authenticated at a client tier. Then the identity is sent to an application tier for its own authorization and auditing. The application in turn generates the requests to query a back-end database potentially using yet another identity that is suitable to access a relational database on end-user's behalf.

The scenarios of such identity propagation differ by a degree of knowledge that the relational database has about the end-users. For example, an end-user may be mapped to a *system identity* - or the identity that is used to represent the application tier. This *many-to-one* mapping greatly simplifies the deployment topology, but effectively places all of the security controls such as authorization and audit on the application tier, as the relational database has no knowledge about the identity of the end-user.

On the other hand, an end-user may be mapped to a *functional identity* which can represent user's role or group in the organization. This *many-to-many* mapping allows placing some of the security controls on the relational database for the price of complicating the deployment topology. In this case, however, the database still does not know who exactly the end-user is. Therefore full *end-to-end* audit trail is impossible which can cause problems with regulatory compliance.

Lastly, an end-user may be mapped to her corresponding identity in the database. This *one-to-one* mapping allows the database perform authorization and audit using end-user's identity. This scenario requires the most difficult deployment topology, but allows placing most of the security controls on the relational database. Since this allows for full end-to-end audit trail, it greatly simplifies regulatory compliance of the enterprise system.

The proposed migration involves shifting from many-to-one to one-to-one identity mapping during the enterprise request processing. What makes this challenging is that the complication of the deployment topology that comes with one-to-one mapping is undesirable. Moreover, the identity propagation mechanism would require an identity provisioning component responsible for keeping identities in the database consistent with identities administered at the application. An optimal solution for handling the identity propagation would need to solve the above problems.

2.2   Policy Propagation

Application-to-database policy propagation is challenging because the underlying security mechanisms at the application are conceptually different from those at the database. Application-level authorization engine can use either proprietary or standards-driven approach such as OASIS XACML [12] to specify, distribute and enforce *high-level* attribute-based security policies. The flexibility of such policies comes from the ability to define the security rules based on attributes of users, resources and environment. For example, a single rule might permit access for all professors in the department of computer science to files pertaining to courses offered in that department. The access decision is made during the request execution time when the rule attributes are retrieved: the authorization IDs of current CS professors and table names of currently offered CS courses.

Relational database *low-level* security enforcement mechanisms are based on the assignment of privileges to users or roles. The privileges can be granted or revoked on complete tables and views. For example, a single database authorization might permit access for professor JohnDoe to table CS101. Clearly, such discretionary database policies are less expressive than attribute-based application policies.

The proposed migration would involve linking the security mechanisms at the application and at the database. A consistent mapping should be established between the security policies, such that the database privileges conform to semantics specified by permissions in the application. Mapping of a single application policy would require the understanding how application-level users and resources identified by their attributes translate to database-level authorization IDs and tables. Mapping of multiple application policies would require the understanding of policy combining and conflict resolution algorithms that are used at the application.

It is important to realize that updates to application security policies or to user and resource attributes may alter the established policy mapping in a nontrivial way. Some of the database permissions would need to be granted, while others would need to be revoked. The solution for policy propagation would need to provide the mechanism that will efficiently maintain the consistent relationship between the database permissions and the policies administered in the application.

3   Implementation

In this section we show how the proposed migration can be implemented in an industrial setting in MDM Server. MDM Server is a large scale enterprise application that runs on WebSphere Application Server (WebSphere) and uses IBM DB2 relational database (DB2) to store master data. First, we describe how the identity of an end-user of MDM Server can be efficiently propagated from WebSphere down to DB2. Next, we talk about how the attribute-based MDM Server security policies can be mapped to discretionary DB2 permissions.

Finally, we show how to efficiently maintain the consistency of DB2 permissions when MDM Server policies or attributes are updated.

3.1   MDM Server Identity Propagation and DB2 Trusted Context

In MDM Server access control checks are implemented at the application tier, and the application itself is able to access the whole master database by using an administrative system identity. The goal of our migration is to place most of the security controls on DB2, instead of relying on MDM Server to perform the security checks. To make an access decision, the database needs to know the identity of an end-user issuing the request that is effectively "masked" by the administrative system identity of the MDM Server application. One way to solve this problem is to modify the MDM Server application logic in such a way that each database connection is established using the end-user's identity who issued the request instead of the administrative system identity. However, this approach would involve the significant modification of the MDM Server code base, which is undesirable. Another, more efficient approach is to plug in a Java Authentication and Authorization Service (JAAS) login module to propagate the end-user identity to the database server. This approach maintains the end-user identity, but does not support connection pooling that is normally leveraged by MDM Server when many-to-one administrative system identity is used.

In our implementation, we utilize *trusted connections* instead of the manual mapping or a JAAS mapping. Trusted connections support client identity propagation and can also use connection pooling to reduce the performance penalty of closing and reopening connections with a different identity. Trusted connections leverage the DB2 trusted context object. The DB2 trusted context is an object that the database administrator defines and that contains a system authorization ID and a set of trust attributes. The trust attributes identify characteristics of a connection that are required for the connection to be considered trusted. The relationship between a database connection and a trusted context is established when the connection to the DB2 server is created. After a trusted context is defined, and an initial trusted connection to the DB2 database server is made, the application server can use that database connection from a different user without a full re-authentication. The database authenticates the end-user and then verifies the user authorization to access the database prior to allowing any database requests to be processed on behalf of that user.

When software stack is set up for use of the trusted context, the authorization of the end-user requests is carried out as follows. First, end-user is authenticated to MDM Client ①, and her identity is sent over to WebSphere which is runnung MDM Server ②. WebSphere initially establishes the connection to DB2 using underpriviliged "connection-only" user ③. However, on query execution time, the user identity is switched to the end-user who executes the query ④. Therefore, the query is executed and audit trail is logged at DB2 with the identity of the end-user ⑤.

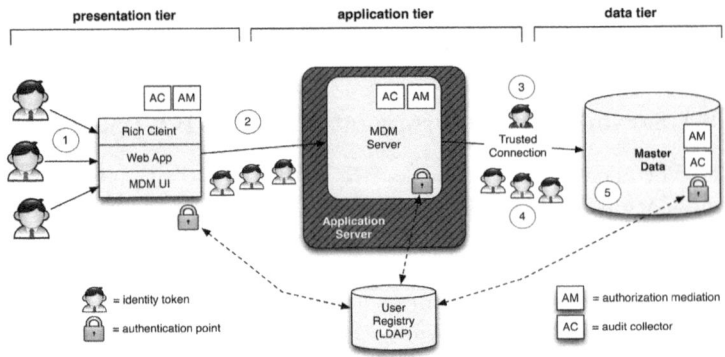

Fig. 1. Identity Propagation in MDM Server

To provision one-to-one identity propagation, database-level authorization IDs should be kept consistent with identity information stored in MDM Server according to the established one-to-one mapping. We solve this problem by using an LDAP [13] registry common to both MDM Server and DB2. Thus, the use of a trusted context and a centralized user registry allows us to establish an efficient one-to-one identity propagation that does not complicate the deployment topology of the enterprise system.

3.2 MDM Server Policy Propagation

To establish application-to-database policy propagation in MDM Server we need to provide a consistent mapping between MDM application security policies and their DB2 counterpart.

In DB2, we consider object-level and content-based authorizations that are granted as primary, secondary or public permissions. Such *discretionary* permissions are represented as access control lists (ACLs), where each object is associated with a list of subjects coupled with a set of actions they can perform on the object. The object, in this case, can be a table or a view, while the subjects are DB2 users (authorization IDs), user groups, user roles and public (all users). The access operations that subjects are allowed to invoke on objects are SQL operations that can be executed on tables: select, insert, delete, and update.

On the other hand, at the application tier, we have MDM *entitlements*, which are managed by Rules of Visibility and Persistency Entitlements components of MDM Server. In an entitlement, an *accessor*, which may be a user or a user group, is entitled to take an *action*, for example adding, updating, or viewing, on set of *business objects* or their *elements* which are identified by the *data association* resource attribute in the entitlement.

To establish a policy mapping, one should understand the semantics of MDM entitlements in the context of master data stored in the relational database.

In MDM, business objects map to database tables and elements that further refine those business objects correspond to the columns of its underpinning database tables. To enable full end-to-end audit, we establish one-to-one mapping between MDM accessors and database users and groups. Finally, MDM data actions are mapped to DB2 data manipulation operations (i.e. update, select, insert and delete).

For example, consider a pair of MDM entitlements shown in Fig. 2. First entitlement permits a particular student to view resources with attribute Data Association = {Course, Grades}. Second entitlement permits all professors updating resources attributed to Grades only. Through corresponding associated objects, COURSE database table is assigned Data Association = {Course, Grades}, and GRADES table has Data Association = {Grades}. According to our policy mapping, the first entitlement is associated with DB2 authorizations $\boxed{1}$, $\boxed{2}$ and $\boxed{3}$, while $\boxed{4}$ and $\boxed{5}$ are associated with the second entitlement.

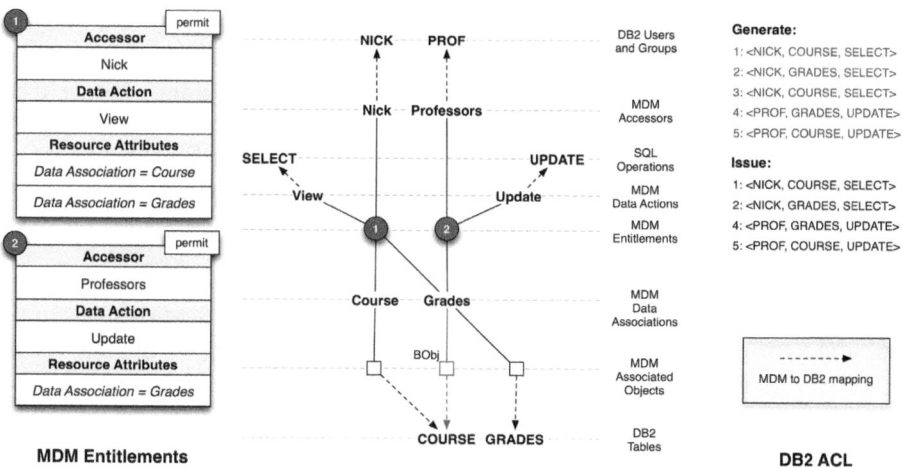

Fig. 2. Policy Mapping in MDM Server

To accommodate the propagation of MDM entitlements down to DB2 we modify MDM Server application logic by introducing Policy Propagation Point (PPP) component (shown in Fig. 3). Policy propagation is carried out in three steps. First step performs entitlement parsing: PPP interacts with existing MDM Server components (we logically group them as *MDM Policy Administration Point (PAP)*) to obtain the list of current entitlements. Each fetched entitlement is described by an accessor, a data action and a data association attribute.

Second step performs attribute extraction: PPP interacts with another group of MDM Server components (*Policy Information Point (PIP)*) to extract the resources that correspond to the attribute values of the parsed entitlements. In this step, data associations are associated with the list of associated objects they contain.

Third, and last, step builds DB2 authorizations and sends them to the database. For each associated object in an entitlement a single DB2 permission is generated in accordance to the established policy mapping. An accessor is mapped to a DB2 user or a group, a data action is mapped to a DB2 data manipulation operation and, lastly, an associated object is mapped to a DB2 table. The generated DB2 permissions are combined according to MDM Server policy combining and conflict resolution algorithms. In MDM Server a *union* approach between the entitlements is assumed (i.e. if a user belongs to a group which is allowed to see a data then, even though that user belongs to the other groups that are not allowed to see that data, the access will still be granted). No conflict resolution is required because the entitlements can only be granted and cannot be restricted (i.e., there are no *negative* entitlements) and the *default* policy is to restrict access unless it is specifically granted. Since DB2 employs the same policy combining and conflict resolution, we combine the generated DB2 permissions simply by issuing them directly to the database.

For each generated database permission, PPP logs all relevant information into the propagation log. This log contains, for each DB2 authorization, the reference to its parent entitlement, accessor, data action, data association and associated object. Information about the status of each DB2 authorization, whether it was issued to the database or was found duplicate, is also stored in this log. For example, for MDM entitlements that are shown in Fig. 2, PPP will generate and log five DB2 authorizations, however, only four of them will be issued to the database as one of the authorizations is a duplicate (assuming that no authorizations were issued for other MDM entitlements before).

Our solution also maintains the consistent relationship between MDM entitlements and DB2 permissions in case when either entitlements or their attributes are updated. A naïve approach to handle this problem would involve manual

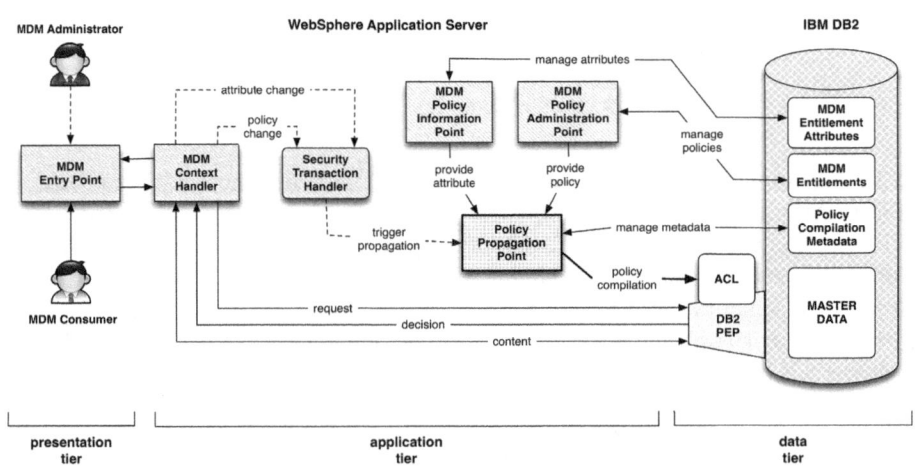

Fig. 3. Policy Propagation in MDM Server

periodical rebuilding of all DB2 permissions. Our *incremental propagation* approach aims to rebuild only a subset of DB2 permissions that is affected by the entitlement or the attribute update.

All administrative MDM transactions that deal with entitlements and their attributes are handled by MDM component called Security Transaction Handler (STH) (shown in Fig. 3). We modify this component in such a way that whenever it receives a transaction that deals with either entitlements or their attributes, it triggers PPP to *immediately* update the affected database permissions.

To illustrate how incremental propagation works consider the following example. Suppose that security administrator updates the data association attribute of the associated object BObj so that it no longer belongs to Grades (Fig. 2). STH notifies PPP that BObj has been updated. Then, PPP extracts from the transaction log all *old* DB2 permissions that were built from BObj. In this case, those are [1] and [5]. Next, PPP checks the propagation log to see if there are unissued permissions that are duplicate to [1] and [5] and were not affected by the update. In this example, it finds permission [3]. Therefore, only [5] is revoked and [3] is unmarked as being duplicate. Finally, PPP interacts with PIP and PAP to generate and issue to the database any *new* permissions resulted from the update.

4   Discussion

In this section we discuss the lessons that we learned from the implementation of the proposed security enforcement migration in MDM Server.

4.1   Integration Footprint

Industrial-scale enterprise applications have hundreds of thousands, sometimes even millions lines of source code. Due its large size a significant recoding of an initial implementation in order to integrate the proposed security enforcement migration is not feasible.

Our identity propagation implementation relies on trusted connections and trusted context features of DB2 and WebSphere. Trusted connections support client identity propagation and can also use connection pooling to reduce the performance penalty of closing and reopening connections with a different identity. Moreover, when WebSphere and DB2 are configured for use of trusted connections, WebSphere, seamlessly for the enterprise application, will propagate end-user identity to DB2 through the trusted data source. Therefore, the existing application code that uses WebSphere APIs to interact with DB2 through configured data sources does not need any modification. Thus, given a properly configured software stack, almost no initial code modifications are needed to implement the identity propagation in our approach, and it also doesn't incur noticeable performance degradation.

Our policy propagation mechanism is encapsulated in a single additional module which integrates seamlessly into the existing application security core architecture. It reuses existing application interfaces to obtain enterprise security policies and attributes. Minor modifications to the initial code base were needed to enable propagation triggering mechanisms in existing application components.

Therefore, our approach has a small integration footprint on the existing code base, which makes it feasible to implement in a large-scale industrial-grade enterprise application such as MDM Server.

4.2   Policy Granularity

We can classify attribute-based security policies into two types based on the *granularity* of the data to which they apply. Specifically, if policy describes *object-level* authorizations, i.e. its resources are mapped to whole database tables, then we call such policy *coarse-grained*. For example, coarse-grained policy allows students to see *all course grades*, when course grades are contained in one, or several, database tables. On the other hand, if a policy describes *content-based* authorizations, i.e. its resources are mapped to specific rows of a table, then we call such policy *fine-grained*. For example, fine-grained policy allows students to see *only their own grades* that are contained in specific tuples of a course grades database table.

In our approach, we compile coarse-grained application policies into a set of database permissions. However, a different approach is needed to handle those application policies that are fine-grained, since individual tuples cannot be specified as objects in database authorizations. In theory, SQL standard provides a way to deal with fine-grained access control through view creation. For example, it is possible to create views for specific users, which allow those users access to only selected tuples of table. However, this approach is not scalable, and essentially impractical in systems with thousands or millions of users, since it would require millions of views to be manually created by a policy administrator.

For this reason, for the past several years, there have been much research on alternative solutions for fine-grained access control for databases. Implementations include *query rewriting* [14], *parameterized view creation* [15,16] and *extensions* to SQL [17–19]. *Hippocratic* databases use meta data stored in the database to make access decisions and are able to enforce cell-level policies. However, neither of those implementations are part of current publicly available commercial SQL servers.

The only currently publicly available implementation of fine-grained database access control (FGAC) is Oracle Virtual Private Database (VPD). This approach is based on Truman Models [15] and attaches predicates to user submitted queries to enforce row-level policies. However, in our implementation, we were limited to IBM software stack and thus were unable to use Oracle VPD. Use of Oracle VPD by switching to Oracle software stack, or a similar IBM offering when it becomes available, to map and synchronize fine-grained policies is a natural next step in our work.

4.3   Security vs. Performance

One of the aspects of policy propagation is the amount of time it takes the change in the enterprise policy, its attribute or attribute relationship to propagate from the application to the database tier. If this time interval is relatively long, then the system security may be compromised. This time period is called the *window of access vulnerability* [20] which is caused by policy or attribute updates.

The duration of the window of vulnerability is determined by the implementation of policy synchronization engine. Generally, updates to database policies can be either *immediate* or *deferred*. In the immediate mode, whenever enterprise policies are updated, the updates to corresponding database policies are handled with the highest priority with no or relatively short delay. On the other hand, in the deferred mode, the updates to database policies are handled with lower priority and, therefore, the delay can be longer. The trade-off between immediate and deferred modes is the performance overhead imposed by the policy synchronization on the enterprise system.

In MDM domain, the confidential nature of master data has required us to handle the policy propagation in the immediate mode, however the implemented mechanism can be configured to operate in the deferred mode in order to help security administrators achieve a balance between security and performance of the system.

4.4   Mapping Correctness

It is important to establish application-to-database security policy mapping *correctly*, such that the permissions that are given in the database as a result of the policy propagation are indeed the permissions intended by a security administrator who defined them in the application. The understanding of privacy semantics of the application security policies is required in order to establish the correct mapping. Specifically, one needs to know both: how to map the *individual* policies and how to correctly *combine* them later.

In order to map the individual attribute-based policy to a set of database authorizations one needs to understand how application-level accessors, resources and data actions correspond to database users, database objects (such as tables and views) and SQL operations. Next, to combine the database authorizations that were produced by the compilation, one needs to use the *policy combining* and *conflict resolution* algorithms that were attached to the corresponding individual application-level policies.

In our implementation, policy combining was performed according to the union approach between MDM entitlements and MDM default no-access policy, hence no conflict resolution was required. Therefore, in MDM Server, it was sufficient to verify the mapping of individual application policies (which was correct given the object-relational design of MDM Server) in order to establish the correctness of the application-to-database policy mapping.

4.5   End-to-End Audit

An *audit trail* [11] is a process of secure recording of key system events and who initiated them in a chronological log. Auditing is used to reconstruct who-did-what after the fact. Typically, for a complete picture of a transaction on a data, audit information must be collected at every enterprise system component along the transaction path - it is called an *end-to-end* trail. The ability of an enterprise system to provide *end-to-end* auditing facilities is essential for regulatory compliance.

When the enterprise security is enforced at the application tier, the requests to a database are made using a *system* identity - the identity that is used to represent the application layer, instead of the end-user identity. Since the end-user identity is not sent to the database it may be problematic to provide an audit trail at the database tier.

Following the proposed migration, the end-user identity is propagated down to the database tier and all the database requests required to perform a transaction are made using this propagated identity. This allows us to easily reconstruct the events after the fact by using native database auditing facilities instead of programming the application logic to collect the audit trail.

5   Migration Methodology

As we mentioned before, many application developers today choose to bypass database-level access control due to the lack of a native database ABAC support. Therefore, we believe that there are many enterprise applications out there which may benefit from our proposed application-to-database security enforcement migration.

Based on our case study, we propose a sound, methodological approach by which organizations can tackle migration projects. To keep our methodology as general as possible, we base it on a well-known industry standard for specifying and managing attribute-based policies - OASIS XACML. In addition to the policy language specification, XACML standard includes the description of the policy management architecture and its data flows (presented in Fig.4a). In our methodology, we propose the necessary changes to XACML reference architecture in order to accomodate the proposed migration and enable database-level enterprise security enforcement (Fig.4b).

We believe that ideal migration plan should be broken down into three phases. Known as analysis, implementation and integration, we define these phases in detail below.

In the analysis phase one should identify if consistent mapping can be established between enterprise attribute-based policies that are consumed by the PDP at the application tier and identity or role-based policies that are consumed by security mechanisms at the relational database. Similar to the migration in MDM Server, this phase might include establishing relations between enterprise data actions and database permission types, enterprise resources (such as business objects, their associations and hierarchies) and database tables, columns and rows, enterprise accessors (such as users, user groups, roles and role hierarchies) and database users, groups and roles.

Then, the core of the PPP is implemented. Given the attribute-based policies, enterprise accessors and resources that correspond to the policy attributes,

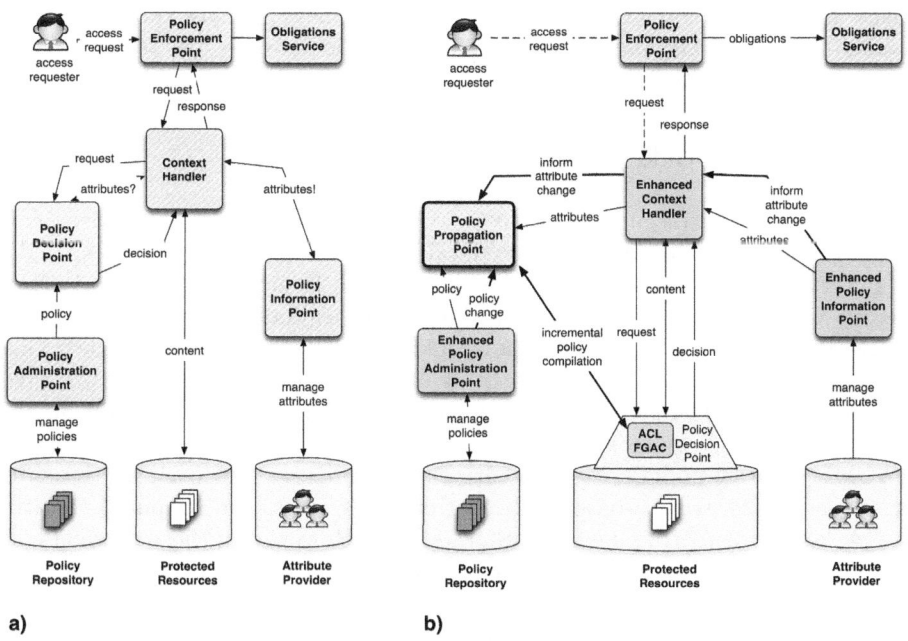

Fig. 4. Extended XACML Architecture

this component performs policy compilation and incremental updates according to the established mapping. It is important to realize that MDM Server's fairly straightforward and coarse-grained policy mapping and lack of negative authorizations allowed for relatively unproblematic implementation. However, finer-grained policies and presence of negative authorizations may considerably complicate this phase as necessary inconsistency and conflict resolution would have to be implemented.

Finally, the implemented engine is integrated into the enterprise system. Similar to the migration in MDM Server, two optimizations might be performed in order to keep the integration footprint small. First, PIP and PAP interfaces that supplied PDP with enterprise policies and evaluated the attributes are reused with PPP. Second, subject identity propagation is implemented with the help of a novel trusted context feature of an underlying software stack. Such feature are already supported by IBM and Oracle application servers and relational databases, with more vendors surely to follow. Finally, existing PIP and PAP are modified to enable PPP notification mechanisms necessary for timely incremental policy propagation.

The proposed migration has the following business benefits. First, it ensures better data protection in the enterprise as it eliminates the risks that are associated with using common application layer's authorization ID by providing an additional layer of security at the database tier. Second, by complying with the least common mechanism principle of protection of information in computer systems our migration eases end-to-end enterprise audit and, consequently, it eases the organization's regulatory compliance. Finally, it may lower the overall transaction processing times due to efficiency of the database-tier security engine when compared to the application-tier one.

6   Related Work

In this section we discuss the related works in the areas of enterprise access control systems and attribute-based database access control techniques.

Enterprise access control enforcement has been interpreted in several efforts. Tivoli Access Manager [21] developed by IBM attempts to decouple the authorization logic from the application logic to allow security policy externalization by using a proxy that sits in front of the application server. IBM Tivoli Security Policy Manager [22] and Axiomatics Policy Server [23] build on that work by enabling enterprise architects to centrally define, manage and enforce XACML security policies for applications and data resources within the enterprise. Ladon [24] is Java-based API that enforces access control policies written in XACML by rewriting incoming SQL queries. However, in these products, the security policies are still enforced at the application level and inherit all disadvantages of this approach.

Little work has been done on the database-level attribute-based policy verification. Our work follows the recent work [20] which aims to compile attribute-based XACML policies into ACLs that are supported by conventional databases. The authors develop a toy prototype based on MySQL relational database to show that such compilation significantly improves attribute-based database access time with a price of reasonable off-line compilation time. Inspired by this work, we realized that such compilation can be used to provide an enhanced security, end-to-end audit and simplified compliance in many existing enterprise systems. Our work builds on this observation by describing the security enforcement migration in existing industrial-grade system and providing standards-driven migration methodology based on the enterprise software stack that enables database-level attribute-based policy enforcement in existing enterprise systems.

7   Conclusions

In this work, we presented a case study of migration of access control enforcement from the application to the database tier. This migration aims to overcome the security and audit concerns that are associated with the application-tier security mechanisms in an existing enterprise application. We considered our approach in the context of master data management where data security, privacy and audit are subject to regulatory compliance. We implemented the proposed migration in an industrial-grade MDM system to prove the applicability and usefulness of our idea. Finally, we proposed a general standards-driven and vendor-independent methodology that is designed to tackle similar security enforcement migration projects.

The proposed migration has two positive effects on the overall security and regulatory compliance of the enterprise system. First, our approach enhances the security by eliminating the vulnerability to privilege escalation attacks. Second, our approach eases the enterprise end-to-end audit by enabling native database-tier auditing facilities. Further, our approach has a small integration footprint by effectively reusing existing enterprise security components and novel features of an underlying software stack. Our reference model seamlessly integrates into the existing enterprise security architecture by requiring only minor modifications to the initial enterprise system code base, which makes it feasible to implement in a large-scale enterprise application.

There are two areas of further interest to us in this project. First, as we mentioned in Section 4.2, our approach handles only coarse-grained table-level policies. This is caused by the limitations of an underlying relational database, which in our case was IBM DB2. In the future, we plan to experiment with Oracle software stack to extend the policy propagation engine to handle fine-grained access control (FGAC). We envision that this can dramatically improve the enterprise transaction processing times since the selectivity of FGAC-transformed query is higher than that of the original query due to introduction of policy related predicates. Second, we plan to perform an evaluation of our approach in an actual production enterprise environment. The organization of the protected data and its accessors and the patterns in which the policies or their attributes are changed would help us understand the real effects our approach has on the enterprise system. This information will be invaluable in determining further optimizations to our engine.

References

1. Scott Graham, G., Denning, P.J.: Protection: Principles and Practice. In: Proceedings of the Spring Joint Computer Conference, AFIPS 1972, May 16-18, pp. 417–429. ACM, New York (1972)
2. Jajodia, S., Sandhu, R.: Toward a Multilevel Secure Relational Data Model. SIGMOD Rec. 20, 50–59 (1991)
3. Sandhu, R.S., Coyne, E.J., Feinstein, H.L., Youman, C.E.: Role-based Access Control Models. Computer 29(2), 38–47 (1996)
4. Wang, L., Wijesekera, D., Jajodia, S.: A Logic-based Framework for Attribute Based Access Control. In: Proceedings of the 2004 ACM Workshop on Formal Methods in Security Engineering, FMSE 2004, pp. 45–55 (2004)
5. Pfleeger, C.P., Pfleeger, S.L., Safari Tech Books Online: Security in Computing, vol. 604. Prentice Hall (2007)
6. Kc, G.S., Keromytis, A.D., Prevelakis, V.: Countering Code-injection Attacks with Instruction-set Randomization. In: Proceedings of the 10th ACM Conference on Computer and Communications Security, pp. 272–280. ACM (2003)
7. United States Code. Sarbanes-Oxley Act of 2002, PL 107-204, 116 Stat 745 (2002)
8. Security Standards Council. PCI DSS v2.0 (2010)
9. Allender, M.: HIPAA compliance in the OR. Aorn Journal (2002)
10. Saltzer, J.H., Schroeder, M.D.: The Protection of Information in Computer Systems. Proceedings of the IEEE 63(9), 1278–1308 (1975)

11. Dreibelbis, A., Hechler, E., Milman, I., Oberhofer, M., van Run, P., Wolfson, D.: Enterprise Master Data Management: An SOA Approach to Managing Core Information. IBM Press (2008)
12. Organization for the Advancement of Structured Information Standards (OASIS), http://www.oasis-open.org/
13. Zeilenga, K., et al.: Lightweight directory access protocol (ldap): Technical specification road map. Technical report, RFC 4510 (June 2006)
14. Franzoni, S., Mazzoleni, P., Valtolina, S., Bertino, E.: Towards a Fine-Grained Access Control Model and Mechanisms for Semantic Databases. In: IEEE International Conference on Web Services, ICWS 2007, pp. 993–1000 (2007)
15. Rizvi, S., Mendelzon, A., Sudarshan, S., Roy, P.: Extending Query Rewriting Techniques for Fine-grained Access Control. In: Proceedings of the 2004 ACM SIGMOD International Conference on Management of Data, SIGMOD 2004, pp. 551–562 (2004)
16. Roichman, A., Gudes, E.: Fine-grained access control to web databases. In: Proceedings of the 12th ACM Symposium on Access Control Models and Technologies, SACMAT 2007, pp. 31–40 (2007)
17. Stoller, S.D.: Trust Management and Trust Negotiation in an Extension of SQL. In: Kaklamanis, C., Nielson, F. (eds.) TGC 2008. LNCS, vol. 5474, pp. 186–200. Springer, Heidelberg (2009)
18. De Capitani di Vimercati, S., Jajodia, S., Paraboschi, S., Samarati, P.: Trust management services in relational databases. In: Proceedings of the 2nd ACM Symposium on Information, Computer and Communications Security, pp. 149–160. ACM (2007)
19. Chaudhuri, S., Dutta, T., Sudarshan, S.: Fine grained authorization through predicated grants. In: IEEE 23rd International Conference on Data Engineering, ICDE 2007, pp. 1174–1183. IEEE (2007)
20. Jahid, S., Gunter, C.A., Hoque, I., Okhravi, H.: MyABDAC: Compiling XACML Policies for Attribute-based Database Access Control. In: Proceedings of the First ACM Conference on Data and Application Security and Privacy, pp. 97–108. ACM (2011)
21. Karjoth, G.: Access Control with IBM Tivoli Access Manager. ACM Transactions on Information and System Security (TISSEC) 6(2), 232–257 (2003)
22. IBM. Tivoli Security Policy Manager (2011), http://www-01.ibm.com/software/tivoli/products/security-policy-mgr/
23. Axiomatics. Axiomatics Policy Server (2011), http://www.axiomatics.com/products/axiomatics-policy-server.html
24. SourceForge. Ladon - XACML enforcement for DB2 (2009), http://xacmlpep4db2.sourceforge.net/

XSS-Dec: A Hybrid Solution to Mitigate Cross-Site Scripting Attacks

Smitha Sundareswaran and Anna Cinzia Squicciarini

College of Information Sciences and Technology
The Pennsylvania State University
{sus263,acs20}@psu.edu

Abstract. Cross-site scripting attacks represent one of the major security threats in today's Web applications. Current approaches to mitigate cross-site scripting vulnerabilities rely on either server-based or client-based defense mechanisms. Although effective for many attacks, server-side protection mechanisms may leave the client vulnerable if the server is not well patched. On the other hand, client-based mechanisms may incur a significant overhead on the client system. In this work, we present a hybrid client-server solution that combines the benefits of both architectures. Our Proxy-based solution leverages the strengths of both anomaly detection and control flow analysis to provide accurate detection. We demonstrate the feasibility and accuracy of our approach through extended testing using real-world cross-site scripting exploits.

1 Introduction

Some of the most well-known and significant vulnerabilities of a Web application are related to cross-site scripting (XSS) [15]. XSS vulnerabilities enable an attacker to inject malicious code into Web pages from trusted Web servers. Typically, when the client receives the document, it cannot distinguish between the legitimate content provided by the Web application and the malicious payload inserted by the attacker. Since the malicious content is handled as the content from the trusted servers, it has the privileges to access the victim users' private data or take unauthorized actions on the user's behalf.

XSS vulnerabilities have been analyzed by a number of researchers and practitioners. One of the most common defense mechanism currently deployed consists of input validation at the server-end, wherein the untrusted input is processed by a filtering module that looks for scripting commands or meta-characters in untrusted inputs. The filtering module then filters any such content before these inputs get processed by the Web application. However, proper input validation is challenging; XSS attacks can be crafted so as to bypass the input sanitization steps. Further, input validation adds a significant burden on the server end, and leaves clients defenseless in case of unprotected sites. Consequently, recent efforts have shifted their attention on client-end solutions, to protect client systems against servers that failed to filter untrusted content [13]. Unfortunately, neither one of these approaches are able to withstand against all forms of XSS attacks. For example, server-side solutions require good control and knowledge of the server's source code, and therefore fare well with attacks that target servers, or that a reflected

N. Cuppens-Boulahia et al. (Eds.): DBSec 2012, LNCS 7371, pp. 223–238, 2012.
© IFIP International Federation for Information Processing 2012

off Web servers [11,8,4]. Client-side solutions instead are most effective against attacks that are perpetrated through attacks which malicious code is contained in the client side page [13,3]. In this paper, we propose a novel approach to combine the benefits of both server-side and client-side defense mechanisms. We leverage the information obtained from both the client and the server-side using a simple, yet effective, security-by-Proxy approach. Our design allows us to uphold users' browsing activities while thoroughly monitoring the sites' vulnerabilities before any attack is carried out. Specifically, the Proxy develops an anomaly-based detection mechanism, enriched with detailed control flow analysis. These two techniques combined together enable early detection of subtle attacks that may involve obfuscation attempts. In addition, control flow analysis helps us validate any redirections and minimizes the leakage of the victim's information through malicious links.

The architecture includes a Plug-in on the client-end. The Plug-in is responsible for ensuring that any Web site visited by the user is checked by the Proxy. Based on the input of the Proxy, the Plug-in also deploys the actual protection. In either case, we effectively stop the attack from being successfully carried out and affect the user's system. The Plug-in is carefully designed so as to maintain limited amount of information of the user's browsing history.

We extensively test our solution over a large number of actual XSS vulnerabilities. Our evaluation results show not only that we are able to protect against all types of XSS attacks, but also that our approach is efficient and does not impose a significant burden either on the client or on the server. In summary, the key benefits of our solution are:

1. **User Friendly**. Our approach does not require any significant level of human involvement. It is based on a simple Plug-in that interacts with the user to inform him of possible attacks and stop them from being carried out.
2. **High Accuracy**. Our approach can detect all known types of XSS script injections, by providing different levels of protection, that include selectively blocking portions of the sites being infected and preventing the site from being accessed.
3. **Acceptable Overheads**. Our approach does not impose any burden on Web application performances. The overhead at the client-side is minimal, most of the computation is carried out by a Proxy. The Proxy is also very efficient, and therefore it can be used to protect multiple users at the same time.

The rest of the paper is organized as follows. Next section provides an overview of XSS attacks. Section 3 provides an overview of our solution, the XSS-Dec. In Section 4 we describe the Proxy's architecture. In Section 5, we discuss the Plug-in. In Section 6 we present our evaluation results. Section 7 analyzes existing body of work in this area, and Section 8 concludes the paper.

2 XSS Attacks and Common Solutions

XSS attacks are a class of code injection attacks caused by the server's lack of input validation and are typically the result of insecure execution of JavaScript, although non-JavaScript vectors, such as Java, ActiveX, or even HTML, may also be used to mount the attack. XSS attacks can be segregated into the following classes:

DOM-based attacks: The attacker sends a specially crafted URL to the victim, altering the DOM structure of the Web page once it's loaded in the browser. The actual source code is not changed. It is often launched using the `document.location` DOM and then used to populate the page with dynamically generated content.

Reflected XSS attacks: The attack code is "reflected" from a Web server. The attacker inserts malicious JavaScript into some form, which typically reflects the string back to the trusted site, using the content inserted to generate a response on the fly. The attack code, which is treated as belonging to the same domain as the rest of the site, is then executed. This is the most common form of attack.

Stored XSS attacks: In this type of attack (also known as HTML injection attack), the payload is stored by the system, and may later be embedded by the vulnerable system in an HTML page provided to a victim. The attack is carried out when a victim visits this page, or the part of the page on which the payload is stored.

Usually, to prevent untrusted code from gaining access to content on other domains, protection mechanisms such as sandboxing are applied or the same origin policy is enforced. However, XSS attacks bypass the same origin policy to gain access to objects stored on different domains by luring the victim to download or execute malicious code from a trusted site. Beyond sandboxing, the most commonly employed defense against XSS attacks is input validation [3,19,8,11]. This approach uses a server-side filtering module that searches scripting command or meta-characters in untrusted input, and filters any such content. If the server fails to filter the input, however, the client is left defenseless. Other popular input validation techniques include dynamic tainting and untrusted information tracking. As highlighted by some recent work, these solutions correctly track whether a filter routine is called before untrusted information is output, but they do not reason about the correctness of employed filters, and fail to consider the Web application output [4]. Further, there are many scenarios where filtering is difficult to carry out correctly, especially when content-rich HTML is used. For example, attacks that are launched by scripts located at multiple locations in a Web application may succeed. A single filter function may not be sufficient if it looks for scripting commands, as injected input may be split across the output statements. In this case, every character in the HTML character set is legal, which implies that the filter cannot reject any individual character that may result in script content. Unauthorized scripts can be obfuscated by entering it within pre-existing execution environments, allowing it to escape the filters. That is, the attacker may embed an environment variable in between two existing tags. Hence, one should check the alteration of the execution flow to identify such hidden attacks.

3 Our Approach: The XSS-Dec

Among the most popular techniques for Web vulnerabilities, anomaly detection and control flow analysis have gained popularity in the recent years. When considered alone, neither approach is however sufficient for effective detection of XSS attacks. First, the complexity of XSS attacks prohibits any approach that solely relies on anomaly detection [17]. Anomaly detection is in fact unable to detect most XSS subtle attacks, that are often deployed by exploiting obfuscation techniques. For example, XSS attacks are

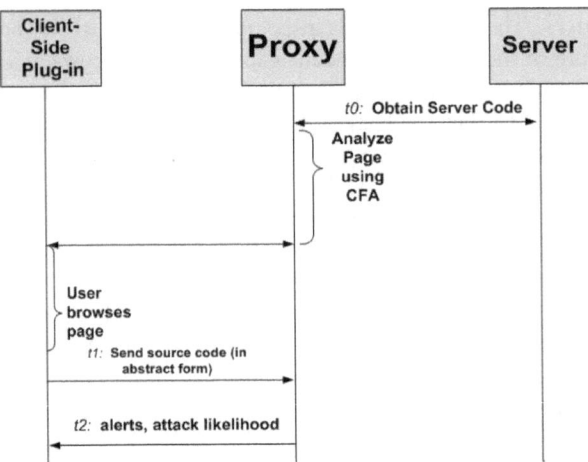

Fig. 1. XSS-Dec main flow

written using JavaScript or ActionScript, where the output of the script is dynamic in nature, thus obfuscating the attack. Second, control flow analysis is effective for detecting subtle attacks, but it is very inefficient for real time detection [10] as it is slow and results in a high number of false positives. Real time detection is important with XSS attacks since the output of the attack script is often developed on the fly.

In order to overcome these limitations, we have devised a hybrid solution that combines the benefits of control flow analysis and anomaly detection to protect client systems against XSS attacks. Specifically, we suggest a security-by-Proxy solution, referred to as XSS-Dec. XSS-Dec relies on a Proxy component for vulnerability analysis and detection. The Proxy acts as a middle-man between the servers of the sites visited by the client, and a client-side Plug-in. This design upholds users' browsing activities while thoroughly monitoring the sites' vulnerabilities *before* any attack is carried out.

An overview of the XSS-Dec main functionality is reported in Figure 1. As shown, a first bootstrapping step is executed at time t_0. The server (or servers, if more than one is in fact connected to the Proxy) sends the encrypted copies of the source files of its Web pages to the Proxy. Subsequently, as the source of a Web site is updated or changed at the server end, more updates are sent to the Proxy by the server (possibly before the newer site is launched to the public). We assume that the Web sites' source codes collected by the Proxy at this point of time are valid, that is, there has been no chance for an attacker to insert any malicious script. The Proxy generates an abstract and accurate representation of the site, using control flow analysis, and stores it for later use. When the user starts browsing (at any point of time $t_1, t_1 > t_0$), the client-side Plug-in deploys the actual defense mechanism. Precisely, the Plug-in keeps a local record of the pages visited by the user. Further, it communicates to the Proxy the client input and the source code of the Web -page as it appears to the client. Upon receiving the client-end input, which is again encoded using control flow techniques, the Proxy detects whether there exists any features indicative of malicious code or script, at time t_2, using both anomaly-based and signature-based detection.

Using signature-based detection, the Proxy searches and extracts features, which it uses to calculate the likelihood of an actual attack taking place. This attack likelihood is used to drive the Plug-in to either work pro-actively by blocking certain user actions and sites, or reactively by waiting for the attack to actually take place before notifying the user. This information regarding the attacks is sent back to the Plug-in at time t_2. The Plug-in, using the information obtained from the Proxy, deploys the actual defense mechanism, by either stopping the attack or preventing it from being executed (time t_3).

Note that the server and client side representations of the page being compared are different: the server source code is free of any injected malicious code, while the input received from the client-side may include malicious content. Although the client's actual input actions may differ from those simulate on the server side, injection of malicious scripts always result in a particular set of code features, like certain HTML tags being manifested in a compromised site. These features are the ones analyzed by the Proxy.

It is worth noting that our security-by-Proxy design assumes that the Proxy is resistant to basic attacks. The servers of the sites frequented by the client are assumed to be semi-trusted, and able to send to the Proxy non-corrupted data. That is, we trust the server to send the source code of its Web sites to the Proxy before the malicious scripts are injected. In line with current solutions based on client systems (e.g. [13]), we also assume that the Plug-in is not compromised.

4   The Proxy

The core algorithms behind our defense mechanism reside at the Proxy. The Proxy is composed of two logically distinct modules, Calculator and Analyser. These two modules serve the complementary tasks of analysis and detection.

4.1   The Calculator

The Calculator is in charge of computing a normalized and detailed view of the server's pages' content. The source code for both client and server's pages are modeled through a control flow graph (CFG), for accurate and efficient computation. The CFG is an abstract representation of the source code of the Web page, including any possible redirections for the URLs contained in it, and execution paths for active components, such as JavaScript or Flash components. In the context of our system, the CFG is represented as a directed graph. The nodes represent either HTML tags or actual program statements and variables. The edges represent the paths of execution, while the directions are dictated by the loops and the conditions present in the code.

CFGs have been often successfully used in static analysis [13,5]. However, given the complexity of certain pages, CFGs can be computationally expensive to generate, and hard to navigate. This makes it very difficult to construct the dynamic CFGs for such pages on the fly, which is essential to identify the possible malicious effects of any code that has been added to the page. Our challenge is then to compute CFGs that are both accurate and efficient for the scope of our detection. To cope with these limitations, we construct different types of CFGs, based on the specific Web page structure and of its content, as specified below.

1. *Page with no active components:* If a page has no active components, its CFG is derived from the control flow information available from the design model of the Web page such as its HTML or XML [18]. Specifically, the Abstract Syntax Tree (AST) [18] is first created, and then any flow information between the nodes is added. In what follows, this CFG is also referred to as model-based.

2. *Pages with active components:* If a page has a lot of active components, the CFG is again derived from the control flow information available from the design model of the Web page, and it is then augmented with the control flow information available from the actual code about active nodes. In particular, Flash-based elements and JavaScript components are expanded to uncover potentially obfuscated attack code. That is, when a ⟨script⟩ tag, or a ⟨∗.swf⟩ file is encountered in a node of the model-based CFG, the node is further expanded based on the component's source code (i.e., the JavaScript, or the ActionScript respectively), and a new sub-CFG is obtained. The new CFG is constructed by representing each command in ActionScript in the code as a node. The flow from one statement to the next is given by directed edges. Notice that for the construction of this sub-CFG (i.e. the one containing expanded active nodes), we do not consider the user's inputs. Instead, we construct all possible execution paths based on all the possible inputs. Therefore, the CFG shows the call relations at the basic block level, while also containing all the possible nodes and edges. An example of a portion of an enriched CFG is given in Fig 2.

3. *JavaScript Rich Pages:* If active components are only JavaScripts, a simpler form of CFG is generated, to save both space and time complexity. Instead of generating the augmented dynamic control flow graph as described above, the JavaScript elements are rendered as augmented ASTs. The grouping parentheses (such as ⟨script⟩ tag, or a ⟨∗.swf⟩) are still left implicit in the tree structure, and the syntactic representation of any conditional nodes are represented using branches, but the call relationships at the block level are still explicitly shown. Therefore in the augmented AST, the nodes are used to represent the commands in the code like in a simple AST, the loops are simply represented by if-then clauses with a given set of steps repeated in between. Any goto statements are also simplified to if-then-else clauses. Any user actions that can alter the loop (e.g. open a new page, click a link, move mouse over some objects of the page) are represented on the edges.

4. *Access Restricted Pages:* The CFG for a site which is access restricted (requires a login to gain access to portions of the site), is developed using a different methodology. Clearly, the site structure and corresponding CFG depend on whether the CFGs are built before or after the user's login. Further, the CFG for each user will be different as users may have customized Web spaces within the site. In case of such access-restricted sites, only the CFG before login remains the same across all users. The Calculator can easily obtain the CFG for this portion of the site. This CFG is very important, as any injected code on this page can potentially allow the attacker to take over the user accounts. Yet, damaging script can be injected in the pages after login too. To compute the CFG of user-restricted portions of the sites, the Calculator logs in using a test account. Intuitively, this CFG will not contain user-specific information. It is however still useful for attack analysis, in that it gives the actual structure of the pages of the site, thus allowing the Proxy to detect any attacks that are launched by modifying the structure of the site. In particular, it helps

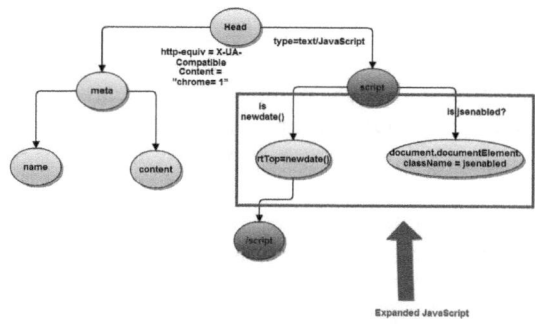

Fig. 2. Portion of the model-based CFG for the Yahoo site

detect any changes to the DOM structure and is therefore useful in detecting Persistent or DOM-based vulnerabilities. However, it cannot detect non-persistent vulnerabilities, which form the most common type of XSS attacks. This is because all the non-persistent vulnerabilities are exploited by data injected in the user-specific pages when the site is a login-based site. To detect the non-persistent vulnerabilities, we depend on the feature extraction capabilities of the Analyzer as explained next. If the Analyzer encounters a JavaScript or ActionScript environment, it requests the Calculator to compute the sub-CFG for that particular portion of the site, to detect possible malicious code injected within these environments.

4.2   The Analyzer

The Analyzer has two main tasks. First, it extracts features indicative of potential exploits. Second, it estimates the likelihood of the attacks being carried out.

Feature Extraction: The Analyzer, upon obtaining a client-side CFG, compares the client-side and the server-side CFGs to extract relevant features that may be indicative of attacks. In the following, we provide a broad classification of the features searched by the Analyzer. The features refer to non-access restricted sites, and are presented in the order of the severity of the attack.

(1) **Redirection to a site not contained in the server-side CFG**: If a CFG generated at the client contains a redirection to some site not contained in the server-side CFG, it likely means the user will be redirected to a site unknown to the original server. This feature, which is the most common and strongest indicator of an XSS attack, is often observed in DOM-based attacks [7].

(2) **SQL Injection Via XSS**: A script capable of inserting input on behalf of the user is potentially indicative of an attack. Specifically, if the script is added to the site without any actual action or permission from the user, and therefore appears to the client-side CF , it may denote a *SQL Injection attack*. The SQL statements are used to commit changes to the database on the victim's behalf. Given below is an example of the code:

```
< TableID = "TNAME" >
< / Table >
< Script Language = "JavaScript" src = "addjscript.js" >
< Script Language = "JavaScript" >
sql("insert + TNAME + values('Victim', 'pwnd', 'again')");
< /Script>
```

This script attempts to insert the values "Victim", "pwnd" and "again" into the table named TNAME. Using such statements, the attacker can change the passwords or other information of the victim.

To identify potentially malicious actions, the Proxy specifically monitors for server-side database actions being committed through SQL commands such as "UPDATE", "DELETE" etc. That is, it scans the JavaScript and ActionScript for any embedded SQL queries as in the above example. The Proxy also checks the CFG for all possible SQL commands including "SELECT", so as to identify a large range of attacks. This feature occurs often, though not exclusively, in stored attacks [20].

(3) **JavaScript based manipulation on the client-side CFG**: If the client side CFG includes nodes with <submit> and < META > tags, forms may be submitted on the user's behalf, or cookies may be manipulated without the user's knowledge. Although each of these tags can occur for legitimate purposes in non-malicious JavaScript, when combined with any of the other features (especially redirection to a site contained on the server-side), these tags are typically representative of an attack. This feature is most often observed with reflected XSS attacks [7].

(4) **Text changed from original server site to the site rendered at the client**: Differences in the way text is rendered on the client's browser versus the way it was stored on the server-side are also to be treated as a warning sign. To check for any changes, the Proxy looks for alterations in the text formatting tags such as the <header> tags, the <para> tags, the use of bold or emphasis tags, etc. In this way, the Proxy can detect subtle attacks, where the attacker simply changes the way a Web site looks with intentions of slander or misrepresentation. Text manipulation can be carried out by any type of XSS attack, but is most commonly observed with DOM-based and reflected XSS attacks [16].

For access restricted sites, extracting the features of a latent attack is more complicated, as comparing the client-side and the server side-side CFGs is not sufficient. This is because the server-side CFG is derived using login information different from the login of the user being monitored. The CFG at the server side, while structurally similar, does not contain the actual information contained in the client-side CFG. For instance, the server-side CFG for a GMail page is constructed using a login different from the login used by the user, and therefore it has the same UML structure as the client-side CFG, but it will differ with respect to the exact content in the CFG.

For such sites, the Analyzer exploits the similar CFG structure of the two versions of the site to identify if the basic representation of the page is altered. In this way, text changes to non-user generated texts such as logos can be detected in the same way as it was for non-access restricted sites. Further, the SQL Injection feature also does not change, as identifying SQL statements which cause actions to be committed on the user's behalf can be detected without the need for comparing the client-side and the server-side CFGs. Yet, the attacker can still inject the malicious script in those portions of the page that are actually user-specific. Referring again to our GMail example, the attacker would insert a URL for redirection in the actual mail content. To address this, the Analyzer checks whether any of the URLs that appear in the user-specific portions of the page link to a potentially malicious page. JavaScript manipulations are also hard to extract for access restricted pages, as the manipulation of the JavaScript can take place within pre-existing environments, which occur only on the user specific portion of the page. For example, the attacker can inject malicious scripts between two $<$ META $>$ tags present in the user's profile on a Facebook page for a link to his personal Web site. Since these $<$ META $>$ tags will vary for each user's Web site, there is no way of identifying whether any code has been injected between them by simply comparing the client-side and the server-side CFGs. Therefore, the Analyzer requests the Calculator to derive the CFG for the user specific parts of the page. Based on this input, the Analyzer checks for any possible malicious execution paths.

Attack Analysis. The Analyzer, upon detecting one or more of these features, computes the likelihood (denoted as α) of an actual attack. Each of the features is assigned a weight. The weight is set to correlate with the amount of past security-relevant information about the feature including the frequency of mentioning in incident reports. For example, the most common condition observed in XSS attacks is redirection or some sort of URL submission [7]. Therefore, the feature of redirection to a site not contained in the server-side CFG is to be assigned a high weight. α is simply calculated using the weighted sum: $\alpha = \sum_{f=1}^{|F|} w_f * n_f$. In the equation, w_f represents weight of the feature f, and n_f indicates the number of times the feature has occurred on the page. $|F|$ is the cardinality of all possible features the Proxy analyses. The equation can be extended to capture additional features, and strings which may be injected by the attacker. For example, a combination of features or a specific pattern of extracted features can be given an additional weight, or an additional attack string can be considered in the equation. For simplicity, however we stick to the formula specified above. As shown in Section 6, it is sufficient to guarantee a very good detection rate.

α is matched against anomaly based thresholds. We consider two threshold levels, a *detection* threshold and a *prevention* threshold. These thresholds are dynamic thresholds in that they are constructed based on the actual set of features which is extracted from a page. The *detection* threshold is indicative of a suspected attack, in that a number of limited features are verified true. It is therefore set based on the total number of features that have been extracted for a given page, and the lowest weight found in the set of features extracted. Using the lowest weight found ensures that the evidence presented to the Plug-in is just enough to register some suspicious but not necessarily harmful activity. The second threshold, which is is instead higher, models cases where there is enough evidence (in terms of features occurrence and importance) to believe that the attack is in fact imminent. This threshold is referred to as *prevention threshold* as it triggers mechanisms to prevent an attack, upon being passed. It is, also set using the total number of features extracted. However, it is based on the highest weight feature that has been extracted for the given page thus making it higher than the detection threshold. For example, page X may contain added redirections to URLs, JavaScript manipulations and text changes from the original page on the server, while page Y may contain redirections, SQL Injections via XSS and JavaScript manipulations. The highest weight extracted for both page X and page Y is the same. However, the lowest weight extracted for page X is the weight associated with text changes, and the one extracted for page Y is the weight associated with JavaScript manipulations.

5   The Client-Side Plug-In

The client-side Plug-in is in charge of providing the Proxy with information about the pages the user is visiting, encoded as a CFG. Further, it deploys protection mechanisms against latent or ongoing attacks, upon being notified by the Proxy.

To complete these tasks, the Plug-in has two main modules: the *Auditor* and the *Detector*. The Auditor obtains information from the Proxy about all the possible attacks on an open page, while the Detector is in charge of stopping or preventing the identified vulnerabilities from being carried out. In order to help identify the possible attack vectors of an open page, the *Auditor* keeps a record of the pages visited by the user, and calculates a model-based CFG for each of such pages. Once created, each CFG is stored in an encrypted form. The CFG is subsequently updated or replaced as needed, according to any changes of the page's code, due to script injections or server-side modifications. The Plug-in sends the latest encrypted CFG to the Proxy, every time the user visits the page, as soon as it is opened on the browser.

The other module of the Plug-in is the *Detector*. The Detector is the component obtaining instructions from the Proxy if the features detected are deemed indicative of a potential attack. Precisely, it receives information about the possible attacks in the form of the attack likelihood α and the specific features extracted by the Analyzer. Each of the features results in a particular type of anomaly. The anomalies consist of execution of a script on a site to which the user has been redirected, personal information of the user being sent to a remote system, and actions such as submission of forms taking place without any corresponding input on the user's part. The first two anomalies are often observed in case of redirection and JavaScript based manipulation, while the third one is often observed with SQL Injection via XSS.

The Detector monitors the client machine for any of such anomaly using dedicated modules. Each module corresponds to a specific monitoring activity and is activated if the Proxy verifies the feature they implement. Specifically, the module checks for specific user actions which actually start the attack based on the features it has detected. If one of the extracted attack features consists of unwanted redirections (see feature (1) in Section 4.2) and the estimated likelihood is high, the Detector prevents the user from being redirected to the targeted page. Otherwise, (i.e. likelihood is below the prevention threshold) the Plug-in simply pops up an alert box to the user when the link is opened. In case an attack presents one or more features (i.e. feature (2), (3) and (4)) beyond redirection, and has an estimated high likelihood value, the Detector prevents the malicious script from being executed. Specifically, it does not render a portion of the page or an entire page and displays an error message to the user. In case of a low likelihood value, it simply pops up an alert to the user before rendering the malicious content. Intuitively, stopping ongoing attacks is less desirable as preventing them. As shown in the next section, the XSS-Dec is efficient enough to stop the attack before any major damage of the client system.

6 Evaluation

We deployed and thoroughly tested a running prototype of the XSS-Dec. Before discussing our evaluation, we briefly describe the prototype developed.

6.1 XSS-Dec Prototype

The Plug-in was developed by logically distinguishing the Auditor from the Detector. Both components were implemented primarily in JavaScript, to guarantee portability. The Auditor uses a separate JavaScript component to construct the CFG based on the HTML and DOM structure of the page. Each of the possible user actions are edges' labels. In case the node contains a URL (i.e., a `href` tag) to some other site or page, this node becomes the second node of the CFG of that page. To reduce the risk of interceptions of the user's browsing history, the Plug-in sends the CFG in an encrypted form. In the current prototype, the Plug-in uses Merkle hash trees [14] for easy graph comparison and reduced graph size. Alternatively, we could serialize the tree and rely on more traditional encryption schemes.

The Detector is organized into four JavaScript modules, one for each of the features possibly extracted by the Proxy. Each module is activated if the Proxy verified true the corresponding feature they implement.

The prototype of the Proxy also has two modules. Both are implemented using Java and JavaScript components. Specifically, the Calculator uses JavaScript to build the nodes for the model-based CFG. Java components are then used to expand the active nodes, i.e. the nodes containing JavaScript or Flash elements. The resulting enriched CFG is built as a serialized tree using the TreeMap Java class. The Calculator stores a number of CFGs of the pages being most often visited by end users. This simplifies both CFG analysis and comparison with Plug-in-received CFGs. The stored CFGs are updated as the pages' content is notified to have changed.

6.2 Experimental Evaluation

The goal of our evaluation was two-fold. First, we aimed at estimating the accuracy of our solution in detecting XSS attacks. Second, we estimated the overhead incurred with our protection mechanism. Estimating the Proxy overhead allows us to make some initial considerations on the scalability of our solution. The Plug-in was tested from a Dell Latitude D630 Laptop, with 2G Ram and a Intel(R) Core(TM)2 Duo CPU T7500 @ 2.20GHz processor. The Proxy was run from an Apache server hosted on the same machine, to maintain a conservative estimate of the system efficiency. The server was running the Apache Web server (version 2.2) and PHP version 5.3.3. Apache was configured to serve requests using threads through its `worker` module. Our tests do not account for any network delays, and are carried out without conducting any fine-tuning or training.

Detection Accuracy

Experimental Settings. Using a trial-and-error approach, we defined two simple anomaly threshold values for assessing whether an attack was latent or not. We express the features' weights of equation in Section 4.2 by means of totally ordered integers, ranging from 1 to k, where $k \leq |TF|$, and $|TF|$ is the possible total number of features (4, in our case). Given a set of extracted features $F = \{f_1, \ldots, f_n\}$, we compute the prevention threshold as follows: $T_{high} = |F| + w_{f_{max}}$, where $w_{f_{max}}$ is the highest weight among all the weights of the extracted features. $|F|$ represents the total number of features extracted (regardless of their actual weight). If a same feature appears more than once, it is counted as a new feature, therefore increasing the overall probability of an attack. Intuitively, from this equation, we can determine whether the feature of highest weight is significant enough to influence α to a point where an attack is most likely to happen. Our detection threshold, referred to as T_{low} is computed in a similar fashion of T_{high}: $T_{low} = |F| + w_{f_{min}}$ and $w_{f_{min}}$ is the lowest weight assigned to the features in $\{f_1, \ldots, f_n\}, n > 1$. When $T_{low} < \alpha < T_{high}$, the Proxy suspects an attack. It sends a warning message to the client Plug-in, providing details about the warning features verified true. When $\alpha > T_{high}$, the Proxy deems that the likelihood of an attack is very high.

Methodology. We evaluated our system on several real-world, publicly available Web applications and on simulated environments. We recorded the number of false positives generated when testing the application with attack-free data and the number of attacks correctly detected when testing the application with malicious traffic. In detecting the attacks we tracked whether they were detected at the time of prevention (i.e. α was above T_{high}) or detection (i.e. α was above T_{low}).

Overall, we ran the XSS-Dec system for a total of 100 pages, in a non-deterministic order. 20 of them were hand-evaluated real-world clean pages. The remaining pages were constructed by us, and contained one or more XSS vulnerabilities. The clean pages were selected from popular Web sites with active components, like MSN, Yahoo, Google, social networking and forum sites. The vulnerable pages were created using the real-world XSS vulnerabilities reported in the security mailing list, Bugtraq [2]. We deployed the given vulnerabilities in similar sites than those listed as vulnerable, and injected the malicious script in the variables described. We constructed 80 sites, and tested 80 different vulnerabilities. Each of these sites hosting the malicious files had benign components. The actual attack code varied for each try, so as to create polymorphic attack code. To create the variations of the attack code, we introduced random NOP blocks in each attack to introduce random delays. Further, we combined one or more attacks with each other, i.e. some vulnerabilities were tested multiple times. Also, the page invoking the malicious content was different for each try. The elements we included in each page consisted of one or more of the following: images, videos, audio components, other benign JARs carried in applets but not embedded in images, text documents, hyperlinks, Java components, JavaScript components, forms, zip files, Microsoft Office Open XML documents, XPI files, benign SWFs and simple games.

Table 1. Evaluation Results

Attack Type	Detected	Prevented	False Positive	False Negative	Total Attacks
Normal XSS	0	All	0	0	15
Image XSS	0	All	0	0	13
HTML entities	0	All	0	0	12
Style-Sheet based XSS	0	All	0	0	13
Flash-Based XSS	5	3	0	2	13
XSS in pre-existing environments	0	All	3	0	14

Results. Table 1 summarizes our results organized according to the classification in the Rsnake Cheat Sheet [7]. The results reported in the table group the different 80 attacks according to the location of the attack vector. As shown, XSS-Dec stopped all but 2 attacks. Both were Flash-Based. Out of the stopped attacks, 94% of them were prevented before being carried out. The remaining 6.2% were stopped at detection time. We reported 3 false positives. A false positive occurred when an attack was detected in a part of the page where there was no attack code.We noticed that false positives were detected on the forums of users sharing coding tips on JavaScript. The code displayed on the pages as part of the discussions was considered as an injection by the XSS-Dec. To improve the false positives on forums, we plan on adding string checking to the Proxy as part of our future work. String checking will help differentiate between the code being discussed in the forums and some malicious script. The sites that were not prevented but only detected were typically sites with a huge number of Flash components. Flash components enable the attacker to hide the consequences of the redirections due to script injection, reducing the overall likelihood of the attack being prevented by the Proxy. We expect that training the model would mitigate these issues. Below, we summarize our results for three of the most challenging categories of attacks: In case of Flash-Based attacks, our approach prevents most of these attacks by not executing the Flash file. For those attacks that are only detected, the file is executed but the user is alerted as soon as some malicious activity is seen on the client end. In case of Cookie stealing XSS attacks, our approach specifically monitors for manipulation of the ⟨META⟩ tags to reset the cookies, and detects all possible instances of this vulnerability. For XSS attacks where the vectors are injected into pre-existing elements (e.g. between pre-existing ⟨script⟩ and ⟨/script⟩ tags), our approach monitors for manipulation of JavaScript and we achieve a 100% prevention rate.

Performance Evaluation

We computed the average time for the most resource consuming activities of our system, i.e. constructing the CFGs and extracting features. Our tests show that the average time for constructing a CFG of level 40 with no dynamic components is 3.25 seconds, and for constructing the CFG with 50% dynamic components is 3.39 seconds. The time grows linearly with respect to the size of the CFG. For these tests, we used CFGs of increasing complexity from 10 to 80 nodes, each corresponding to real sites. The CFGs of level 80 correspond to popular sites, with a large number of active components (3/4 of the nodes), such as Youtube and Bigfish. The complexity of the CFGs increased as the ratio of active nodes to inactive nodes increased. For the simpler CFGs, the ratio of active nodes to inactive node was 1:4, while for the more complex CFGs, the ratio was greater than 3:4. The highest complexity for a CFG of level 80 was 83% of the nodes were active nodes. The time taken for constructing the CFG of level 80, with 83% active nodes was 4.183 seconds, while the time taken to calculate the least complex CFG was 3.2432 seconds. This makes the overhead for the most complex CFG compared to the least complex CFG less than 1 second different. We notice that while this time is not negigible, CFGs are only calculated periodically, and cached for efficient reuse.

The time for extracting features on an average for a CFG of level 40 is 2 seconds, while the maximum time for a level 80 CFG is 2.673 seconds. Since the CFGs are computed at the Proxy, these results confirm that the Proxy is indeed scalable. In real-world settings, the Proxy would be hosted on a dedicated server with larger processing power than our system. Further, we notice that since most of the pages maintain a similar structure the Proxy can improve the size of the cached directory of model-based CFGs, for similar Web sites. This would likely improve the performance substantially. Finally, the Proxy in the real world would not be running in parallel with the Plug-in as was the case for our system.

7   Related Work

XSS attacks have been identified as a threat since the 1990s. Since then, various solutions to detect and prevent these attacks have been explored. Traditional solutions focus on sanitizing the input at the server side, but recently client side approaches have also been proposed. There also exist Proxy-based solutions which aim to protect Web applications by analyzing the HTTP requests. Despite these efforts, XSS attacks still remain on the top of Web security attacks in the OWASP lists [15].

Server-side solutions or Proxy-based solutions are commonly used for Web -based attacks, since they enable users' inputs sanitization [3,19,11,22]. In particular, Scott and colleagues proposed an interesting Proxy-based solution [19]. The Proxy is similar to an application firewall; it enforces pre-written security policies. Their proposed mechanism requires that all Web applications patch themselves to prevent an XSS attack. In case a Web application is not patched, the end user is left defenseless. Our focus, on the other hand, is how to ensure that any malicious script does not affect the user. Consequently, it does not require any patching from either the user or the server. Similar to the above is a commercial product called AppShield [1]. AppShield also inspects the HTTP messages to prevent application level attacks. While it is similar to our system in inspecting the HTML of the pages outbound from the server, it does not specifically look for any code injection. Hence, Appshield can recognize attacks based on the (proprietary) rules that it uses to validate the HTTP requests. Wurzinger also propose a proxy-based solution to detect HTML responses and any injected scripts [23]. To identify malicious scripts, any legitimate script calls in the original web page are changed into unparsable identifiers called script IDs. Therefore, if any unparsed script is found, it is assumed ti be indicatory of an attack. This system focuses on stored and reflected XSS but not on DOM Based attacks. Further, the parsing of a script may be a significant bottleneck of the system.

Bisht et al. [4], propose to remove any server side script in the output of a Web application, when the script is not originally inserted by the application itself. This approach is complementary to ours in that we focus on preventing the attacks at the client-end, rather than relying on servers' filters only. Further, as any other server-based solution, Bisht's approach relies on the server ability to patch and remove server side scripts. The fact that a solution focusing on protecting the servers may leave end-users vulnerable has inspired some interesting client-oriented solutions. One of these is the Noxes system, proposed by Kirda et al [13]. Noxes is a Web firewall aiming at protecting the client from XSS attacks. Noxes' detection is based on the analysis of the server-side scripts. In XSS-Dec, we also use server-side scripts. However, our detection of code injection relies on a detailed comparison of the server-side scripts with client-side scripts. Kirda and colleagues instead choose to rely on validating the HTTP referrer headers. The HTTP headers, however, do not represent a useful indicator in case the attacks come from trusted sites. Further, the information leaked via embedded URLs is contained by limiting the information sent through each.

We borrowed the idea of using control flow analysis from some recent interesting work [9,21,5,6]. The Swaddler system [9], for example, focuses on detecting any violations in the workflow of a stateful application or input violations by users. We differ from this work both in scope and in the detection mechanism: our focus is on script injections rather than state violations. Further, our solution accounts for both stateful and stateless applications. Bonfante et al. in [5] used control flow graphs for extracting malware signatures. The authors present a system for extracting signatures of malware by using CFGs composed at the assembly language instruction set level. While similar to our approach in spirit, our CFGs are derived based on high level languages. We employed control-flow analysis in our previous work, the DeCore [21]. The DeCore is aimed at detecting content repurposing attacks, from the client-side end, and therefore focuses on a different set of attacks. Close to the notion of control flow analysis is script analysis, which has been leveraged to detect XSS vulnerabilities. A specific example of this approach is the Pixy tool proposed by Jovanovic et al.[12]. We take a complementary approach, in that we analyze JavaScript, ActionScript and HTML. Further, the Pixy tool relies on taint analysis of the data whereas we leverage the notion of control flow analysis by using CFGs. The CFGs allows the XSS-Dec to detect any malicious script injection using any type of script, while the taint analysis in the Pixy tool helps detect any input violation.

8   Conclusion

In this paper, we presented XSS-Dec, a security-by-Proxy approach to protect end-users against XSS attacks. Our solution combines the benefits of both server-side and client-side protection mechanisms. We leverage the information obtained from both the client and the server-side to provide an anomaly based detection approach complemented by control flow analysis. In the future, we will study whether a server can use the Proxy features without having the server's sending pages beforehand. Finally, we will test the scalability of the XSS-Dec in distributed settings.

References

1. Appshield, Sanctum Inc. (2004)
2. Security focus-bugtraq (2010), http://www.securityfocus.com/archive/1

3. Bates, D., Barth, A., Jackson, C.: Regular expressions considered harmful in client-side XSS filters. In: 19th International Conference on World Wide Web, WWW 2010, pp. 91–100. ACM (2010)
4. Bisht, P., Venkatakrishnan, V.N.: XSS-GUARD: Precise Dynamic Prevention of Cross-Site Scripting Attacks. In: Zamboni, D. (ed.) DIMVA 2008. LNCS, vol. 5137, pp. 23–43. Springer, Heidelberg (2008)
5. Bonfante, G., Kaczmarek, M., Marion, J.-Y.: Control Flow Graphs as Malware Signatures. In: International Workshop on the Theory of Computer Viruses, TCV 2007, Nancy, France (2007)
6. Chen, S., Meseguer, J., Sasse, R., Wang, H.J., Wang, Y.-M.: A systematic approach to uncover security flaws in gui logic. In: IEEE Symposium on Security and Privacy, pp. 71–85. IEEE Computer Society (2007)
7. ComputerWeekly.com. Hackers broaden reach of cross-site scripting attacks (2007)
8. Cook, S.: A Web developer's guide to cross-site scripting. t. r, SANS institute (2003)
9. Cova, M., Balzarotti, D., Felmetsger, V., Vigna, G.: Swaddler: An Approach for the Anomaly-Based Detection of State Violations in Web Applications. In: Kruegel, C., Lippmann, R., Clark, A. (eds.) RAID 2007. LNCS, vol. 4637, pp. 63–86. Springer, Heidelberg (2007)
10. Earl, C., Might, M., Horn, D.V.: Pushdown control-flow analysis of higher-order programs. In: The 2010 Workshop on Scheme and Functional Programming (2010)
11. Gundy, M.V., Chen, H.: Noncespaces: Using randomization to enforce information flow tracking and thwart cross-site scripting attacks. In: Annual Network & Distributed System Security Symposium (2009)
12. Jovanovic, N., Kruegel, C., Kirda, E.: Pixy: A static analysis tool for detecting web application vulnerabilities. In: IEEE Symposium on Security and Privacy, pp. 258–263. IEEE Computer Society (2006)
13. Kirda, E., Kruegel, C., Vigna, G., Jovanovic, N.: Noxes: a client-side solution for mitigating cross-site scripting attacks. In: 2006 ACM Symposium on Applied Computing, SAC 2006, pp. 330–337. ACM (2006)
14. Munoz, J.L., Forne, J., Esparza, O., Soriano, M.: Certificate revocation system implementation based on the merkle hash tree. International Journal of Information Security 2, 110–124 (2004), 10.1007/s10207-003-0026-4
15. OWASP. Top 10 2010 - the open web application security project (2007), http://www.owasp.org
16. OWASP. DOM based XSS (2011), https://www.owasp.org/index.php/DOM_Based_XSS
17. Raman, P.: JaSpin: JavaScript Based Anomaly Detection of Cross-Site Scripting Attacks. Master's thesis, Carleton University, Ottawa, Ontario (2008)
18. Schwartz, N.: Steering clear of triples: Deriving the control flow graph directly from the Abstract Syntax Tree in C programs. Technical report, New York, NY, USA (1998)
19. Scott, D., Sharp, R.: Abstracting application-level web security. In: Proceedings of the 11th International Conference on World Wide Web, pp. 396–407. ACM (2002)
20. SpiderLabs. Analysis of lizamoon: Stored XSS via SQL injection (2011), http://blog.spiderlabs.com/2011/04/analysis-of-lizamoon-stored-xss-via-sql-injection.html
21. Sundareswaran, S., Squicciarini, A.C.: DeCore: Detecting Content Repurposing Attacks on Clients' Systems. In: Jajodia, S., Zhou, J. (eds.) SecureComm 2010. LNICST, vol. 50, pp. 199–216. Springer, Heidelberg (2010)
22. Wassermann, G., Su, Z.: Static detection of cross-site scripting vulnerabilities. In: 30th International Conference on Software Engineering, pp. 171–180. ACM (2008)
23. Wurzinger, P., Platzer, C., Ludl, C., Kirda, E., Kruegel, C.: Swap: Mitigating XSS attacks using a reverse proxy. In: Proceedings of the 2009 ICSE Workshop on Software Engineering for Secure Systems, IWSESS 2009, pp. 33–39. IEEE Computer Society, Washington, DC (2009)

Randomizing Smartphone Malware Profiles against Statistical Mining Techniques

Abhijith Shastry, Murat Kantarcioglu, Yan Zhou, and Bhavani Thuraisingham

Computer Science Department
University of Texas at Dallas
Richardson, TX 75080
{abhijiths,muratk,yan.zhou2,bxt043000}@utdallas.edu

Abstract. The growing use of smartphones opens up new opportunities for malware activities such as eavesdropping on phone calls, reading e-mail and call-logs, and tracking callers' locations. Statistical data mining techniques have been shown to be applicable to detect smartphone malware. In this paper, we demonstrate that statistical mining techniques are prone to attacks that lead to random smartphone malware behavior. We show that with randomized profiles, statistical mining techniques can be easily foiled. Six in-house proof-of-concept malware programs are developed on the Android platform for this study. The malware programs are designed to perform privacy intrusion, information theft, and denial of service attacks. By simulating and tuning the frequency and interval of attacks, we aim to answer the following questions: 1) Can statistical mining algorithms detect smartphone malware by monitoring the statistics of smartphone usage? 2) Are data mining algorithms robust against malware with random profiles? 3) Can simple consolidation of random profiles over a fixed time frame prepare a higher quality data source for existing algorithms?

1 Introduction

Compared to conventional mobile phones, smartphones are built to support more advanced computing needs modern custom software demands. An unpleasant byproduct of the ongoing smartphone revolution is its invitation to malicious exploits. As smartphone software grows more complex, more malware programs will be created to attempt to exploit specific weaknesses in smartphone software [4,6]. Smartphones of end users all together constitute a large portion of the powerful mobile network. Having access to the enormous amount of personal information on this network is a great incentive for the adversary to attack the smartphone mobile world.

Malicious activities on mobile phones are often carried out through lightweight applications, scrupulously avoiding detection while leaving little trace for malware analysis. Over the years many malware detection techniques have been proposed. These techniques can be roughly divided into two groups: static analysis and dynamic analysis. Static analysis techniques discover implications of

N. Cuppens-Boulahia et al. (Eds.): DBSec 2012, LNCS 7371, pp. 239–254, 2012.

unusual program activities directly from the source code. Although static analysis is a critical component in program analysis, its ability to cope with highly dynamic malware is unsatisfactory. A number of obfuscation techniques have been shown to easily foil techniques that rely solely on static analysis [14]. Dynamic analysis (also known as behavioral analysis) identifies security holes by executing a program and closely monitoring its activities [24]. Information such as system calls, network access, and files and memory modifications is collected from the operating system at runtime [18]. Since the actual behavior of a program is monitored, threats from dynamic tactics such as obfuscation are not as severe in dynamic analysis. However, dynamic analysis can not guarantee a malicious payload is always activated every time the host program is executed.

We follow a similar perspective of dynamic analysis by analyzing real-time collections of statistics of smartphone usage. Metrics of real-time usage are recorded for analysis. We choose the Android platform in our study. Android is open source and apparently has a solid customer base given that many devices are using this platform. For the convenience of security analysis on this platform, we developed custom parameterized malware programs on the Android platform. These malware programs can target the victim for the purpose of denial of service attacks, information stealing, and privacy intrusion. Our second contribution is the empirical analysis of the weaknesses of data mining techniques against mobile malware. We demonstrate that a malware program with unpredictable attacking strategies is more resilient to commonly used data mining techniques.

The rest of the paper is organized as follows. Section 2 presents the related work in the field of malware detection. Malware detection techniques developed for general-purpose use and those designed specifically for mobile phones are discussed in this section. Section 3 presents six malware programs and their tuning parameters. Section 4 discusses the experimental setup and the data collected for analysis. Experimental results are also presented. Conclusions are presented in Section 5.

2   Related Work

We first give a broad overview of malware detection techniques in general since those techniques share common roots with techniques specifically developed for smartphone malware. In the second part of this section, we discuss work directly related to mobile malware detection.

2.1   Malware Detection Technique

Techniques used for malware detection can be categorized broadly into two categories: anomaly-based detection and signature-based detection. Anomaly-based detection techniques use the knowledge of what constitutes normal behavior to decide the maliciousness of a program. For example, a rule-based system decides whether a program is benign or malicious based on a pre-defined set of rules. Signature-based detection, on the other hand, makes use of static characterization of known malicious software [24]. Detection techniques generally follow three different approaches: static, dynamic, or hybrid analysis. A static approach typically attempts to detect malware without executing the program under inspection, while a dynamic approach attempts to detect malicious behavior during program execution or after program execution. Thus the dynamic approach is often referred to as behavior based analysis. Hybrid techniques leverage the advantages of the previous two approaches by combining them as described in [17].

Lee and Stolfo [11] propose to use association rules and frequent episodes in intrusion detection systems. The association rules and frequent episodes can be collectively referred to as a rule set. Rule sets are created for various security-critical aspects of the target host. These rule sets serve as the knowledge of what activities are considered as normal on the host.

Hofmeyr et al. [8] propose a technique that monitors system call sequences in order to detect maliciousness. Initially profiles representing the normal behavior of a system are developed. The behavior is characterized by sequences of system calls. Hamming distance is used to determine how closely a system call sequence resembles another. Thresholds are used to determine whether a process is anomalous. Okazaki et al. [15] also propose a detection method based on the frequency of system calls.

Static anomaly-based detection techniques use the characteristics of the file structure of the program to identify malicious code. A major advantage of static anomaly-based detection is that it is possible to detect malware without having to execute the program containing the malware. Stolfo et al. [21] describe fileprint (n-gram) analysis as a means for detecting malware. Many other existing anomaly-based malware detection mechanisms use a hybrid approach [17].

2.2   Malware Detection in Mobile Phones

Malware detection techniques developed for use on the computer platform cannot be directly used in a mobile environment due to limited resources and processing capabilities of a mobile phone. Many anomaly-based and signature-based detection techniques, mostly using a dynamic approach, have been proposed to detect malware on mobile phones.

Zhou et al. [23] present permission-based behavioral footprinting and heuristic-based filtering techniques for identifying both known and unknown malware in the Android family. They first filter out Android apps based on the permissions required to grant wrongdoings on the phone, and then define suspicious behaviors of malicious apps and use them to detect zero-day malware.

Yap and Ewe [22] propose a behavior checker solution that detects malicious activities in a mobile system. A proof of concept scenario using a Nokia mobile phone on the Symbian operating system is provided. Bose et al. [1] propose a behavioral detection framework that employs a temporal logic approach to detect malicious activities over time. An efficient representation of malware behaviors is proposed based on a key observation that the logical ordering of an application's actions over time often reveals malicious intent even when each action alone may appear harmless.

Kim et al. propose a detection mechanism based on power signatures [10]. The technique can detect and analyze previously unknown energy depletion threads based on a collection of power signatures. Moreau et al. [13] use artificial Neural Networks (ANNs) to detect anomalous behavior indicating a fraudulent use of the operator services. An example of such behavior is unusually high call rate. Cheng et al. [3] propose SmartSiren, a virus detection and alert system for smartphones. SmartSiren was evaluated by detecting SMS viruses by monitoring the amount of SMSs sent by a single device.

Schmidt et al. [19] extract features representing device state from a smartphone running the Symbian OS. These extracted features are used for anomaly detection to distinguish between normal and abnormal behavior. The processing of the extracted features was performed on a remote server. Dixon and Mishra [5] propose a rootkit and malware detection mechanism for smartphones in which processing is performed on a computer which is connected to the mobile device. An implementation on the Android platform is also provided.

Shabtai et al. propose a behavioral malware detection framework for android devices [20]. The framework includes a host-based malware detection system that continuously monitors various features and events obtained from the mobile devices, and then applies machine learning anomaly detectors to classify the collected data as benign or malicious. They develop four malicious applications on the Android platform and evaluate the proposed framework. They show that such a behavioral malware detection scheme is able to detect unknown malware programs.

3   Malware Setup

We developed six different parameterized malware programs on the Android platform. These malware programs perform privacy intrusion, information theft attacks, and denial of service attacks. By varying the parameters of the malware programs, different profiles of the same malware can be generated. Moreover, the parameters themselves can be randomized (with an expected mean value) rather than being a fixed value. By randomizing the parameters, interesting malware profiles can be prepared for further analysis.

We assume that either through a direct installation or an indirect installation (through the payload of a benign application), the victim's mobile phone is infected with the developed malware. All malware programs were developed and tested on a Samsung Captivate smartphone running on the Android platform. One important thing to know about the Android framework is that applications run in sand boxes (virtual machines), and therefore do not impact other applications in the system. Moreover, all permissions required by the application running on the Android platform (such as Internet access, microphone access) have to be declared, which is prompted to the user when the application is installed. Again the assumption we are leaning on allows us to get away from any practical difficulties of installing the developed malware programs in a furtive manner.

3.1   Call Recorder

The *Call Recorder* malware performs eavesdropping on incoming and outgoing phone calls. Both incoming and outgoing calls are recorded. The recorded file is kept locally on the phone. A configuration option is provided to upload the recorded file to a server. This malware attempts to compromise the privacy of the person using the infected mobile phone. Parameters of this malware include:

- MAX_DURATION—maximum duration a phone call is recorded
- MAX_FILESIZE—maximum size of the recorded file
- NUM_SKIPPED_CALLS—specifying that only every $(n+1)^{th}$ phone call is recorded, where n is the value of NUM_SKIPPED_CALLS.
- INTERVAL_RECORD—specifying the length of every recording (after each sleep) during a phone call
- INTERVAL_SLEEP—specifying the duration in which the malware sleeps (stops recording)
- SHOULD_UPLOAD—uploading the recorded content to a server
- DELETE_LOCAL—deleting the local copy of the recording output

3.2   DoS Malware

Dos Malware performs a Denial of Service (DoS) attack. Upon loading this application, it spawns many threads. Each thread performs a large number of multiplications between two randomly generated numbers. As the number of threads increases, the phone starts becoming unresponsive. When the number of threads spawned is above 200, the phone hangs and has to be rebooted. Thus, this malware paralyzes the device by driving the CPU beyond its limit. Parameters of this malware are:

- MAX_THREADS—number of threads spawned by the malware
- NUM_MULTIPLICATIONS—number of multiplications performed by each thread
- INTERVAL_RESTART—duration after which all the spawned threads are killed before new ones are spawned
- INTERVAL_SLEEP—duration in which the malware sleeps before new threads are spawned

3.3   Mass Uploader

As the name suggests, this malicious application uploads the contents of the memory device of the mobile phone to a server. Thus, it is designed to steal information from the device. Other than uploading, this malware can also download content from a server. When this application is started, it begins the process of uploading and downloading content to/from a server. Parameters of this malware are:

- UPLOAD/DOWNLOAD_BW—the upload/download bandwidth limits for the malicious application

- UPLOAD/DOWNLOAD_INTERVAL—the   duration   after   which   an
  upload/download is performed by the malicious application
- UPLOAD/DOWNLOAD_INTER_LIM—specifying the limit of the amount
  of data sent (burst) in one upload/download attempt

Note that memory private to an application is protected by linux permissions
on Android. Therefore normally other service cannot access it. We assume a
root exploit has enabled the application to elevate to root and steal sensitive
data.

3.4   Smart Recorder

Smart Recorder performs eavesdropping on incoming and outgoing phone calls
from specific phone numbers. These specific phone numbers are read from a
server whenever a phone call is made. The specific phone numbers can be changed
at run time. After recording a phone call, the recorded file is uploaded to a server.
This malware gives the attacker more control over the recorded phone conversations. Specific phone calls can be targeted as the attacker tries to compromise
the privacy of the person using the infected mobile phone. Parameters of this
malware are:

- MAX_DURATION—maximum duration that the phone call is recorded.
- MAX_FILESIZE—maximum size of the recorded file
- INTERVAL_RECORD—length of every recording (after each sleep) during
  a phone call
- INTERVAL_SLEEP—duration in which the malware sleeps (stops recording)

3.5   Spy Camera

Spy Camera can spy on the unsuspecting user by taking snap shots from the
mobile phone camera every few seconds. These snapshots can be uploaded to
a server. Thus, this malware compromises the privacy of the user. When the
malware takes a snap from the mobile phone, the user is not notified in any way
(by sound or other notifications) that a picture has been taken from the mobile
phone camera. Parameters of this malware are:

- SNAP_INTERVAL—duration after which a snap is taken from the camera
  on board the mobile device and stored locally on the phone
- PIC_DSAMPLE_RATIO—specifying the down sample ratio that impacts the
  quality of the pictures taken form the camera
- PIC_COMP_QUALITY—specifying the amount of compression the raw im-
  age is subjected to before saving the image
- SHOULD_UPLOAD—uploading the picture to a server
- DELETE_LOCAL—deleting the local copy of the pictures taken

3.6   Spy Recorder

Spy Recorder can remotely turn on the microphone of the mobile phone and start recording any voice input. The recording is initiated and terminated by a phone call from a specific number that a spy has registered. The microphone is turned on when a phone call from a specific number is made to the victim's phone. The call is automatically rejected afterwards. Similarly, making another phone call from a specific number will turn off the microphone. The entire process is completed without the user's attention. This spyware can be used to record conversations duration important meetings. The recorded file can be stored on the mobile device for later retrieval or can be uploaded to a server. Parameters of this malware are:

- MAX_DURATION—maximum duration that the conversation is recorded
- MAX_FILESIZE—maximum size of the recorded file
- INTERVAL_RECORD—length of every recording (after each sleep) during a phone call
- INTERVAL_SLEEP—duration in which the malware sleeps (stops recording)
- SHOULD_UPLOAD—uploading recorded conversation to a server
- DELETE_LOCAL—deleting the local copy of the recorded conversation

4   Experimental Analysis

We now discuss the experiments in which we investigate the weaknesses of common data mining detection techniques. We present metrics characterizing the behavior of an application and the data sets used in the experiments data. We also discuss the data mining tools used in our experiments and the evaluation metrics.

4.1   Run-Time Behavior Metrics

The run-time behavior of a program can be defined using the statistics of usage of a smartphone. In our experiments, we record the run-time behavior of an application as a set of pre-defined features while the application is running. The collected feature sets are the data source for the data mining techniques later.

The features used in this study characterize the typical behavior of an application. They include various metrics, such as CPU consumption, network traffic, memory usage, and battery (power) consumption. Table 1 lists all the features that need to be recorded in real-time. A light-weight utility program was developed for collecting these features every five seconds and storing them in a database on the mobile phone. This application runs as a service in the background.

Whenever an application is running, the feature extraction utility collects the features from the running application. Once the features are collected, the next step involves training a classifier on the collected feature vectors. After the classifier is trained, dynamic decisions can be made for a running application by

Table 1. List of features extracted

Feature Category	Feature
CPU	cpu_totutil, cpu_totproccount, cpu_userproccount cpu_load_avg1min, cpu_load_avg5min, cpu_load_avg15min
Memory	mem_tot, mem_free, mem_buffer mem_cached, mem_active, mem_inactive mem_dirty, mem_writeback, mem_pageanon mem_mapped, mem_anon, mem_file
Battery	btr_lvl_rem, btr_status, btr_temp btr_volt, btr_lvl_change
Network	net_cell_upd_tx_pkts, net_cell_upd_tx_bytes net_cell_upd_rx_pkts, net_cell_upd_rx_bytes net_wifi_upd_tx_pkts, net_wifi_upd_tx_bytes net_wifi_upd_rx_pkts, net_wifi_upd_rx_bytes

the classifier using the new collection of the features of the running application. Proper actions can be performed after detection, such as notifying the user when the classifier flags an application as malicious.

4.2 Data Sets

We developed a feature extractor application for collecting data. It runs as a background service on the Android phone. Features of the benign and malicious applications are collected from the Android OS while they are running. Using the feature extractor application, we create data sets from 20 benign applications and 6 malware programs over a period of 10 minutes, resulting 120 feature vectors per application. 10 of the benign applications were tools and the other 10 were games. The applications used in our experiments are listed in Table 2.

Features from malware programs were collected when no limits (such as bandwidth, CPU usage) were imposed on the malware programs. This malware profile is referred to as the general profile. In addition, features of malware programs with randomized profiles were also collected. For each malware program, five different randomized profiles were created. Thus the entire data set consists of feature

Table 2. Applications used in the experiments

Malicious	Benign (tools)	Benign (games)
Call Recorder	MusicPlayer	AngryBirds
DoS Malware	Phone	AirControlLite
Mass Uploader	Browser	SmartTacToe
Smart Recorder	Calculator	Snake
Spy Camera	Youtube	Minesweeper
Spy Recorder	Calendar	3DBowling
	Clock	Solitaire
	Contacts	RacingMoto
	Market	DragonFly
	Memo	DragRacing

vectors from 20 benign applications, 6 malware programs, and 30 randomized profiles of malware programs, in the total of 56 applications. Each of the randomized profile varies from the other with respect to the amount of randomization, i.e., the amount of deviation from the mean value of a parameter. For example, Rand$x$ refers to a profile in which the variance from the mean value of the parameters is x.

4.3   Data Mining Tools

Five data mining algorithms in WEKA [7] are selected to build the classifiers in our study. The algorithms are: decision tree (DT), logistic regression (LR), naïve Bayes (NB), artificial neural network (ANN), and support vector machine (SVM). Classifiers typically operate in two phases: training and testing. During the training phase, a classifier is trained on input vectors with proper class values. The classifier then builds a model that generalizes on data in the entire domain. After the training phase, the classifier can be used to make predictions for unseen instances, and it is known as the test phase.

Decision trees are tree-structured predictive models used in many application domains of Machine Learning. Many different types of decision trees exist. In our study we built a decision tree model using the C4.5 algorithm [16].

Logistic Regression is a type of predictive model that can be used when the target variable is a categorical variable with two categories and thus is suitable for our study. During the training phase, the logistic regression algorithm builds a model that is similar to fitting a polynomial to a set of data values. This model is then used to predict class labels during the test phase.

A naïve Bayes classifier is a probabilistic classifier based on the Bayes theorem with strong (naïve) independence assumptions. During the training phase, a model is built which is described by a set of probabilities. An important limitation of this classifier is that input features must all be discretized. It cannot directly handle continuous values. Continuous valued features can be handled using a mixture of Gaussians [12].

A neural network classifier builds a graph model that consists of a set of inter-connecting artificial neurons in its training phase. The neural network [12] exploits patterns in data by modeling complex relationships between the input and the target output. Typically, training neural network models takes more time than that for other models.

A standard support vector machine (SVM) algorithm builds a predictive model by constructing a high-dimensional hyper-plane that discriminates between two categories of data. Given a set of training examples, a hyper-plane that maximizes the distance to the closest training examples on either side is chosen as the optimal separating hyper-plane. The SVM methods have demonstrated great success in many application domains since it was first introduced to the machine learning research community [2].

4.4   Evaluation

The following standard metrics were used to measure the performance of the selected data mining algorithms:

1. True Positive Rate (TPR): Proportion of positive instances (feature vectors of malicious applications) classified correctly.
2. False Positive Rate (FPR): Proportion of negative instances (feature vectors of benign applications) misclassified.
3. Total Accuracy: Proportion of absolutely correctly classified instances, either positive or negative.
4. Receiver Operating Characteristic (ROC) - Area Under Curve (AUC): The ROC curve is a graphical representation of the trade-off between the TPR and FPR for every possible detection cut-off. AUC is the area under this curve.

4.5   Experiments

We now describe in detail two separate experiments we have performed on the datasets of the benign and malicious applications. We also discuss the experimental results and their implications.

Experiment 1. The first experiment evaluates the performance of the five data mining techniques when the adversary does not spread out the attacks in an unpredictable manner. Table 3 shows the 10-fold cross validation results when the data set consisted of the 20 benign applications and 6 malware programs. Four out of the five algorithms work fairly well with a classification accuracy of more than 95%. Naïve Bayes turned out to be a disappointing exception. Violation of the independence assumption in the data may be the main reason that hampers the performance of the naïve Bayes classifier. This experiment implies that statistical analysis in these data mining algorithms is in general applicable when an attack is not disseminated.

Table 3. 10-fold cross validation results on the 20 benign applications and 6 malware programs

Classifier	Accuracy	TPR	FPR	AUC
DT	99.615	0.988	0.001	0.992
LR	99.231	0.988	0.006	0.995
NB	82.692	0.303	0.016	0.788
ANN	99.808	0.996	0.001	0.999
SVM	95.416	0.832	0.009	0.911

Table 4 and Table 5 present the cross validation results when the data set consists of 10 out of the 20 genuine applications and the 6 malware programs. The genuine applications chosen in Table 4 and Table 5 are tools and games, respectively. All classifiers except naïve Bayes perform well with respect to the cross validation results. When games are used in the data sets instead of tools, the classification accuracy is slightly better for four classifiers. This result is consistent with the observation made in a similar experiment in [20].

Table 4. 10-fold cross validation results on 10 benign applications (tools) and 6 malware programs.

Classifier	Accuracy	TPR	FPR	AUC
DT	99.531	0.993	0.003	0.996
LR	99.167	0.989	0.007	0.996
NB	71.563	0.292	0.030	0.849
ANN	99.792	0.997	0.002	1.000
SVM	98.698	0.986	0.0125	0.987

Table 5. 10-fold cross validation results on 10 benign applications (games) and 6 malware programs.

Classifier	Accuracy	TPR	FPR	AUC
DT	99.948	0.999	0.000	0.999
LR	99.271	0.994	0.008	0.995
NB	74.688	0.339	0.008	0.829
ANN	99.948	1.000	0.000	1.000
SVM	93.490	0.840	0.008	0.916

Experiment 2. This experiment demonstrates the performance of the five data mining algorithms on a non-randomized malware profile (marked as General in the following plots) and five randomized malware profiles, namely Rand0, Rand5, Rand25, Rand50, and Rand75. Each randomized profile is generated by varying the parameters of each malware program as described earlier. Through this experiment, we hope to answer the two questions we raised earlier:

- Are data mining algorithms robust against malware with random attacking activities?
- Can simple consolidation of activities over a fixed time frame prepare a higher quality data source for existing algorithms?

To answer the second question, we experimented on datasets in which instances are consolidated by sequentially averaging over n consecutive samples we have extracted in *Experiment 1*, where $n = 1$, 5, 10, and 50, marked as 5sec, 25sec, 50sec, and 250sec interval respectively in the figures. Figures 1 illustrates the average classification accuracy over all five classifiers on 6x6 training-test data pairs. The training data consists of all the genuine applications and the malware programs with each of the six profiles shown as a label on the x-axis. Each cluster in a plot shows the average classification accuracy on six test sets—the same datasets of six different profiles { *General, Rand0, Rand5, Rand25, Rand50, Rand75*}, labeled with specific bar patterns as shown in the figures.

Due to space limitations, we do not show the figures of the classification accuracy of each individual algorithm. In general, none of the classifiers was able to consistently outperform the others as the training and test data varies according to different random profile configurations. The decision tree algorithm performs best on datasets of all six profiles when the training data is extracted

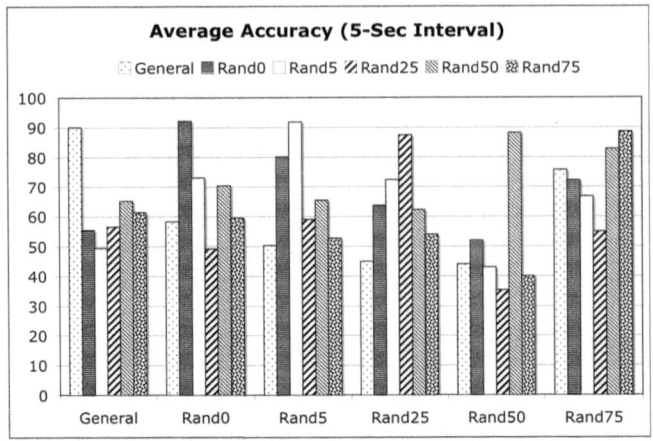

Fig. 1. Average Accuracy over all algorithms on 5sec data profiles

from the Rand75 profile and instances are formed by averaging 50 consecutive samples in each 250-second interval. Logistic regression performs best when the training data comes from the Rand25 profile and all instances in each data set is formed by averaging 10 consecutive samples in each 50-second interval. Naïve Bayes performs best with the training data of the Rand50 profile and instances are formed by averaging 50 consecutive samples. ANN performs best with training data of the Rand75 profile and samples of every 5-second interval. SVM performs best with training data of the Rand75 profile and instances are formed by averaging 50 consecutive samples. As can be seen, behavioral analysis may become very difficult when malware exhibits random behavior. Figure 2 shows the classification accuracy of each algorithm averaged over the 6x6 profile pairs.

Other key observations are: 1.) When the training set consists of benign applications and general malware programs while the test set consists of randomized profiles of malware programs, the classification accuracy is very poor irrespective of the classifiers. In most cases the classification accuracy is below 70%. This has important implications that a behavioral analysis-based malware detection scheme will fail when the training set consists of just general malware programs; 2.) Another observation is that when training set includes a randomized malware profile say, Rand-x and tests are carried out on another randomized malware profile, say Rand-y, classification accuracy is good when x is close to y, in general. For instance, using the decision tree classifier, training on the Rand50 profile and testing on the Rand75 profile gives a classification accuracy of around 86%. Some anomalies exist to this trend such as training on a Rand75 profile using a decision tree classifier.

Figures 3— 5 illustrate performance change as instances are formed by averaging samples in longer durations. As can be observed in Figure 3, the majority of performance change is positive when we average the samples every 25 sec-

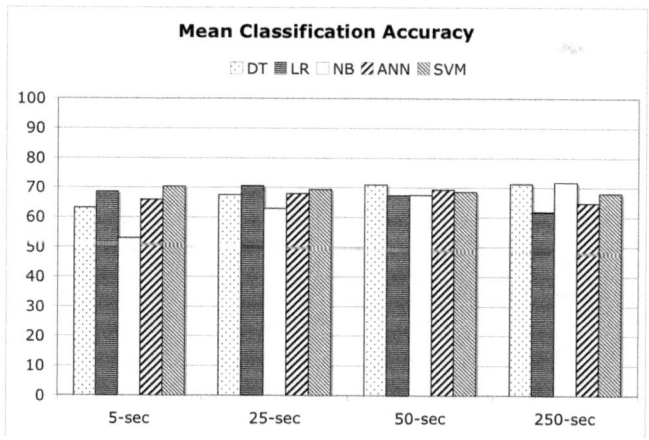

Fig. 2. Mean Accuracy of each algorithm on all data profiles

onds. This implies the mean point of a few consecutive samples serves better as an instance in the data set. Further consolidation using longer durations of 50 seconds and 250 seconds do not appear to improve the performance further except for the Rand50 profile. For individual algorithms, we observe significant performance improvement from the *decision tree* and *naïve Bayes* classifiers. The *Logistic Regression, Artificial Neural Network,* and *Support Vector Machine* classifiers all experienced an initial increase followed by slight decreases in classification accuracy as instances are formed over a longer duration. SVM is the most consistent one among the five classifiers. In general, sample consolidation does seem to improve classifier performance.

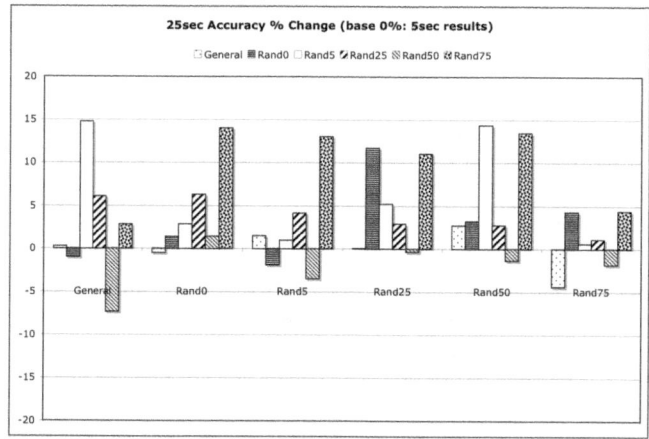

Fig. 3. Average Accuracy% change over all algorithms on 25sec data profiles

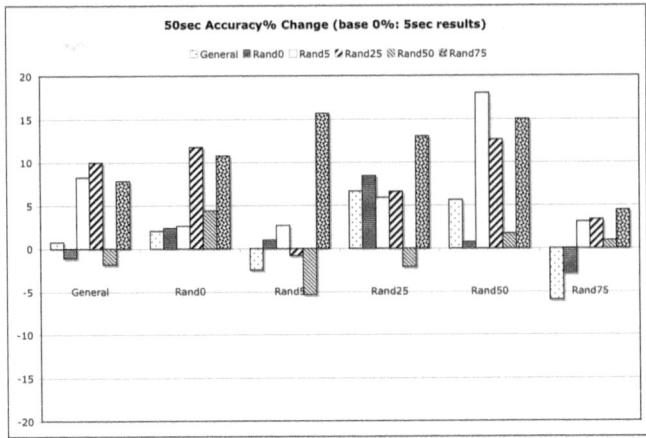

Fig. 4. Average Accuracy% over all algorithms on 50sec data profiles

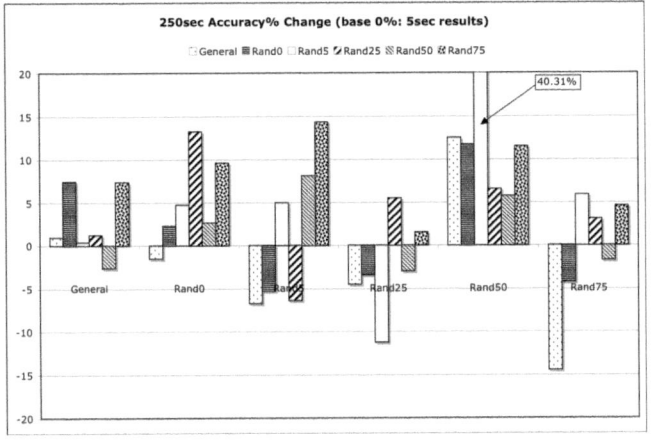

Fig. 5. Average Accuracy% over all algorithms on 250sec data profiles

5   Conclusions

We developed six custom parameterized malware programs on the Android platform. These malware programs can perform a variety of malicious activities on the victim's smartphone. We demonstrate that, although a data mining algorithm can be very successful when the training and the test data follow similar distributions, its performance is unsatisfactory on randomized profiles of the same malware programs. Therefore it is necessary to search for solutions that can better handle random behavioral patterns of malware programs. We also demonstrate that simple consolidation may effectively improve classification performance. In the future, we plan to expand the datasets by developing additional malware applications and including real-world malware, and moreover, search for reliable ways to improve detection in a volatile environment using adversarial classification techniques [9]. We also plan to compare data mining techniques with existing practical techniques for malware detection such as permission-based filtering and behavioral footprint matching methods [23].

Acknowledgments. This work was partially supported by Air Force Office of Scientific Research MURI Grant FA9550-08-1-0265, National Institutes of Health Grant 1R01LM009989, National Science Foundation (NSF) Grant Career-CNS-0845803, and NSF Grants CNS-0964350, CNS-1016343.

References

1. Bose, A., Hu, X., Shin, K.G., Park, T.: Behavioral detection of malware on mobile handsets. In: Proceeding of the 6th International Conference on Mobile Systems, Applications, and Services, MobiSys 2008, pp. 225–238. ACM, New York (2008)
2. Boser, B.E., Guyon, I.M., Vapnik, V.N.: A training algorithm for optimal margin classifiers. In: Proceedings of the 5th Annual ACM Workshop on Computational Learning Theory, pp. 144–152. ACM Press (1992)
3. Cheng, J., Wong, S.H., Yang, H., Lu, S.: Smartsiren: virus detection and alert for smartphones. In: Proceedings of the 5th International Conference on Mobile Systems, Applications and Services, MobiSys 2007, pp. 258–271. ACM, New York (2007)
4. Christodorescu, M., Jhacomputer, S.: Testing malware detectors. In: Proceedings of the 2004 ACM SIGSOFT International Symposium on Software Testing and Analysis, ISSTA 2004, pp. 34–44. ACM Press (2004)
5. Dixon, B., Mishra, S.: On rootkit and malware detection in smartphones. In: 2010 International Conference on Dependable Systems and Networks Workshops (DSN-W), June 28-July 1, pp. 162–163 (2010)
6. Gary McGraw, G.M.: Attacking malicious code: a report to the infosec research council. IEEE Software, 33–41 (2000), magazine article
7. Hall, M., Frank, E., Holmes, G., Pfahringer, B., Reutemann, P., Witten, I.H.: The weka data mining software: an update. SIGKDD Explor. Newsl. 11, 10–18 (2009)
8. Hofmeyr, S.A., Forrest, S., Somayaji, A.: Intrusion detection using sequences of system calls. J. Comput. Secur. 6, 151–180 (1998)
9. Kantarcioglu, M., Xi, B., Clifton, C.: Classifier evaluation and attribute selection against active adversaries. Data Min. Knowl. Discov. 22, 291–335 (2011)
10. Kim, H., Smith, J., Shin, K.G.: Detecting energy-greedy anomalies and mobile malware variants. In: Proceeding of the 6th International Conference on Mobile Systems, Applications, and Services, MobiSys 2008, pp. 239–252. ACM, New York (2008)
11. Lee, W., Stolfo, S.J.: Data mining approaches for intrusion detection. In: Proceedings of the 7th Conference on USENIX Security Symposium, vol. 7, p. 6. USENIX Association, Berkeley (1998)
12. Mitchell, T.M.: Machine Learning. McGraw-Hill, New York (1997)
13. Moreau, Y., Shawe-taylor, P.B.J., Stoermann, C., Ag, S., Vodafone, C.C.: Novel techniques for fraud detection in mobile telecommunication networks. In: ACTS Mobile Summit (1997)

14. Moser, A., Kruegel, C., Kirda, E.: Limits of static analysis for malware detection. In: Twenty-Third Annual Computer Security Applications Conference, ACSAC 2007, pp. 421–430 (2007)
15. Okazaki, Y., Sato, I., Goto, S.: A new intrusion detection method based on process profiling. In: Proceedings of the 2002 Symposium on Applications and the Internet, SAINT 2002, pp. 82–90 (2002)
16. Quinlan, J.R.: C4.5: programs for machine learning. Morgan Kaufmann Publishers Inc., San Francisco (1993)
17. Rabek, J.C., Khazan, R.I., Lewandowski, S.M., Cunningham, R.K.: Detection of injected, dynamically generated, and obfuscated malicious code. In: Proceedings of the 2003 ACM Workshop on Rapid Malcode, WORM 2003, pp. 76–82. ACM, New York (2003)
18. Rieck, K., Holz, T., Willems, C., Düssel, P., Laskov, P.: Learning and Classification of Malware Behavior. In: Zamboni, D. (ed.) DIMVA 2008. LNCS, vol. 5137, pp. 108–125. Springer, Heidelberg (2008)
19. Schmidt, A., Schmidt, H., Clausen, J., Camtepe, A., Albayrak, S.: Enhancing security of linux-based android devices. Image Rochester NY (2008)
20. Shabtai, A., Kanonov, U., Elovici, Y., Glezer, C., Weiss, Y.: "Andromaly": a behavioral malware detection framework for android devices. Journal of Intelligent Information Systems, 1–30 (2011)
21. Stolfo, S.J., Wang, K., Li, W.-J.: Worms 2005 columbia ids lab fileprint analysis for malware detection 1. In: 6th IEEE Information Assurance Workshop (2005)
22. Yap, T.S., Ewe, H.T.: A Mobile Phone Malicious Software Detection Model with Behavior Checker. In: Shimojo, S., Ichii, S., Ling, T.-W., Song, K.-H. (eds.) HSI 2005. LNCS, vol. 3597, pp. 57–65. Springer, Heidelberg (2005)
23. Zhou, Y., Wang, Z., Zhou, W., Jiang, X.: Hey, you, get off of my market: Detecting malicious apps in official and alternative android markets. In: Proceedings of the 19th Network and Distributed System Security Symposium, NDSS 2012 (2012)
24. Zolkipli, M.F., Jantan, A.: Malware behavior analysis: Learning and understanding current malware threats. In: International Conference on Network Applications, Protocols and Services, pp. 218–221 (2010)

Layered Security Architecture for Masquerade Attack Detection

Hamed Saljooghinejad and Wilson Naik Bhukya

Department of Computer and Information Science,
University of Hyderabad, Hyderabad, India
hamed.saljooghinejad@gmail.com,
naikcs@uohyd.ernet.in

Abstract. Masquerade attack refers to an attack that uses a fake identity, to gain unauthorized access to personal computer information through legitimate access identification. Automatic discovery of masqueraders is sometimes undertaken by detecting significant departures from normal user behavior. If a user's normal profile deviates from their original behavior, it could potentially signal an ongoing masquerade attack. In this paper we proposed a new framework to capture data in a comprehensive manner by collecting data in different layers across multiple applications. Our approach generates feature vectors which contain the output gained from analysis across multiple layers such as Window Data, Mouse Data, Keyboard Data, Command Line Data, File Access Data and Authentication Data. We evaluated our approach by several experiments with a significant number of participants. Our experimental results show better detection rates with acceptable false positives which none of the earlier approaches has achieved this level of accuracy so far.

Keywords: Masquerade Detection, Intrusion Detection System, Anomaly Detection, User Profiling.

1   Introduction

Masquerade attacks are ranked second on the top five lists of electronic crimes perpetrated after viruses, worms or other malicious code attacks. The most common information, which can be used to detect masquerade attacks, is contained within the actions a masquerader performs. This set of actions is known as a behavioral profile. Behavior is not something that can be easily stolen. Masquerade detection techniques are based on the premise that when a masquerader attacks the system he will sufficiently deviate from the user's behavior and thus can be recognized using machine learning techniques [9]. In this paper we demonstrate an approach for detecting masqueraders in an efficient manner. We show how multiple layers of user data records together can construct a meaningful behavioral profile in order to have better detection results. The paper is organized as follows: next section introduces the related works on masquerade detection. Then, we describe architecture for the layered approach, which is followed by the experimental designs including data collection, feature extraction, learning and classification phases. Results of several experiments and conclusion are presented in last sections.

N. Cuppens-Boulahia et al. (Eds.): DBSec 2012, LNCS 7371, pp. 255–262, 2012.
© IFIP International Federation for Information Processing 2012

2   Background and Related Work

Masquerade detection was done by observing the command line data and watching a user's behavior and then finding anomalies in their usage. In the category of the command line approaches, the initial activity was done by Schonlau et al. [6] which collected a dataset of Unix command line data from 50 users called the SEA dataset for testing and comparing and was used with different intrusion detection techniques. This research utilized different statistical techniques on the dataset and then compared the results. The NaïveBayes classifier was first applied on Schonlau's dataset by Roy A. Maxion [7]. They extended their previous work by applying the NaïveBayes classifier on Greenberg's enriched command line data [8]. Kim [5] applied SVM with a voting engine on SEA, 1v49 and Greenberg datasets.

However command line data detection mechanisms could not truly detect masquerade attacks in the modern graphical user interface (GUI) systems like windows and variants of Unix like Linux or Mac OSx. Today working with GUI systems is more common and the study of their different aspects is crucial. GUI based data is mostly related to data which comes from the interaction between the user and their mouse and keyboard. Poursa[12] considered Analysis of mouse data which was taken from users who worked with browsers. This approach has its disadvantages mainly because their work focused on the browser data, even though users may be working with applications other than browsers. Later works then focused on comprehensive GUI behavior. For this purpose, [1] developed an active system logger by using C# on a Windows XP System. This logger examined GUI event data captured from users and then useful parameters are extracted to construct the feature vectors. This profiling method, while good, comes with the disadvantage that they only implemented it for Microsoft GUI systems with much of the focus set only on mouse usage. Moreover, their detection rate was not impressive. [2] designed a logger in the KDE environment. The disadvantages of their work were that they collected data only from a single window and did not consider the complexity of profiling multi applications. It was [4] who later showed the advantages of user profiling across multiple applications. Other activity regarding GUI based detections can be found in [3] which does not use mouse movements or keystroke dynamics but rather profiles how the user manipulates the windows, icons and menus. They do not appear to consider time as a factor, which is crucial for intrusion analysis.

3   Our Layered Approach

We propose a new approach for detecting a masquerader in a system. A layered approach is introduced to collect comprehensive data across multiple applications. Fig.1 shows an overview of the architecture which will be discussed in this paper. As it is shown, the training phase is based on collecting the genuine user data from which a behavioral profile will be created for each user and started with an event logger tool which is designed and implemented to collects all events during a session. Then a feature extraction tool is designed to generate the useful features. It constructs feature vectors which are generated from different layers; namely layer-1 as GUI data contains window data, mouse data, keyboard data, layer-2 as command line data, layer-3 as file access data and layer-4 as authentication data. This will then be used in the detection phase. In the detection phase the new user profile will be generated from the new user records and compared to the genuine profile. Any significant deviation between them can be recognized as an attack. This task can be possible with the help of a binary classifier which we can call the detection engine. We took the help of two well known classifiers for this task namely SVM and NaïveBayes. More details regarding the experiment will be addressed in later sections.

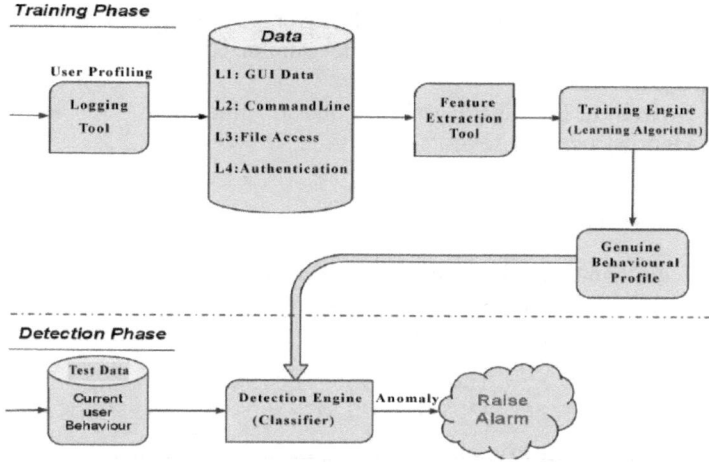

Fig. 1. Architecture of the Layered Approach

4   Data Collection and Calculation of Features

In the absence of a real-world data set for the study of masquerade attacks, we developed our own data collection project. In this section we described the data collection and feature extraction phases. For collecting the data our own logger was developed to collect the information from system. Since we needed data passing through Window Manager, we chose C as the programming language using X library which helped us communicate with Window Manager for capturing the events. The details of each event were logged to a file for processing in the next phase. The collected event details include identification of the window, the name of that window and time of occurrence along with different attributes of that particular event. At the preprocessing phase, by analyzing and observing the impact of each feature on detection rate and false positive rates, it was concluded that the following features were to be considered in the training phase. The features are generated from 6 categories including window data, keyboard data, mouse data as GUI data, command line data, file access data and authentication data as below:

4.1   Window Data(9features)

Users in the GUI environment try to interact with a particular window such as Maximizing, Minimizing, Opening, Closing and switching between windows. In total 9 features were extracted. Here is a window event sample:

```
Event Occurred at: Tue Feb 8 10:03:13 2011
WID=65011715-WName=Openoffice msg:The active window  changed
from previous WID=69206091-Wname=Firefox
```

Window Coordination and Size (4). The average number of times that the user changes the x, y coordinates of a window or width or height of it per user session.
Window Maximize, Minimize (4). The average number of times that the user minimizes, maximizes, restores from the minimized or maximized position of a window per user session.
Window Switching (1). The average number of times that a user switches between different windows per user session.

4.2   Mouse Data(6featurs)

In this category mouse-related user activities like mouse click, mouse right click, mouse movement, etc. were captured for every application. In total 6 features were extracted. Here is a mouse event sample:

```
Event Occurred at: Tue Feb 8 09:58:22 2011
WID=650117—WName=Firefox--msg:Mouse left button clicked
```

Mouse Enter and Exit (2). The average number of times that the mouse enters to a window or exit from a window for each application per user session.
Mouse Clicks (2). The average number of left and right mouse clicks for each application per user session.
Mouse Scroll Up and Down (2). The average number of times that the mouse scrolls up or scroll down for each application per user session.

4.3   Keyboard Data(5features)

All the keyboard events are logged and stored separately for each application. The different keyboard events are key press, key release, and shortcut key (Ctrl, Alt, shift modifiers). In total 5 features were extracted. Here is a keyboard event sample:

```
Event Occurred at: Tue Feb 8 10:00:33 2011
WID=692060--WName=Firefox--msg:Shortcut key Pressed—Ctrl+z
```

Key Pressed (1). The average number of keys pressed in each application per user session.
Shortcut Key Pressed (1). The average number of shortcut keys pressed in each application per user session.
Ctrl, Alt, Shift Modifier (3). The average number of times that Ctrl, Alt or Shift modifiers pressed for each application per user session.

4.4   Command Line Data(2features)

All the commands which are entered in command prompt are logged. They are divided into 2 major categories, Normal and Critical. Normal commands are those which are harmless and can be used with normal user privileges such as ls, clear, cd, etc. Critical commands are those used by administrators and contain any commands that are critical for the system and need root privileges such as su, sudo, etc.

Normal and Critical Commands (2). Number of normal and critical commands that the user enters to the command prompt.

4.5   File Access Data(38 features)

In this layer, users' accesses to crucial files were logged. Attackers usually try to access the victim's data. Analyzing this pattern of behavior can be helpful in order to determine when such an act is being conducted by a masquerader within the system. For example attackers try to access password files to obtain the users' password. Another example can be seen by the attempt to access log files by the attacker in order to delete any evidence they may have left behind. We used the Auditd tool to log the user access to specific files or folders. It is part of the Linux kernel auditing toolkit and captures auditing trails created by the kernel auditing facility from /proc/audit. Using the feature extraction tool, data was preprocessed and a total of 38 features are created. The first 19 features are successful attempts to access specific files or folders and the rest of them are features regarding failed access.

Successful and Unsuccessful Access (38). Number of successful attempts to access particular files and folders include password file, log folder, etc folder, var folder, home directory, proc and bin folder as well as the number of unsuccessful attempts.

4.6   Authentication Data(2features)

Normal users rarely perform actions of administrative domain, so when unusual numbers of authentication occur, it can be suspected as malicious activity. From this 2 features are extracted, the first one is the total number of authentication request and the second one is the total number of failed authentication request.
Successful and Unsuccessful Authentication (2). Number of successful and unsuccessful authentications.

5   Learning and Classification

We collected the data from 16 distinct users who were the students of a particular course in our department. For each user an account was created on a shared computer in our lab and they were given an individual choice of operating conditions and applications. The data collection phase took about two months and the data was collected for multiple sessions, each lasting about 10 minutes. Roughly 3140 minutes of user data was collected and profiled which was significant due to the considerable number of users.

The dataset contains the different number of sessions for different users including 2 users with 21 sessions, 10 users with 20 sessions, 3 users with 19 sessions, and 1 user with 15 sessions as it is shown in table 1. This data was fed into the parsing engine to sanitize and extract features for each application per session. The methodology to train and test the data is explained below.

Table 1. Data Sets Description

Users	No of Sessions (each 10minutes)
A,B	2*21
C,D,E,F,G,H,I,J,K,L	10*20
M,N,O	3*19
P	1*15
Sum	314 (3140 minutes)

- The Collected feature vectors were then divided into different training and test sets. From these, the training sessions were used for the learning phase and for the test phase due to the lack of real masquerade records the other user's records were treated as non genuine or as masquerade records. Obviously if the real masquerade data was available, the anomaly detection would be easier due to the fact that masqueraders tend to put more effort into changing the system state than a normal user would.

- We started with the proportion of 50 percent for the training set vs. 50 percent of the testing set. To show the effect that the number of training sets has in relation to accuracy we repeated the experiments, increasing the proportion of the training set by 5 percent for each case. In total 8 different cases were constructed and it ended up with proportion of 85 percent for the training set and 15 percent for the testing set. Results show a better performance, with an increase in the number of training sets.

- The Collected feature vectors were divided into different training and test sessions per user. We defined a binary classification problem for our data due to the fact that users should be tagged as positive (genuine) or negative (masquerader). SVM [10] [11] and NaïveBayes [14] classifiers are used to classify the data and measure the detection rate, false positive and false negative rates. We used the Weka tool [13] to perform the classification. For improving the performance of classifiers we used the SMOTE filter [15].

6   Results and Discussion

We evaluated the performance of our approach with different experimentations as explained in previous section. Our best result was the detection rate of 97.5% with a false positive rate of 10.18% and false negative rate of 1.5% for SVM. Following sections explain more about the results and comparison between different approaches.

6.1   Detection Rate Evaluation by Different Number of Training and Test Sets

Table 2 shows the average detection rate for different numbers of training sets and test sets using different classifiers. In all cases the detection accuracy of SVM is far better than the other classifier. We can also see an improvement in detection rate by increasing the number of training sets.

Table 2. Average Detection Rate with Different Training and Test sets Size

Classifiers	50% Train 50% Test	55% Train 45% Test	60% Train 40% Test	65% Train 35% Test	70% Train 30% Test	75% Train 25% Test	80% Train 20% Test	85% Train 15% Test
SVM	95.83	96.02	96.59	96.55	96.92	96.89	97.05	97.55
NaïveBayes	91.06	91.2	91.51	91.65	91.61	91.83	91.28	91.34

6.2   ROC Score Evaluation by Different Number of Training and Test Sets

A receiver operating characteristic (ROC), or simply ROC curve, is a graphical plot of the sensitivity, or true positive rate vs. false positive rate with equivalent value between 0 and 1. 1 means we have 100% detection rate with 0 false positive and 0 false negative. ROC analysis provides tools to select the optimal classifier. We calculated the average ROC scores for different numbers of training and test sets. For each user the ROC score is calculated and we then computed the average ROC for each classifier. Table 3 shows the better performance of SVM for our approach. Concretely SVM is famous for its use in this type of problem because it is a maximal-margin classifier as compared to NaïveBayes which is probabilistic and it has been known to be highly effective in text classification and providing better classification results with less training data.

Table 3. Average ROC Score with Different Number of Train and Test sets Size

Classifiers	50% Train 50% Test	55% Train 45% Test	60% Train 40% Test	65% Train 35% Test	70% Train 30% Test	75% Train 25% Test	80% Train 20% Test	85% Train 15% Test
SVM	0.894	0.905	0.914	0.914	0.919	0.924	0.932	0.941
NaïveBayes	0.848	0.851	0.861	0.863	0.868	0.883	0.895	0.902

6.3   Comparison With Other Approaches

To show the advantages of the layered approach, a brief comparison between the detection rate of our approach vs. the previous research works is been shown. As indicted in table 4, layered approach shows better detection rate than other methods either GUI [1] [2] [4] or Command line [5] [7] [8] approaches.

Table 2. Comparison of detection rate results achieved by different approaches

Metrics	Layered Approach	GUI Data			Command Line Data		
		[1]	[2]	[4]	[7]/[8]	[5]	[5]
Number of Users/Dataset used	16	3	8	3	Greenburg	Greenburg	SEA/1vs49
Detection Rate	97.55	91.41	94.88	91.7	70.9/82.1	87.3	80.1/94.8

7   Conclusion

In this paper, we described and developed a new framework for constructing comprehensive feature vectors in different layers for the improvement of masquerade detection accuracy. After capturing the events of a user we went through preprocessing and then extracted relevant features for each application to then construct the relevant feature vectors. We considered six different layers to collect the data, consisting of window data, mouse data, keyboard data, command line data, file access data and authentication data. These feature vectors are classified for masquerade detection using two classifiers namely SVM and NaïveBayes. Our experiments show that this layered approach is well classified using SVM and thus provides better masquerade detection capabilities than single layer approaches. Moreover we observed the impact of increasing the number of training sets which led to an improvement of the detection rate.

References

1. Garg, A., Rahalkar, R., Upadhyaya: Profiling Users in GUI Based Systems for Masquerade Detection. In: Proc. of 2006 IEEE Information Assurance Workshop (IAW), New York (2006)
2. Bhukya, W., Kommuru, S., Negi, A.: Masquerade Detection Based Upon GUI User Profiling in Linux Systems. In: Cervesato, I. (ed.) ASIAN 2007. LNCS, vol. 4846, pp. 228–239. Springer, Heidelberg (2007)
3. Imsand, E.S., Hamilton Jr., J.A.: GUI Usage Analysis for Masquerade Detection. In: Proceedings of 2007 IEEE, Information Assurance Workshop (IAW 2007), New York (2007)
4. Saljooghinejad, H., Rathore, W.N.: Multi Application User Profiling for Masquerade Attack Detection. In: Abraham, A., Lloret Mauri, J., Buford, J.F., Suzuki, J., Thampi, S.M. (eds.) ACC 2011, Part II. CCIS, vol. 191, pp. 676–684. Springer, Heidelberg (2011)
5. Kim, H.S., Cha, S.D.: Empirical evaluation of svm-based masquerade detection using Unix commands. Computers and Security 24(2), 160–168 (2005)
6. Schonlau, M., DuMouchel, W., Ju, W.-H., Karr, A.F., Theus, M., Vardi, Y.: Computer Intrusion: Detecting Masquerades. Statistical Science 16, 58–74 (2001)
7. Maxion, R.A., Townsend, T.N.: Masquerade Detection Using Truncated Command Lines. In: Proceedings of Int. Conf. on Dependable System & Networks (DSN 2002), pp. 219–228 (2002)
8. Maxion, R.A.: Masquerade Detection Using Enriched Command Lines. In: Proceedings of Int. Conference on Dependable Systems and Networks (DSN 2003), CA (June 2003)
9. Lane, T., Brodley, C.E.: An Application of Machine Learning to Anomaly Detection. In: Proceedings of 20th National Information System Security Conf., vol. 1, pp. 366–380 (1997)
10. Joachims, T.: Text Categorization with SVM: Learning with Many Relevant Features. In: Nédellec, C., Rouveirol, C. (eds.) ECML 1998. LNCS, vol. 1398, pp. 137–142. Springer, Heidelberg (1998)
11. Joachims, T.: Transductive Inference for Text Classification Using Support Vector Machines. In: Proc. European Conf. Machine Learning (ECML 1999), June 27-30 (1999)
12. Pusara, M., Brodley, C.: User Re-authentication via mouse movements. In: Proceedings of the ACM Workshop on Visualization and Data Mining for Computer Security, USA (2004)
13. http://www.cs.waikato.ac.nz/ml/weka/
14. McCallum, A., Nigam, K.: A comparison of event models for naivebayes text classification. In: Learning for Text Categorization, AAAI Workshop, Wisconsin, July 27, pp. 41–48 (1998)
15. Chawla, N.V., Hall, L.O., Bowyer, K.W.: SMOTE: Synthetic Minority Oversampling Technique. Journal of Artificial Intelligence Research 16, 321–357 (2002)

k-Anonymity-Based Horizontal Fragmentation to Preserve Privacy in Data Outsourcing

Abbas Taheri Soodejani, Mohammad Ali Hadavi, and Rasool Jalili

Data and Network Security Laboratory,
Department of Computer Engineering, Sharif University of Technology
{a.taheri@cert.,mhadavi@ce.,jalili@}sharif.edu

Abstract. This paper proposes a horizontal fragmentation method to preserve privacy in data outsourcing. The basic idea is to identify sensitive tuples, anonymize them based on a privacy model and store them at the external server. The remaining non-sensitive tuples are also stored at the server side. While our method departs from using encryption, it outsources all the data to the server; the two important goals that existing methods are unable to achieve simultaneously. The main application of the method is for scenarios where encrypting or not outsourcing sensitive data may not guarantee the privacy.

Keywords: Data outsourcing, privacy, horizontal fragmentation, *k*-anonymity.

1 Introduction

In fragmentation-based approach, some data columns are separated from each other to hide their sensitive associations, called vertical fragmentation; and also some sensitive tuples are separated from non-sensitive tuples, called horizontal fragmentation.

From the owner involvement in data storage view, fragmentation-based methods fall into two categories: (1) **Partial-outsourcing** methods [1-3], which store a portion of data at the owner side, (2) **Full-outsourcing** methods [4-6], which completely outsource the data to the external server. Partial-outsourcing methods get involved the owner in data storage and consequently data management, which is largely in contradiction with the goal of outsourcing, i.e., outsourcing data management. On the other hand, full-outsourcing methods outsource all data exploiting data encryption. Consequently, they suffer from the same disadvantages as the encryption-based approach.

Our method is based on horizontal fragmentation. The main idea is to identify and *k*-anonymize the sensitive tuples. Anonymized tuples are stored as a fragment and non-sensitive tuples are stored as a logically separate fragment, both at the server side.

This paper proposes a full-outsourcing method that unlike the existing full-outsourcing methods does not use encryption but anonymization to provide privacy. The method is appropriate for scenarios that even encrypting or even not outsourcing the sensitive data cannot guarantee the privacy. In addition, our method inherits some benefits of the horizontal fragmentation in [2] such as is consistency with database normalization techniques, content-aware fragmentation, introducing a logical formalism for our fragmentation and controlling inference using data dependencies.

N. Cuppens-Boulahia et al. (Eds.): DBSec 2012, LNCS 7371, pp. 263–273, 2012.

The rest of this paper is organized as follows. Section 2 introduces the basic concepts and definitions. Section 3 presents an algorithm for our fragmentation method. Finally, section 4 concludes the paper.

2   Basic Concepts

Our view of relational databases is the formalism of first-order predicate logic with equality. We describe how to formalize some relational concepts with this formalism. **Relational instance:** we view an instance I of a relational database schema $\mathcal{R}$ as a set of expressions of the form $R(a_1, a_2, \dots, a_n)$, where R is an n-ary relation name in $\mathcal{R}$ and the a_i's are constants. Such expressions are called *ground tuples*.

Illness		Treatment		Info		
Name	**Disease**	**Name**	**Medicine**	**Name**	**DoB**	**ZIP**
Andy	Hypertension	Andy	Med A	Andy	1981/01/03	94142
Alice	Obesity	Alice	Med B	Alice	1953/10/07	86342
Bob	Aids	David	Med C	Bob	1952/02/12	79232
David	Heart disease	Bob	Med D	David	1999/01/20	20688
Linda	Cholera	Bob	Med E	Linda	1989/01/03	94139
Sara	Flu	Tom	Med X	Sara	2000/01/20	40496
Tom	Viral disease	Tom	Med Y	Tom	1970/11/01	23567

Fig. 1. An instance (I) of a database schema

EXAMPLE 1. Consider the database instance I in Figure 1 (our running example throughout the paper). The schema is $\mathcal{R} = \{Illness, Treatment, Info\}$, consisting of three relations. The instance I of $\mathcal{R}$ is a set of ground tuples *Illness*(Andy, Hypertension), *Treatment*(Andy, Med A), *Info*(Andy, 1981/01/03 , 94142), and the rest.   □

Database Dependency: Database dependencies are domain-specific declarations reflecting the intended meaning of the stored data in a database. Along with the definition of a schema $\mathcal{R}$, a set of data dependencies $\mathcal{D}$ is defined that consists of tuple-generating dependencies (TGDs) and equality-generating dependencies (EGDs).

Definition 1 (Tuple-generating dependency). *A* tuple-generating dependency (TGD) *is a closed formula of the form* $\forall x \ (\phi(x) \longrightarrow \exists y \ \psi(x, y))$, *where* x *and* y *are vectors of variables;* $\phi(x)$ *is a (possibly empty) conjunction of atomic formulas, all with variables among the variables in* x; $\psi(x, y)$ *is a conjunction of atomic formulas, all with variables among the variables in* x *and* y.

Definition 2 (Equality-generating dependency). *An* equality-generating dependency (EGD) *is a closed formula of the form* $\forall x \ (\phi(x) \longrightarrow \psi(x))$, *where* x *is a vector of variables;* $\phi(x)$ *is a conjunction of atomic formulas, all with variables among the variables in* x; $\psi(x)$ *is a conjunction of formulas of the form* $x = x'$, *where* x *and* x' *are distinct variables in* x.

For a dependency, we call φ the *body* and ψ the *head* of dependency, respectively.

EXAMPLE 2. In our example, set of dependencies $\mathcal{D}$ can contain the following formulas:

d_1: ∀n, d (*Illness*(n, d) ⟶ ∃b, z *Info*(n, b, z))

d_2: ∀n, m (*Treatment*(n, m) ⟶ ∃d *Illness*(n, d))

d_3: ∀n (*Treatment*(n, Med D) ∧ *Treatment*(n, Med E) ⟶ *Illness*(n, Aids))

d_4: ∀n, d, z, d′, z′ (*Info*(n, d, z) ∧ *Info*(n, d′, z′) ⟶ (d = d′ ∧ z = z′))

The TGD d_1 states that if a person is ill, his/her personal information must be available; that is, for each tuple for a person, say 'n', in the relation *Illness*, there must be a tuple for him/her in the relation *Info*. The TGD d_2 states that if a person takes a medicine, there must be a disease s/he contracted. The TGD d_3 states that if someone takes two medicines 'Med D' and 'Med E', s/he is certainly contracted disease 'Aids'. The EGD d_4, states that personal information of each person is unique. □

For a database schema $\mathcal{R}$, a set of dependencies $\mathcal{D}$, and an instance *I* of $\mathcal{R}$, there must not exist a dependency $d \in \mathcal{D}$ that is violated by *I*. The meaning of dependency violation is captured by Definition 3.

Definition 3 (Dependency violation). *Let* I *be a database instance. The dependency* d *is said to be violated by* I *if:*

- *There exists a vector of constants **a** such that the instantiation of the body* φ(**x**)[**a**/**x**] *of variables **x** with constants **a** holds in* I: I ⊨ φ(**x**)[**a**/**x**]
- *but the instantiated head ∃**y** ψ(**x**, **y**) of variables **x** with constants **a** is false in* I: I ⊭ ψ(**x**, **y**)[**a**/**x**] *if* d *is a TGD (similarly,* ψ(**x**) *is false in* I: I ⊭ ψ(**x**)[**a**/**x**] *if* d *is an EGD).*

In other words, a dependency *d* is said to be violated if there exists a set of ground tuples in *I* from which the body of *d* can be instantiated but no tuples exist in *I* that the head of *d* can be fully instantiated from.

If the instance *I* violates some dependencies in $\mathcal{D}$, a procedure called *Chase* [7, 8] is run on *I*. Running the chase on *I*, fixes the violated dependencies. In other words, the result of the chase is an instance, called *chased instance*, that satisfies all dependencies in $\mathcal{D}$. We refer the reader to [2, 9] for more details about the chase procedure.

2.1 Syntax of Privacy Constraints

Data owner's privacy requirements are modeled through privacy constraints:

Definition 4 (Privacy constraint). *Given a database schema* $\mathcal{R} = \{R_1, R_2, \dots, R_m\}$, *a privacy constraint c is a closed formula of the form ∃**x** α(**x**), where **x** is a vector of variables; α(**x**) is a conjunction of positive atomic formulas, all with variables among the variables in **x** and constants from the domains of attributes.*

In the above definition, α(**x**) is a conjunction of partial instantiation of R_i's in $\mathcal{R}$. Definition 4 states that if there exist tuples in the database instance that yield an instantiation of the formula $\alpha(\boldsymbol{x})$, then there is a privacy violation, we say *a sensitive knowledge is disclosed*. The set of tuples that violate a privacy constraint is called the *violation set* denoted by V. Note that we restrict the syntax of privacy constraints to formulas without negation and with only conjunction as logical connective.

EXAMPLE 3. For the database schema $\mathcal{R}$ in our example, the set of privacy constraints, denoted by $\mathcal{C}$, may contain the following formulas:

c_1: ∃n *Illness*(n, Aids)
c_2: ∃d *Illness*(Sara, d)
c_3: ∃n, n′ (*Illness*(n, Cholera)   *Treatment*(n′, Med X)   *Treatment* (n′, Med Y))

where c_1 states that if there exists a tuple with value 'Aids' for attribute *Disease*, a privacy violation occur. In this case, the name of the person who contracted Aids is the sensitive knowledge. Similar interpretation holds for c_2. The constraint c_3 states that if there exist three tuples, one from relation *Illness* with name 'n' and disease 'Cholera' and the other two from relation *Treatment* with the same name 'n′' but one with medicine 'Med X' and the other with medicine 'Med Y', then these tuples together violate the privacy. Here, the sensitive knowledge is a sensitive association between n and $n′$, e.g., $n′$ will be contracted Cholera (say, because Cholera is an infectious disease and all persons in the database live in the same place).   □

2.2   *k*-Anonymity

In some scenarios, encrypting or not outsourcing sensitive data may not guarantee the privacy. For example, consider a healthcare database of patients' records. Let the records of the patients that contracted Aids be sensitive. To provide privacy, one may encrypt or not outsource the records of the patients with disease Aids. Alice has disease Aids, hence she has a record in the database. The attacker Bob knows that Alice has a disease but not exactly which kind of disease. He also knows that records with disease Aids are sensitive. He examines the data and observes that there is no record for Alice in the database. Therefore, he can infer that Alice has disease Aids. There are other scenarios that encrypting or suppressing data do not guarantee the privacy. In these scenarios, we exploit the *k*-anonymity concept to provide privacy.

Our method, inspired from the *k*-anonymity concept [10], aims to provide privacy of degree *k*. To this aim, some fake tuples is produced in a way that for each sensitive knowledge at least *k* - 1 sensitive but fake knowledge can be inferred from database instance. Thus, the probability of a sensitive knowledge being real is equal to or less than $1/k$.

Let *c* be a violated constraint and *V* be its violation set. To provide privacy with degree *k* for *c*, we produce *k* - 1 (possibly overlapping) violation sets, all violating the same constraint   *c*. Thus, for *c*, we have *k* violation sets and the sensitive knowledge of *c* is anonymized with *k* - 1 fake knowledge.

For the sake of simplicity, for each violated constraint we assume that the number of sensitive tuples in its violation set is one (generalizing the concepts to violation sets with more than one tuple is straightforward). Also, we assume that the sensitive knowledge is the value of an attribute *S*, called *sensitive attribute*. With this assumption, the *k*-anonymity concept in our method is captured by the following principle:

Definition 5 (*k*-anonymity principle). *For a violation set of a constraint* c *with sensitive attribute* S, *there must be at least* k - 1 *fake violation sets all instantiated from* c, *and the set of* k *violation sets contains at least* k *distinct values for attribute* S.

EXAMPLE 4. Two tuples *Illness*(Bob, Aids) and *Illness*(Sara, Flu) violate the constraints c_1 and c_2, respectively, from Example 3. Let the attribute *Name* for c_1 and *Disease* for c_2 be the sensitive attributes. We produce tuples *Illness*(Jim, Aids) and *Illness*(Sara, Influenza) to 2-anonymize these two violation sets, respectively.   □

2.3   Syntax of Anonymization Rules

As mentioned in the previous section, the sensitive knowledge inferable from those tuples violating a privacy constraint should be anonymized. For each privacy constraint we define an anonymization rule that states which attribute or combination of attributes (that are sensitive) in which tuples should be anonymized.

Definition 6 (Anonymization rule). *An* anonymization rule *for a privacy constraint* c *of the form* $\exists \boldsymbol{x}\ \alpha(\boldsymbol{x})$, *is a formula of the form* $\exists \boldsymbol{x}\ \alpha(\boldsymbol{x}) \xrightarrow{k-1} \exists \boldsymbol{y}\ \beta(\boldsymbol{y})$, *where* $\boldsymbol{x}$ *and* $\alpha(\boldsymbol{x})$ *are those defined in Definition 4;* $\boldsymbol{y}$ *is a (possibly empty) vector of variables;* $\beta(\boldsymbol{y})$ *is a conjunction of positive formulas all with free variables, bounded variables, and constants; and* k *is the k-anonymity parameter.*

In Definition 6, $\alpha(\boldsymbol{x})$ and $\beta(\boldsymbol{y})$ are called the *body* and *head* of the rule, respectively. According to the head we determine which tuples and, more specifically, which attributes should be anonymized. The above formula states that if there exist tuples that instantiate the body $\alpha(\boldsymbol{x})$, consequently violating a constraint, then there must exist *k* - 1 fake instantiations for the head $\beta(\boldsymbol{y})$. In fact, for a violated constraint, we apply an anonymization rule to achieve the *k*-anonymity principle in Definition 5. In $\beta(\boldsymbol{y})$, constants are values of the attributes, such as name, that their values can identify a sensitive knowledge; free variables represent a set of attributes, such as disease, that their values are sensitive; and bounded variables represent the attributes that have no role in the identification and have no association with the sensitive knowledge.

EXAMPLE 5. The set of anonymization rules, denoted by $\mathcal{A}$, corresponding to the privacy constraints in Example 3 contains the following rules:

r_1: $\exists$n *Illness*(n, Aids) $\xrightarrow{k-1}$ *Illness$_{FS}$*(n, Aids)

r_2: $\exists$d *Illness*(Sara, d) $\xrightarrow{k-1}$ *Illness$_{FS}$*(Sara, d)

r_3: $\exists$n, n$'$ (*Illness*(n, Cholera) *Treatment*(n$'$, Med X) *Treatment*(n$'$, Med Y)) $\xrightarrow{k-1}$ (*Treatment$_{FS}$*(n$'$, Med X) *Treatment$_{FS}$*(n$'$, Med Y))

Relation names above with subscripted *FS* indicate relations in the fragment F_S. The rule r_1 states that to anonymize a tuple that instantiates *Illness*(n, Aids), violating

c_1, k - 1 fake tuples should be produced that are instantiations of $Illness_{FS}(n, Aids)$. In the set of k - 1 fake tuples, the attribute *Disease* should take the value 'Aids' and the attribute *Name* should be anonymized by taking k - 1 distinct values. Similar interpretation holds for r_2. The rule r_3 states that if there exist tuples that instantiate c_3, the sensitive knowledge can be anonymized by generating $2(k$ - 1) fake tuples; k - 1 tuples that instantiate $Treatment_{FS}(n', Med X)$, and k - 1 tuples that instantiate $Treatment_{FS}(n', Med Y)$. For the attribute *Name* in the latter k - 1 tuples, we use the same values used in the former k - 1 tuples.   □

3   Fragmentation

In this section, we first define the requirements of a correct horizontal fragmentation. Then, we introduce an algorithm to produce a correct fragmentation.

3.1   Fragmentation Correctness

A fragmentation is correct if it satisfies three requirements: **completeness**, **non-redundancy**, and **privacy**. The completeness requirement states that we must be able to reconstruct the original database instance from its corresponding fragments. The non-redundancy requirement states that the fragments should not have common tuples and/or attributes (depending on the fragmentation method). The privacy requirement states that the fragmentation must satisfy the privacy constraints so that no sensitive knowledge can be inferred from the fragments. The completeness and privacy are two mandatory requirements that any fragmentation method must satisfy them, while the non-redundancy requirement can be considered optional as it is not applicable to all methods. We define fragmentation correctness in our method as follows:

Definition 7 (Fragmentation correctness). *Let $\mathcal{R}$ be a database schema with a set of dependencies $\mathcal{D}$. Let $\mathcal{C}$ be a set of privacy constraints and* k *be the k-anonymity parameter. Also let $\mathcal{F} = \{F_{NS}, F_S\}$ be a fragmentation for instance I of $\mathcal{R}$, where F_S and F_{NS} are sensitive and non-sensitive fragments, respectively. $\mathcal{F}$ is a correct fragmentation with respect to $\mathcal{C}$ iff both the following conditions hold:*

1. $I \cap (F_{NS} \cup F_S) = I$ (Completeness requirement),
2. $\forall c_i \in \mathcal{C}: (F_{NS} \cup F_S) \cup \mathcal{D} \not\models c_i$, otherwise $P(c_i) \leq 1/k$ (Privacy requirement).

According to Definition 7, a fragmentation is complete if taking the union of the fragments and removing the fake tuples, yields the original instance *I*. Also, it preserves privacy if applying the dependencies, as deduction rules, to the tuples in F_{NS} and F_S, as ground tuples, does not imply a privacy constraint. Otherwise, the probability of the knowledge inferable from the constraint should be equal to or less than $1/k$.

3.2   *k*-Anonymity-Based Horizontal Fragmentation Algorithm

Based on Definition 7, we present an algorithm that produces a correct fragmentation. The general scheme of our method is showed in Figure 2. First, we identify those tuples in I that instantiate some privacy constraints (step 1). These tuples are called *explicitly sensitive tuples* and move them to F_S (step 2). We also identify the tuples in I that do not explicitly violate any constraint but implicitly violate some constraints, i.e., by applying dependencies. These tuples are called *implicitly sensitive tuples* and move them to F_S (step 2). By this step, all explicitly and implicitly sensitive tuples are moved to F_S but there may be some tuples that can be used to gain extra information about the sensitive tuples. Consider the tuples that their existence depends *only* on the existence of other tuples, i.e., produced by only applying dependencies to other tuples. These tuples are called *dangling tuples* and move them to F_S (step 2). After step 2, all remaining tuples in I are non-sensitive (because all sensitive tuples have been moved to F_S) and move them to F_{NS} (step 3). For the explicitly sensitive tuples in step 1, we produce some fake tuples (k - 1 violation sets) to k-anonymize their respective sensitive knowledge and place them in F_S (step 4). In addition to the fake violation sets, some other fake tuples is produced and placed in F_S (step 4), c.f. Section 3.3. Subsequently, we chase F_S to satisfy all dependencies (step 5, not shown in Figure 2).

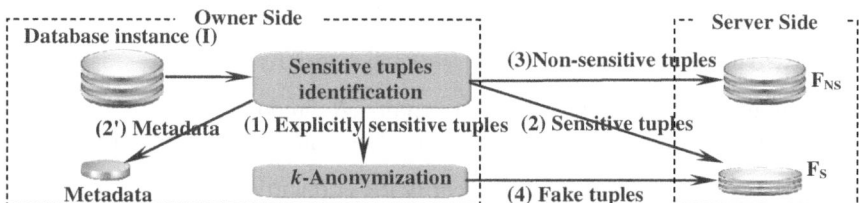

Fig. 2. General scheme of the proposed method

For each sensitive tuple in step 2 a metadata, e.g., a tuple id, is produced and placed in F_S along with that tuple (step 2′). These metadata are also stored at the owner side and used to recognize the real tuples in the future accesses to F_S.

3.3   Cascading Tuples

Consider tuples $t_1, t_2, \ldots, t_n$, where tuple t_n is generated by applying dependency d_{n-1} to tuple t_{n-1} and t_{n-1} itself is generated by applying dependency d_{n-2} to tuple $t_{n-2}, \ldots$, finally tuple t_2 is generated by applying dependency d_1 to tuple t_1. We call this phenomenon *cascading tuples*. Let tuple t_n be explicitly sensitive. Therefore, tuples $t_1, t_2, \ldots, t_{n-1}$ are implicitly sensitive. According to the previous section, t_n is anonymized with some fake tuples and moved to F_S. Tuples $t_1, t_2, \ldots, t_{n-1}$ are also moved to F_S. The attacker observes that $t_1, t_2, \ldots, t_{n-1}$ are not explicitly sensitive but they are in F_S and if dependencies $d_1, d_2, \ldots, d_{n-1}$ be applied consecutively to these tuples, then t_n will be generated. Thus, s/he infers that $t_1, t_2, \ldots, t_{n-1}$ are implicitly sensitive tuples and consequently *real* tuples. Based on this observation, s/he infers that t_n is also a *real* tuple with probability 1 that is greater than value $1/k$ in Definition 7, violating the privacy requirement. To thwart this inference, we propose the following solution:

For a set Σ of TGDs we construct a directed graph, called *dependency graph*, denoted by DG, as follows:

- For each TGD $\forall x \; (\phi(x) \longrightarrow \exists y \; \psi(x, y))$: for the set of atoms $a_1, a_1, \ldots, a_n$ in ϕ, add a new node to DG, if it does not already exist.
- For each TGD $\forall x \; (\phi(x) \longrightarrow \exists y \; \psi(x, y))$: add an edge from the node corresponding to ϕ to the node that contain atom b of ψ.

Let t_n be anonymized with k - 1 fake tuples $f_1, f_2, \ldots, f_{k-1}$. For each f_i ($i=1, \ldots, k$ - 1) we find a node in DG from which f_i can be instantiated. Then, proceed downwards to a leaf node and instantiate with fake values (values from domains of attributes) all the nodes on the path to that leaf. In this way, the probability of the tuples $f_1, f_2, \ldots, f_{k-1}$, and t_n and their respective implicitly sensitive tuples being real, is equal to or less than $1/k$. Therefore, for each explicitly sensitive real tuple we produce some fake tuples in the above way to prevent from inferences about the real tuples.

3.4   *k*-Anonymity-Based Horizontal Fragmentation Algorithm

The proposed algorithm consists of the following steps:

Step 1. Initially, fragments F_S and F_{NS} are empty. A temporary set *fake_tuples* is considered that will contain the fake tuples used to anonymize the sensitive tuples.

Step 2. The purpose of this step is to identify the explicitly sensitive tuples and move them to F_S. For each constraint $\alpha(x)[a/x]$ in C, we identify vectors of constants a such that the instantiation $\alpha(x)[a/x]$ of variables x with constants a holds in the input instance I, i.e., $I \vDash \phi(x)[a/x]$. This implies that I explicitly violates the constraint (as a result, a sensitive knowledge can be inferred from I). To prevent this explicit inference, we remove from I and move to F_S the tuples participating in the instantiation of $\alpha(x)[a/x]$, i.e., the violation set of the constraint. Also, we anonymize these tuples with some fake tuples. The fake tuples are moved to *fake_tuples*.

Step 3. The purpose of this step is to identify the implicitly sensitive and dangling tuples and move them to F_S. By removing the explicitly sensitive tuples from I, some dependencies may be violated by I, helping the attacker to gain extra information about the sensitive tuples. To prevent this, for each TGD $\forall x \; (\phi(x) \longrightarrow \exists y \; \psi(x, y))$ we identify those tuples of constants a, b such that the instantiations $\alpha(x)[a/x]$ and $\psi(x, y)[a/x, b/y]$ of variables x, y with constants a, b hold in the union of the input instance and sensitive fragment. If both of the body and head are only in I or only in F_S, no inference can be done. Otherwise, the following inferences may be possible:

Inference type 1: The TGD body is in I and the TGD head is in F_S:
$$I \vDash \phi(x)[a/x], I \nvDash \psi(x, y)[a/x, b/y], F_S \vDash \psi(x, y)[a/x, b/y]$$

In this case, an attacker can apply the TGD, produce its head, and implicitly infer about some sensitive tuples, i.e., the head of TGD, in F_S. To prevent this implicit inference, all parts of the TGD body must be moved to F_S.

Inference type 2: The TGD body is in F_S and the head is in I:
$$F_S \vDash \phi(x)[a/x] \, , F_S \nvDash \psi(x, y)[a/x, b/y], I \vDash \psi(x, y)[a/x, b/y]$$

In this case, an attacker observes that there exist some tuples that are generated by the TGD in I but the tuples in the TGD body are missing in I (dangling tuples), because they are moved to F_S. To prevent this inference, all parts of the TGD head must be moved to F_S.

Inference type 3: Some parts of the body are in I and some parts of the head are in
$$F_S: \quad I \cup F_S \vDash \phi(x)[a/x], I \cup F_S \vDash \psi(x, y)[a/x, b/y]$$

This case is a general type of the two previous types. In this case, all parts of the body and all parts of the head that are in I must be moved to F_S.

If any tuple is moved to F_S in this step, we will repeat this step until there exists no tuple that can be used for one or more of the above three types of inferences.

Step 4. By this step, there is no tuple in I that can be used for inference. Now, we move the remaining tuples, i.e., non-sensitive tuples, to F_{NS}.

Step 5. We add the fake tuples produced in Step 1, i.e., tuples in *fake_tuples*, to F_S. The newly added tuples may cause F_S violating some dependencies. In this case, we chase F_S with the dependencies and produce a chased fragment satisfying all dependencies. In the newly generated tuples, we replace the "labeled nulls" (see [9]) with some fake values (values from domains of attributes). This prevents the attacker from knowing that they are produced from some fake tuples.

Step 6. Finally, fragmentation $\mathcal{F}$ consisting of F_S and F_{NS} is returned as the result of the fragmentation algorithm.

Illness_$_{NS}$	
Name	**Disease**
Andy	Hypertension
Alice	Obesity
David	Heart disease
Linda	Cholera

Treatment_$_{NS}$	
Name	**Medicine**
Andy	Med A
Alice	Med B
David	Med C

Info_$_{NS}$		
Name	**DoB**	**ZIP**
Andy	1981/01/03	94142
Alice	1953/10/07	86342
David	1999/01/20	20688
Linda	1989/01/03	94139

Illness_$_S$	
Name	**Disease**
Bob	Aids
Jim	Aids
Sara	Influenza
Sara	Flu
Tom	Viral disease
Mary	**Viral disease**

Treatment_$_S$	
Name	**Medicine**
Bob	Med D
Bob	Med E
Tom	Med X
Tom	Med Y
Mary	Med X
Mary	Med Y

Info_$_S$		
Name	**DoB**	**ZIP**
Bob	1952/02/12	79232
Sara	2000/01/20	40496
Tom	1970/11/01	23567
Jim	*1959/02/15*	*78452*
Mary	**1972/04/03**	**63234**

Fig. 3. A fragmentation for instance I in Figure 1

EXAMPLE 6. Consider the instance I, the set of dependencies $\mathcal{D}$, the set of privacy constraints $\mathcal{C}$, and the set of anonymization rules $\mathcal{A}$, from Examples 1-5, respectively. We fragmented I using our algorithm. The result is shown in Figure 3. Attribute values in bold represent the tuples produced for cascading tuples. Attribute values in italic represent the labeled nulls replaced with fake values; the chase applied the dependency d_3 to the fake tuple *Illness*(Jim, Aids) and generated *Info*(Jim, labeled_null$_1$, labeled_null$_2$) and replaced them with values 1959/02/15 and 78452, respectively. □

4   Conclusion

We presented a horizontal fragmentation that outsources all data but do not use encryption to provide privacy. Also, an algorithm presented that produces a correct fragmentation. The main application of the method is for scenarios where even encrypting or suppressing the sensitive data cannot guarantee the privacy.

Several issues remain: our method, to achieve k-anonymity, produces $k - 1$ fake violation sets for each violated constraint. This may result in high storage and bandwidth overheads for large volumes of sensitive data and large values of k. The adoption of other ways, such as a probabilistic approach, to provide privacy with degree k but with less than $k - 1$ fake violation sets, can be investigated as a future work. In this paper, we employed a version of the chase, called *standard chase*, which put some restrictions on the dependencies and constrains, such as being positive and conjunctive. Investigating the applicability of other versions of the chase in the method can be studied further. The anonymity principle of our method has some similarities to the l-diversity privacy model [11]. Investigating other privacy models, such as t-closeness [12], to provide a stronger privacy model for the proposed method can be valuable.

References

1. Ciriani, V., De Capitani di Vimercati, S., Foresti, S., Jajodia, S., Paraboschi, S., Samarati, P.: Enforcing Confidentiality Constraints on Sensitive Databases with Lightweight Trusted Clients. In: Gudes, E., Vaidya, J. (eds.) Data and Applications Security 2009. LNCS, vol. 5645, pp. 225–239. Springer, Heidelberg (2009)
2. Wiese, L.: Horizontal Fragmentation for Data Outsourcing with Formula-Based Confidentiality Constraints. In: Echizen, I., Kunihiro, N., Sasaki, R. (eds.) IWSEC 2010. LNCS, vol. 6434, pp. 101–116. Springer, Heidelberg (2010)
3. Ciriani, V., De Capitani di Vimercati, S., Foresti, S., Jajodia, S., Paraboschi, S., Samarati, P.: Keep a Few: Outsourcing Data While Maintaining Confidentiality. In: Backes, M., Ning, P. (eds.) ESORICS 2009. LNCS, vol. 5789, pp. 440–455. Springer, Heidelberg (2009)
4. Ciriani, V., De Capitani di Vimercati, S., Foresti, S., Jajodia, S., Paraboschi, S., Samarati, P.: Combining fragmentation and encryption to protect privacy in data storage. ACM Transactions on Information and System Security 13, 1–33 (2010)

5. Aggarwal, G., Bawa, M., Ganesan, P., Garcia-molina, H., Kenthapadi, K., Motwani, R., Srivastava, U., Thomas, D., Xu, Y.: Two can keep a secret: A distributed architecture for secure database services. In: Second Biennial Conference on Innovative Data Systems Research, pp. 186–199 (2005)
6. Foresti, S.: Preserving privacy in data outsourcing. Springer-Verlag New York Inc. (2011)
7. Beeri, C., Vardi, M.Y.: A Proof Procedure for Data Dependencies. J. ACM 31, 718–741 (1984)
8. Maier, D., Mendelzon, A.O., Sagiv, Y.: Testing implications of data dependencies. ACM Trans. Database Syst. 4, 455–469 (1979)
9. Fagin, R., Kolaitis, P.G., Miller, R.J., Popa, L.: Data exchange: semantics and query answering. Theoretical Computer Science 336, 89–124 (2005)
10. Samarati, P., Sweeney, L.: Generalizing data to provide anonymity when disclosing information (abstract). In: Proceedings of the Seventeenth ACM SIGACT-SIGMOD-SIGART Symposium on Principles of Database Systems, p. 188. ACM, Seattle (1998)
11. Machanavajjhala, A., Kifer, D., Gehrke, J., Venkitasubramaniam, M.: l-diversity: Privacy beyond k-anonymity. ACM Trans. Knowl. Discov. Data 1, 3 (2007)
12. Ninghui, L., Tiancheng, L., Venkatasubramanian, S.: t-Closeness: Privacy Beyond k-Anonymity and l-Diversity. In: 23rd International Conference on Data Engineering, pp. 106–115 (2007)

Reconstruction Attack through Classifier Analysis

Sébastien Gambs[1,2], Ahmed Gmati[1], and Michel Hurfin[2]

[1] Université de Rennes 1,
Institut de Recherche en Informatique et Systèmes Aléatoires,
Campus de Beaulieu, Avenue du Général Leclerc, 35042 Rennes Cedex, France
{sebastien.gambs,ahmed.gmati}@irisa.fr
[2] Institut National de Recherche en Informatique et en Automatique,
INRIA Rennes - Bretagne Atlantique, France
michel.hurfin@inria.fr

Abstract. In this paper, we introduce a novel inference attack that we coin as the reconstruction attack whose objective is to reconstruct a probabilistic version of the original dataset on which a classifier was learnt from the description of this classifier and possibly some auxiliary information. In a nutshell, the reconstruction attack exploits the structure of the classifier in order to derive a probabilistic version of dataset on which this model has been trained. Moreover, we propose a general framework that can be used to assess the success of a reconstruction attack in terms of a novel distance between the reconstructed and original datasets. In case of multiple releases of classifiers, we also give a strategy that can be used to merge the different reconstructed datasets into a single coherent one that is closer to the original dataset than any of the simple reconstructed datasets. Finally, we give an instantiation of this reconstruction attack on a decision tree classifier that was learnt using the algorithm C4.5 and evaluate experimentally its efficiency. The results of this experimentation demonstrate that the proposed attack is able to reconstruct a significant part of the original dataset, thus highlighting the need to develop new learning algorithms whose output is specifically tailored to mitigate the success of this type of attack.

Keywords: Privacy, Data Mining, Inference Attacks, Decision Trees.

1 Introduction

Data mining and Privacy may seem *a priori* to have two antagonist goals: Data Mining is interested in discovering knowledge hidden within the data whereas Privacy seeks to preserve the confidentiality of personal information. The main challenge is to find how to extract useful knowledge while at the same time preserving the privacy of sensitive information. *Privacy-Preserving Data Mining* (PPDM) [14,1,3] addresses this challenge through the design of data mining algorithms providing privacy guarantees while still ensuring a good level of utility on the output of the learning algorithm.

N. Cuppens-Boulahia et al. (Eds.): DBSec 2012, LNCS 7371, pp. 274–281, 2012.

In this work, we take a first step in this direction by introducing an inference attack that we coined as the *reconstruction attack*. The main objective of this attack is to reconstruct a probabilistic version of the original dataset on which a classifier was learnt from the description of this classifier and possibly some auxiliary information. We propose a general framework that can be used to assess the success of a reconstruction attack in terms of a novel distance between the reconstructed and original datasets. In case of multiple releases of classifiers, we also give a strategy that can be used to merge the different reconstructed datasets into a single one that is closer to the original dataset than any of the simple reconstructed datasets. Finally, we give an instantiation of this reconstruction attack on a decision tree classifier that was learnt using the algorithm C4.5 and evaluate experimentally its efficiency. The results of this experimentation demonstrate that the proposed attack is able to reconstruct a significant part of the original dataset, thus highlighting the need to develop new learning algorithms whose output is specifically tailored to mitigate the success of this type of attack.

The outline of the paper is as follows. First, in Section 2, we describe the notion decision tree that is necessary to understand our paper and we review related work on inference attacks. In Section 3, we introduce the concept of reconstruction attack together with the framework necessary to analyze and reason on the success of this attack. Afterwards, in Section 4, we describe an instantiation of a reconstruction attack on decision tree classifier and evaluate its efficiency on a real dataset. Finally, we conclude in Section 5 by proposing new avenues of research extending the current work.

2   Background and Related Work

Decision tree. Decision tree is a predictive method widely used in data mining for classification tasks, which describes a dataset in the form of a top-down taxonomy [4]. Usually, the input given to a decision tree induction algorithm is a dataset $\mathcal{D}$ composed of n data points, each described by a set of d attributes $\mathcal{A} = \{A_1, A_2, A_3, \ldots, A_d\}$. One of these attributes is a special attribute A_c, called the *class attribute*. The output of the induction algorithm is a rooted tree in which each node is a test on one (or several) attribute(s) partitioning the dataset into two disjoint subsets (*i.e.*, depending on the result of the test, the walk through the tree continues either by following the right or the left branch if the tree is binary). Moreover in a rooted tree, the root is a node without parent and leaves are nodes without children. The decision tree model outputted by the induction algorithm can be used as a classifier $\mathcal{C}$ for the class attribute A_c that can predict the class attribute of a new data point $x_?$ given the description of its non-class attributes. The construction of a decision tree is usually done in a top-down manner by first setting the root to be a test on the attribute that is the most discriminative according to some splitting criterion that varies across different tree induction algorithms. The path from the root to a leaf is unique and it characterizes a group of individuals at the same time by the class at the leaf but also by the path followed. In his seminal work, Ross Quinlan has introduced in 1986 an induction tree algorithm called ID3 (*Iterative Dichotomiser 3*) [11]. Subsequently, Quinlan has developed an extension to this algorithm called C4.5 [12], which incorporates several extensions such as the possibility to handle continuous attributes or missing attribute values. However, both C4.5 and ID3 rely on the notion of *information gain*, which is directly based on the Shannon entropy [13], as a splitting criterion.

Inference attacks. An *inference attack* is a data mining process by which an adversary that has access to some public information or the output of some computation depending on the personal information of individuals (plus possibly some auxiliary information) can deduce private information about these individuals that was not explicitly present in the data and that was normally supposed to be protected. In the context of PPDM, a *classification attack* [8] and *regression attack* [9] working on decision trees were proposed by Li and Sarkar. The main objective of these two attacks is to reconstruct the attribute class of some of the individuals that were present in the dataset on which the decision tree has been trained. This can be seen as a special case of the *reconstruction attack* that we propose in this work that aims at reconstructing not only the class attribute of a data point but also the other attributes. It was also shown by Kifer that the knowledge of the data distribution (which is sometimes public) can help the adversary to cause a privacy breach. More precisely, Kifer has introduced the *deFinetti attack* [5] that aims at building a classifier predicting the sensitive attribute corresponding to a set of non-sensitive attributes. Finally, we refer the reader to [7] for a study evaluating the usefulness of some privacy-preserving techniques for preventing inference attacks.

3   Reconstruction Attack

3.1   Reconstruction Problem

In our setting, the adversary can observe a classifier $\mathcal{C}$ that has been computed by running a learning algorithm on the *original dataset* $\mathcal{D}_{orig}$. The main objective of the adversary is to conduct a *reconstruction attack* that reconstruct a *probabilistic version of this dataset*, called $\mathcal{D}_{rec}$, from the description of the classifier $\mathcal{C}$ (and possibly some auxiliary information Aux) that is as close as possible from the original dataset $\mathcal{D}_{orig}$ according to a distance metric Dist that we defined later.

Definition 1 (Probabilistic dataset). *A probabilistic dataset $\mathcal{D}$ is composed of n data points $\{x_1, \ldots, x_n\}$ such that each datapoint x corresponds to a set of d attributes $\mathcal{A} = \{A_1, A_2, A_3, \ldots, A_d\}$. Each attribute A_k has a domain of definition $\mathcal{V}_k$ that includes all the possible values of this attribute if this attribute is* categorical *or corresponds to an interval $[min, max]$ if the attribute is* numerical. *The knowledge about a particular attribute is modeled by a probability distribution over all the possible values of this attribute. If a particular value of the attribute* *gathers all the probability mass (i.e., its value is perfectly determined), then the attribute is said to be* deterministic. *By extension, a probabilistic dataset whose attributes are all deterministic (i.e., the knowledge about the dataset is perfect) is called a* deterministic dataset.

In this work, we assume that the original dataset $\mathcal{D}_{orig}$ is deterministic in the sense that it contains no uncertainty about the value of a particular attribute and no missing values. From this dataset $\mathcal{D}_{orig}$, a classifier $\mathcal{C}$ is learnt and the adversary will reconstruct a probabilistic dataset $\mathcal{D}_{rec}$. For the sake of simplicity in this paper, we also assume that the adversary has no prior knowledge about some attributes being more likely than others. Therefore, if for a particular attribute A_k of a datapoint x, the adversary hesitates between two different possibles values then both values are equally probable for him (*i.e.*, uniform prior). In the same manner, if the adversary knows that the value of a particular attribute belongs to a restricted interval $[a, b]$ then no value within this interval seems more probable to him than other. Finally, in the situation in which the adversary has absolutely no information about the value of a particular attribute, we use the symbol "$*$" to denote this absence of knowledge (*i.e.*, $A_k = *$ if the adversary has no knowledge about the value of the kth, attribute or even $x = *$ if the adversary has no information at all about a particular data point).

3.2   Evaluating the Quality of the Reconstruction

In order to evaluate the quality of the reconstruction, we define a distance between two datasets that quantifies how close these two datasets are. We assume that the two datasets are of same size and that before the computation of this distance they have been *aligned* in the sense that each data point of one dataset has been paired with one (and only one) data point of the other dataset.

Definition 2 (Distance between probabilistic datasets). *Let D and D' be two probabilistic datasets each containing n data points (i.e., respectively $D = \{x_1, \ldots, x_n\}$ and $D' = \{x'_1, \ldots, x'_n\}$) such that each datapoint x corresponds to a set of d attributes $\mathcal{A} = \{A_1, A_2, A_3, \ldots, A_d\}$. The* distance *between these two datasets* $\mathsf{Dist}(D_1, D_2)$ *is defined as*

$$\mathsf{Dist}(D_1, D_2) = \frac{1}{nd} \sum_{i=1}^{n} \sum_{k=1}^{d} \frac{H(V_k(x'_i) \cup V_k(x_i))}{H(V_k)}, \tag{1}$$

for which $V_k(x'_i) \cup V_k(x_i)$ corresponds to the union of the values for the kth attribute of x_i and x'_i, V_k is all the possible values of this k^{th} attribute (or all the discretized values in case of an interval) and H denotes the Shannon entropy.

Basically, this distance quantifies for each data point and each attribute, the uncertainty that remains about the particular value of an attribute if the two knowledges are pooled together. In particular, this distance is normalized and will be equal to zero *if and only if* it is computed between two copies of the same deterministic dataset (e.g., $\mathsf{Dist}(\mathcal{D}_{orig},\mathcal{D}_{orig})= 0$). On the other extreme, let D_* be a probabilistic dataset in which the adversary is totally ignorant of *all the attributes of all the data points* (i.e., $\forall k$ such that $1 \leq k \leq d$, $\forall i$ such that $1 \leq i \leq n$, $V_k(x_i) = *$). In this situation, $\mathsf{Dist}(D_*,D_*)= 1$ as the distance simplifies to $\mathsf{Dist}(D_*, D_*) = \frac{1}{nd}\sum_{i=1}^{n} \sum_{k=1}^{d} \frac{H(|V_k|)}{H(|V_k|)} = \frac{nd}{nd}$. For a reconstructed dataset $\mathcal{D}_{rec}$, the computation of the distance between this dataset and itself returns a value between 0 and 1 that quantifies the level of uncertainty (or conversely the amount of information) in this dataset.

While Definition 2 is generic enough to quantify the distance between two probabilistic datasets, in our context we will mainly use it to compute the distance between the probabilistic dataset $\mathcal{D}_{rec}$ and the deterministic dataset $\mathcal{D}_{orig}$. More precisely, we will use the value of $\mathsf{Dist}(\mathcal{D}_{rec},\mathcal{D}_{orig})$ as the *measure of success of a reconstruction attack*.

3.3   Continuous Release of Information

In this work, we are also interested in the situation in which a classifier is re-leased on a regular basis (i.e., not just once), after the additions of new data points to the dataset. We now define the notion of *compatibility* between two probabilistic datasets, which is in a sense also a measure of distance between these two datasets.

Definition 3 (Compatibility between probabilistic datasets). *Let D and D' be two probabilistic datasets each containing n data points (i.e., respectively $D = \{x_1, \ldots, x_n\}$ and $D' = \{x'_1, \ldots, x'_n\}$) such that each datapoint x corresponds to a set of d attributes $\mathcal{A} = \{A_1, A_2, A_3, \ldots, A_d\}$. The compatibility between these two datasets $\mathsf{Comp}(D_1, D_2)$ is defined as*

$$\mathsf{Comp}(D_1, D_2) = \frac{1}{nd}\sum_{i=1}^{n} \sum_{k=1}^{d} \frac{H(V_k(x'_i) \cap V_k(x_i))}{H(V_k)}, \tag{2}$$

for which $V_k(x'_i) \cap V_k(x_i)$ corresponds to the intersection of the values for the kth attribute of x_i and x'_i, V_k is all the possible values of this k^{th} attribute (or all the discretized values in case of an interval) and H denotes the Shannon entropy.

Note that the formula of the compatibility between two datasets is the same as for the distance with the exception of using the intersection rather than the union when pooling together two different knowledges about the possible values of the kth attribute of a data point x. The main objective of the compatibility is to measure how much the uncertainty is reduced by combining the two different datasets into one. Suppose for instance, that $\mathcal{D}$ and $\mathcal{D}'$ are respectively the reconstruction obtained by performing a reconstruction attack on two different classifiers $\mathcal{C}$ and $\mathcal{C}'$. .

Merging reconstructed data sets. Let us consider that a first classifier $\mathcal{C}$ has been generated at some point in the past. Later, in the future, new records have been added to the dataset and another classifier $\mathcal{C}'$ is learnt on this updated version of the dataset. We assume that an adversary can observe the two classifiers $\mathcal{C}$ and $\mathcal{C}'$ and apply a reconstruction attack on $\mathcal{C}$ and $\mathcal{C}'$ to build respectively two probabilistic datasets $\mathcal{D}$ and $\mathcal{D}'$. In order to merge these two datasets $\mathcal{D}$ and $\mathcal{D}'$ To merge the two probabilistic datasets $\mathcal{D}$ and $\mathcal{D}'$ into one single probabilistic dataset, denoted $\mathcal{D}_{rec}$, the adversary can adopt the following strategy.

1. Apply the reconstruction attack on the classifiers $\mathcal{C}$ and $\mathcal{C}'$ to obtain respectively the reconstructed datasets $\mathcal{D}$ and $\mathcal{D}'$ (we assume without loss of generality that the size of $\mathcal{D}'$ is smaller or equal to the size of $\mathcal{D}$).
2. Pad $\mathcal{D}'$ with extra data points that have perfect uncertainty (*i.e.*, $x = *$) until the size of $\mathcal{D}'$ is the same as the size of $\mathcal{D}$.
3. Apply the *Hungarian algorithm* [6,10] in order to align $\mathcal{D}$ and $\mathcal{D}'$. Defining an alignment amounts to sort one of the datasets such that the ith record x_i of $\mathcal{D}$ corresponds to the ith record x'_i of $\mathcal{D}'$. The Hungarian method solves the alignment problem and finds the optimal solution that maximizes the compatibility $\mathsf{Comp}(\mathcal{D}, \mathcal{D}')$ between two sets of n data points.
4. Merge $\mathcal{D}$ and $\mathcal{D}'$ into a single reconstructed dataset $\mathcal{D}_{rec}$ by using the alignment computed in the previous step. For each attribute A_k, the domain of definition the merged point is made of the intersection of $V_k(x) \cap V_k(x')$ if this intersection is non-empty and set to the default value $*$ otherwise.
5. Compute the distance metric $\mathsf{Dist}(\mathcal{D}_{rec}, \mathcal{D}_{orig})$ for evaluating the success of the reconstruction attack.

4   Reconstruction Attack on Decision Tree

Let $\mathcal{C}$ be a classifier that has been computed by running a C4.5 algorithm on the original dataset $\mathcal{D}_{orig}$. This decision tree classifier is the input of our reconstruction algorithm. For each branch of the tree, the sequence of tests composing this branch form the description of probabilistic data points that will be reconstructed out of this branch. The reconstruction algorithms follows a branch either in a top-down manner and refines progressively the domain of definition $V_k(x)$ for each attribute A_k of a probabilistic data point x until the leaf is reached. As we have run a version of C4.5 in which each leaf also contains the number of data points for each class, we can add the corresponding number of probabilistic data points of each class with the refined description to the probabilistic dataset $\mathcal{D}$ under construction. The algorithm explores all the branches of tree to reconstruct the whole probabilistic dataset $\mathcal{D}$.

To evaluate the success of this reconstruction attack on a decision tree classifier, we have run an experiment on the "Adult" dataset from UCI Machine Learning Repository [2]. This dataset is composed of $d = 14$ attributes such as age or marital status, including the income attribute, which is either "> 50K" or "<= 50" and that we have used as class attribute during the construction of the decision tree. To construct the C4.5 classifiers, we have used the WEKA software [15]. Moreover, for each attribute A_k, we have computed its domain of

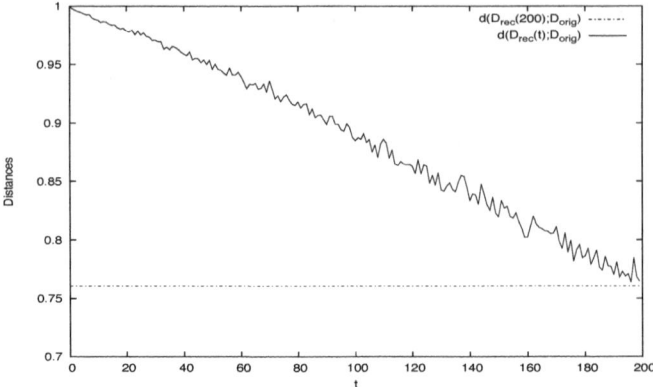

Fig. 1. Distance between a reconstructed dataset D' and $\mathcal{D}_{orig}$ when the reconstruction attack is run on a decision tree learnt on a number of data points varying between 1 and 200 (the size of $\mathcal{D}_{orig}$)

definition V_k, which is defined by the finite set of possible values. For continuous attribute such as age, the extremal values observed the complete database were used to determine the minimal and maximal possible values. The experimentations were performed in a random subset of 200 records of the original "Adult" dataset and not on the complete database. We denote this subset of 200 records by $\mathcal{D}_{orig}$ and the reconstruction attack aims at reconstructing this particular dataset. The metric Dist is used to evaluate the success of the reconstruction attack. The smaller this distance, the more accurate the reconstruction is. Figure 1 displays the result of our experiments obtained when computing the distance between a reconstructed learnt on a dataset whose number of points varies between 1 to 200. Not surprisingly, we can see from these results that a reconstruction attack performed on a classifier that contains more information about the original dataset leads to a reconstruction that is more accurate (*i.e.*, closer to the original dataset).

We have also conducted several experiments in which two reconstructed datasets learnt from classifiers released at different time where merged using the algorithm described in Section 3.3. Our main finding is that it is possible to obtain a limited gain in the order of 0.01 or 0.02 when combining the two datasets (we leave the details of these experiments for the full version of the paper due to lack of space).

5   Conclusion

In this paper, we have introduced the concept of reconstruction attack whose aim is to reconstruct a probabilistic version of the original dataset from the description of a classifier. We have also proposed a novel distance based on information entropy that measures the closeness between the original and the reconstructed datasets and can be used to assess the success of the attack. Moreover, we have design a specific instance of a reconstruction attack and demonstrate his efficiency on a real dataset coming from the UCI repository. The current work is only the first step towards the development of a framework for evaluating the impact of releasing a classifier for the privacy of the dataset. As future works, we want to design reconstruction attack for other types of classifiers such as neural networks or ensemble methods such as boosting. We also want to develop a method for merging several reconstructed datasets into a single coherent one in case of multiple releases.

References

1. Aggarwal, C.C., Yu, P.S. (eds.): Privacy-Preserving Data Mining - Models and Algorithms. Advances in Database Systems, vol. 34. Springer (2008)
2. Asuncion, A., Frank, A.: UCI machine learning repository (2010), http://archive.ics.uci.edu/ml
3. Bertino, E., Lin, D., Jiang, W.: A survey of quantification of privacy preserving data mining algorithms. In: Aggarwal, Yu (eds.) [1], pp. 183–205
4. Duda, R.O., Hart, P.E., Stork, D.G.: Pattern Classification, 2nd edn. Wiley, New York (2001)
5. Kifer, D.: Attacks on privacy and definetti's theorem. In: Çetintemel, U., Zdonik, S.B., Kossmann, D., Tatbul, N. (eds.) SIGMOD Conf., pp. 127–138. ACM (2009)
6. Kuhn, H.W.: The hungarian method for the assignment problem. Naval Research Logistics Quarterly 2, 83–97 (1955)
7. Li, C., Shirani-Mehr, H., Yang, X.: Protecting Individual Information Against Inference Attacks in Data Publishing. In: Kotagiri, R., Radha Krishna, P., Mohania, M., Nantajeewarawat, E. (eds.) DASFAA 2007. LNCS, vol. 4443, pp. 422–433. Springer, Heidelberg (2007)
8. Li, X.B., Sarkar, S.: Against classification attacks: A decision tree pruning approach to privacy protection in data mining. Operations Research 57(6), 1496–1509 (2009)
9. Li, X.B., Sarkar, S.: Protecting privacy against regression attacks in predictive data mining. In: Galletta, D.F., Liang, T.P. (eds.) ICIS, pp. 1–15. Association for Information Systems (2011)
10. Munkres, J.: Algorithms for the assignment and transportation problems. Journal of the Society for Industrial and Applied Mathematics 5, 32–38 (1957)
11. Quinlan, J.R.: Induction of decision trees. Machine Learning 1(1), 81–106 (1986)
12. Quinlan, J.R.: C4.5: Programs for Machine Learning. Morgan Kaufmann (1993)
13. Shannon, C.E.: A mathematical theory of communication. The Bell Systems Technical Journal 27, 379–423, 623–656 (1948)
14. Verykios, V.S., Bertino, E., Fovino, I.N., Provenza, L.P., Saygin, Y., Theodoridis, Y.: State-of-the-art in privacy preserving data mining. SIGMOD Record 33(1), 50–57 (2004)
15. Witten, I.H., Frank, E.: Data Mining: Practical Machine Learning Tools and Techniques with Java Implementations. Morgan Kaufmann (1999)

Distributed Data Federation without Disclosure of User Existence

Takao Takenouchi[1,2], Takahiro Kawamura[2], and Akihiko Ohsuga[2]

[1] Knowledge Discovery Research Laboratories, NEC Corporation
takenouchi@bu.jp.nec.com
[2] Graduate School of Information Systems,
The University of Electro-Communications

Abstract. Service providers collect user's personal information relevant to their businesses. Personal information stored by different service providers is expected to be combined to make new services. However, specific user records risk being identified from the combined personal information, and the user's sensitive information may be revealed. Also, personal information collected by a service provider must not be disclosed to other service providers because of security issues. Thus, several researchers have been investigating distributed anonymization protocols, which combine the personal information stored by the providers and sanitize it to ensure an anonymity policy with minimum disclosure. However, when providers have different sets of the users, there is a problem that the existence of users in either service provider may be revealed. This paper introduces a new notion, *δ-max-site-presence*, which indicates the probability of the existence of users being revealed in a distributed environment and a new distributed anonymization protocol for hiding the existence of users. Our evaluation results show that the proposed protocol can anonymize users in accordance with the policy of hiding their existence and user anonymity without too much information loss.

Keywords: Distributed Anonymization, Privacy Preserving Data Publishing, k-anonymity.

1   Introduction

Service providers have recently started providing applications on cloud platforms, on which they collect vast amounts of users' personal information for their businesses. Personal information stored by different service providers is expected to be combined to make new services. For example, we expect a usage case in which an online video service (Provider A) and a finance company (Provider B) cooperate. In this case, Provider A has the information of the video titles rented by its customers and the times they watch them, and Provider B has the information of the customers' incomes. They then combine the three types of information and send it all to an advertising agency (Provider C) that performs segmentation analyses for targeting advertisements. In this case, Provider C will find three clusters: "daytime viewers", "high-income nighttime viewers",

N. Cuppens-Boulahia et al. (Eds.): DBSec 2012, LNCS 7371, pp. 282–297, 2012.
© IFIP International Federation for Information Processing 2012

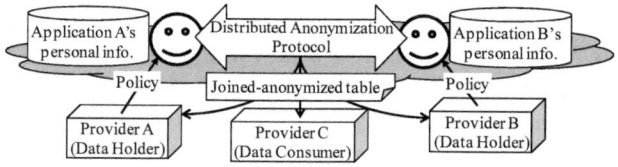

Fig. 1. Agents and distributed anonymization protocol for data federation

and "low-income nighttime viewers". However, if Provider C cannot obtain the income information, they will find only two clusters: "daytime viewers" and "nighttime viewers".

Although combining the personal information will generate new beneficial information, it may cause two problems. First, a specific user record could be identified from the combination of personal information, and the user's sensitive information may be revealed. Second, personal information stored by a service provider must not be disclosed to other service providers, because the personal information on cloud platforms is managed separately and protected by a security policy. Also, because the personal information is an asset of the service providers, the providers need to combine it with minimum disclosure.

Thus, several researchers have been investigating distributed anonymization protocols [4], which combine the personal information stored by different providers and sanitize it to ensure an anonymity policy with minimum disclosure. We expect that distributed anonymization protocols will be used for data federations between the applications of the service providers on a cloud platform. By using distributed anonymization protocols, the providers can create a joined-anonymized table, which is a combination of the tables of Providers A and B that is sanitized while ensuring the anonymity policy, and send it to other providers (Fig. 1).

However, when the service providers have different sets of users, there is a problem that the existence of a user in the data of each service provider may be revealed. The information of the existence of a user is sensitive information for that user. For example, if a user's information is stored by a financial service, then some people may guess that the user has debts.

In this paper, we consider the problem of revealing the existence of a user on each provider in distributed anonymization. This problem occurs when the attributes of the personal information are vertically partitioned, which means that the providers have different attributes of the user, and the providers have different sets of users. This paper consists of two main contributions. First, this paper propose a new notion, named δ-max-$site$-$presence$, which indicates the probability of the existence of the users being revealed in a distributed environment. This is an extension of an existing notion, named δ-$presence$, which indicates the probability of the existence of a user being revealed in a local environment. Second, this paper introduces a new distributed anonymization protocol that hides the existence of a user in each database.

The rest of this paper is organized as follows. In Section 2, we discuss related works. Section 3 describes distributed anonymization and the problems of hiding the users' existence. Section 4 proposes the new notion, named δ-*max-site-presence*. In Section 5, we propose a new anonymization protocol that hides the existence. Next, we evaluate the utility of the proposed protocol in Section 6 and evaluate the security in Section 7. Then, in Section 8, we conclude the paper in Section 8.

2   Related Works

Anonymization is a method to sanitize personal information to prevent a specific user being identified. In this paper, we named a set of attributes that may uniquely identify a user *quasi-identifier*. Also, we named the attributes that users would not like to be revealed *sensitive attributes*. There is a well-known notion, named *k-anonymity* [12,13], which indicates the anonymity of a user in a table. If a table satisfies the condition that the number of records identified by each value of the *quasi-identifier* is at least k, then the table satisfies *k-anonymity*.

Distributed anonymization is a method to join and anonymize the table stored by some providers [4]. Distributed anonymization can be categorized into two types: vertically partitioned data and horizontally partitioned data. Vertically partitioned data means that each provider has different types of attributes from the other provider. On the other hand, horizontally partitioned data means that each provider has different users from the other provider.

There are some distributed anonymization methods in vertically partitioned data [10,14,6]. Mohammed et al. [10,14] used the top-down approach and some secure computation protocols [9,15] in order to join the tables in multiple providers. The top-down approach is an algorithm to *specialize* a value *qid* of a *quasi-identifier* in the tables step by step. At the start, all *qid*s in the tables are generalized as top level, such as "*". The term *specialize* here means dividing a group identified by a *qid* into two groups on a *division point*. After the division, the set of the identifiers of the users of the two divided groups is sent to the other provider. Then, the providers continue dividing the tables as long as the tables satisfy *k-anonymity*. Finally, the providers join the divided tables in order to create the joined-anonymized table. Some secure computation protocols are used in order to calculate a heuristic function which decides the division point. By using secure computation protocols, the providers can calculate the heuristic function without sharing of the providers' local data. Jiang and Clifton [6] used the bottom-up approach to join the tables in multiple providers. In this approach, all providers anonymize their tables locally and join with each other by checking anonymity securely by using cryptographic technology.

Jurczyk and Xiong [7] proposed horizontal distributed anonymization. They mentioned a new privacy problem in which the location of a data holder is revealed by differences in providers' data types. They used some secure computations and extended Mondrian[8], which is a well-known top-down approach algorithm. They also proposed a new notion, named ℓ-*site-diversity*.

Also, there is a method to hide the existence of a user in a local environment. Nergiz et al. [11] proposed δ-*presence* and the algorithm in order to satisfy this. However, this algorithm is not for a distributed environment. Therefore, we propose a new notion, named δ-*max-site-presence*, and a new distributed anonymization protocol for a distributed environment.

3   Problem of Distributed Anonymization Protocol

3.1   Distributed Anonymization

In this paper, we assumed that Providers A and B have vertically partitioned tables T_A and T_B and create joined-anonymized table T^*:

$$T_A(UID, QID_A)\,,\quad T_B(UID, QID_B, SA)\,,\quad T^*(QID_A, QID_B, SA)$$

where QID_A, QID_B is the *quasi-identifier*, SA is a *sensitive attribute*, and UID is a common identifier of a user in Providers A and B. T_A and T_B are joined by UID and anonymized into table T^*. Also, UID is a unique identifier of the record in T_A and T_B. In distributed anonymization, a joined-anonymized table should be created with minimum disclosure of personal information. This is because the providers in real-world businesses have difficulty fully trusting each other.

3.2   Problem of Revealing the Existence of a User

Existing distributed anonymization protocols assume that the sets of the users in Providers A and B are the same [4,10,7,6]. In other words, each user who exists in Provider A also exists in Provider B. However, in the future, many providers are expected to federate with each other. Therefore, it is necessary to support the case in which the sets of the users in Providers A and B are not the same. However, if we use existing distributed anonymization protocols in these cases, the *joined-anonymized table problem* and the *UID sending problem* will occur.

In the *joined-anonymized table problem*, a provider can infer the existence of a user in the other provider by comparing the table of the first provider and the joined-anonymized table. For example, let us assume a case in which Provider A has Table 1(a) as T_A, Provider B has Table 1(b) as T_B, and the joined-anonymized table T^* is divided on *50K* of *income* like Table 1(c). In this case, the number of records where *income* $<50K$ in T^* (Table 1(c)) is two, and also the number of records where *income* $<50K$ in T_A is two: User 1 and User 2. Thus, Provider A can infer that Users 1 and 2 must exist in T^* (Table 1(c)). Furthermore, because T^* contains a common user who exists both on T_A and T_B, Provider A can certainly infer that Users 1 and 2 also exist in Provider B. In contrast, if T^* is divided on *60K* like Table 1(d), Provider A can infer only that two of Users 1, 2, and 3 exist in Provider B.

In the *UID sending problem*, the existence of a user in a provider is revealed to the other provider when *UID*s are sent. For example, if a provider sends the *UID*s that exist in the provider to the other provider, then the receiving provider can easily infer that the received *UID*s must exist in the sending provider. Also, if providers calculate common users beforehand, the existence of a user is revealed.

Table 1. Example of the joined-anonymized table problem

(a) Provider A (T_A)		(b) Provider B (T_B)			(c) Joined-anonymized table (T^*) with disclosure of the existence			(d) Joined-anonymized table (T^*) without disclosure of the existence		
UID	income	UID	time	title	income	time	title	income	time	title
User 1	30K	User 1	16:00	X movie	<50K	16:00-	X movie	<60K	16:00-	X movie
User 2	40K	User 2	17:00	Y sports	<50K	16:00-	Y sports	<60K	16:00-	Y sports
User 3	55K	User 4	17:30	X movie	50K<=	-15:59	X movie	60K<=	-15:59	X movie
User 5	60K	User 5	16:30	Y sports	50K<=	-15:59	Y sports	60K<=	-15:59	Y sports
User 6	65K	User 6	15:00	X movie						
User 8	70K	User 7	12:00	Y sports						
		User 9	14:00	Y sports						
		User 10	14:30	X movie						

4 Proposed Notion: δ-max-site-presence

This section proposes a new notion that indicates the probability of the disclosure of the existence of a user in order to solve the *joined-anonymized table problem*. There is a existing notion, named δ-*presence*[11], which indicates the probability of the disclosure of the existence of a user by comparing two tables in a centralized environment. Thus, by applying δ-*presence* to a distributed environment, we propose a new notion, named δ-*max-site-presence*.

Let us assume that there are Tables T_1 and T_2, which is subset of records of T_1. Let $|T|$ be the number of records in Table T. Nergiz et al. [11] define the probability that a record in T_1 also exists in T_2 is $\frac{|T_2|}{|T_1|}$. We apply this definition of the probability of the existence of a user being disclosed to a distributed environment. For example, we assumed that Provider A's T_A is Table 1(a), Provider B's Table T_B is Table 1(b), and the joined-anonymized table is Table 1(d). In this case, the number of records of *income* <*60K* in T_A is three: Users 1, 2, and 3. Moreover, the number of records of *income* <*60K* in T^*(Table 1(d)) is two. According to the definition mentioned above, the probability that Users 1, 2, and 3 in T_A exist in T_B is $\frac{2}{3}$. Furthermore, because T^* is created from a common user in T_A and T_B, the probability that Users 1, 2, and 3 in Provider A also exist in Provider B is $\frac{2}{3}$.

We define a new notion that indicates the probability of the existence of a user in a provider as δ-*max-site-presence*.

Definition 1. *(δ-max-site-presence) Let T_A and T_B be the tables stored by Providers A and B. Note that T_A and T_B have different attributes. Also, let T^* be the joined-anonymized table, T_n^* be the table that consists of attributes of Provider $n \in \{A, B\}$, and values$_n$ be the set of values of any set of attributes in T_n^*. We represent $T[v]$ as the table that consists of records that are identified by value v in the table T. Then, we define that T^* satisfies δ-max-site-presence if the probability of the existence of all users in other providers being disclosed from the view of Providers A and B is less than δ as follows:*

$$\frac{|T^*[v_{n,i}]|}{|T_n[v_{n,i}]|} \leq \delta \quad \forall v_{n,i} \in values_n \quad \forall n \in \{A, B\} \tag{1}$$

For example, in Table 1(d), the $values_A$ is the set of { $<60K$, $60K\leq$ }. In this example, let us consider $<60K$. When $v_{A,i}$ is $<60K$, the number of records of $income <60K$ in Table 1(d) is two. This means $|T^*[v_{A,i}]| = 2$. Also, the number of records of $income <60K$ in Table 1(a) is three. This means $|T_A[v_{A,i}]| = 3$. Also, let us consider $values_B$. The $values_B$ is the set of {$(16:00\text{-},X$ $movie),(16:00\text{-},Y$ $sports),(\text{-}15:59,X$ $movie),(\text{-}15:59,Y$ $sports)$}. When $v_{B,i}$ is $(16:00\text{-},X$ $movie)$, the number of records of $(16:00\text{-},X$ $movie)$ in Table 1(d) is one. This means $|T^*[v_{B,i}]| = 1$. Also, the number of records of $(16:00,X$ $movie)$ in Table 1(a) is two. This means $|T_B[v_{B,i}]| = 2$. Thus, Table 1(d) satisfies $\frac{2}{3}$-max-site-presence.

5   Proposed Protocol: Dummy User Protocol

This section proposes a new distributed anonymization protocol named *Dummy user protocol*, which does not disclose the existence of a user. The proposed protocol is designed to create T^* that has as much detailed information as possible and satisfies the following requirements:

Requirement 1. T^* must satisfy k-anonymity and δ-max-site-presence.
Requirement 2. The disclosed information that is more detailed than T^* should be minimized.

In addition, every provider behaves semi-honestly. In this trust model, the provider who has received the messages of the protocols will analyze them to gain new knowledge, but the provider must follow the protocol. We assume that the providers partly trust each other in the real-world business. Therefore, we think this trust model is reasonable.

5.1   Dummy User Protocol

To solve the *UID sending problem* (Section 3.2), we introduce *dummy user* and propose *Dummy user protocol*. In this protocol, a provider treats the users who do not exist in the provider as if they actually did. In addition, we call the users who really exist in the provider *existing user*s. By using the dummy user, it will be difficult for a provider to distinguish whether the received *UID* is of an existing or a dummy user.

As the same as the work of Jurczyk and Xiong [7], Dummy user protocol is based on Mondrian [8], which is widely used as a top-down approach algorithm, and consists of two sub protocols: *Dividing protocol* and *Joining protocol*. First, Providers A and B execute Dividing protocol to create an internal anonymized table $T_n^*(n \in \{A, B\})$ locally (Fig. 2(a)). After that, Provider C executes Joining protocol to obtain the joined-anonymized table T^* by joining the T_n^* that Provider n has. Table 2 shows an example of T_n^* and T^*. The combination of income and watching time is a *quasi-identifier* and the title is a *sensitive attribute*.

Dividing Protocol: Step 1 At the first step, Providers A and B assign dummy users in their tables. We assume that each provider knows the parent population

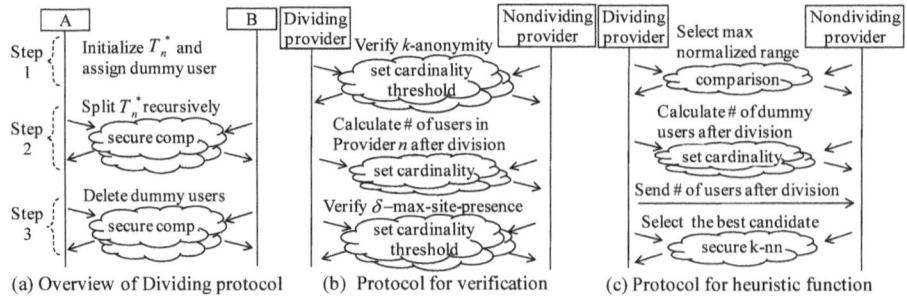

Fig. 2. Protocol sequences of Dividing protocol

---

function split(U_p: a set of *UIDs* including dummy user's *UIDs*)
1:    update the dummy values of all dummy users in U_p
2:    *point* $\Leftarrow$ decide division point by the heuristic function
3:    verify *k-anonymity* and *δ-max-site-presence* of the dividing on the *point*
4:    **if** the verification is failed **then**
5:       give up the split of U_p
6:    **endif**
7:    **if** the *point* is on my side **then**
8:       divide my T_n^* on the *point*, and send the *UIDs* to the nondividing provider
9:    **else**
10:    receive the *UIDs* from the dividing provider, and divide my T_n^*
11:    **endif**
12:    $U_{hi}, U_{low} \Leftarrow$ *UIDs* after the dividing, call split(U_{hi}), split(U_{low}) recursively

Fig. 3. Algorithm of Step 2 of Dividing protocol

U beforehand. Here, let U_A be the set of the existing users in Provider A, U_B be the set of the existing users in Provider B, and U_O be the set of the users who do not exist in Providers A or B. Then, U can represent that $U = U_A \cup U_B \cup U_O$ ($U_O \neq \phi, U_A \cap U_B \neq \phi$). This assumption is satisfied if Providers A and B use the same centralized authentication systems like Facebook Connect. In this case, all users registered in the authentication provider are the parent population U. Then, Provider A assigns $U - U_A$ as a dummy user of Provider A, and Provider B assigns $U - U_B$ as dummy user of Provider B.

Next, Providers A and B initialize T_n^* by generalizing all values in T_n^* into top state (Table 2(a)(b)(Initial)). The internal anonymized tables are T_A^* (*GID*, *userIDs*, *QID_A*), T_B^* (*GID*, *userIDs*, *QID_B*, *userCounts*). *GID* is a sequentially assigned identifier of a record of the T_n^*. *userIDs* is a set of *UIDs* in a record of the T_n^*. *userCounts* is the count of common users in *userIDs* for each value of the *sensitive attribute*. *userCounts* is calculated after all dividing is finished.

Dividing Protocol: Step 2 Providers A and B execute a split function that divides T_A^* and T_B^* by communicating with each other (Fig. 3). First, Providers

A and B assign the *qid* for a dummy user. We call these values *dummy values*. To make it difficult to distinguish between a dummy user and a existing user in a provider from the view of the other provider, the providers assign dummy values in accordance with the distribution of the existing users' *qids*.

Next, the providers decide the division point by using the heuristic function (Section 5.2). After that, the providers verify whether the T^* that is divided on the division point satisfies *k-anonymity* and *δ-max-site-presence*(Fig. 2(b)). In this verification, the providers use *cardinality* protocols and *cardinality threshold* protocols of *secure set intersection*[3]. The *cardinality* protocols can calculate the number of the intersections of two private sets, and the *cardinality threshold* protocols can compare the number of the intersections and a given number.

To verify the *k-anonymity*, the providers use a *cardinality threshold* protocol and check whether the number of the common users in the divided group is larger than k. The input parameters are two sets: the set of *UID*s of the existing users in the group after division by the *dividing provider*, who has an attribute of the dividing point, and the set of *UID*s of the existing users in the group before division by the other provider (*nondividing provider*).

To verify the *δ-max-site-presence*, the providers use a *cardinality* protocol and *cardinality threshold* protocol. By using these protocols, the providers check the following conditional formula for each Provider n ($n \in \{A, B\}$) :

$$|T^*[v_{n,i}]| \leq \delta * |T_n[v_{n,i}]| \quad \forall v_{n,i} \in values_n \quad (2)$$

When Provider n is a nondividing provider, Provider n cannot calculate $|T_n[v_{n,i}]|$ locally, because the nondividing provider does not know the groups after division. Therefore, the provider uses the *cardinality* protocol. The input parameters are the set of *UID*s of the group after division by the dividing provider and the set of the existing users' *UID*s of the group before division by the nondividing provider. Then, the nondividing provider obtains $|T_n[v_{n,i}]|$. Thus, the providers can calculate $\delta * |T_n[v_{n,i}]|$ locally. Then, the providers verify the conditional formula (2) by using the *cardinality threshold* protocol that has the same inputs as verifying *k-anonymity*.

After that, if T^* satisfies *k-anonymity* and *δ-max-site-presence*, the dividing providers divide T_n^*. Table 2(a)(b)(First division) list the results of the first division. In these divisions, the division point is *19:00* of *watching time* in *Provider B*. In this case, Provider B divides T_B^* on the division point locally and then sends the *UID*s of the groups before and after division to Provider A. Then, Provider A divides T_A^* along with the received *UID*s.

Finally, the providers call the split function recursively with the groups after division. Table 2(a)(b)(Second division) list the results of the second division. In this case, the second division point is *40K* of *income* in *Provider A*.

Dividing Protocol: Step 3 After finishing all dividing, Provider B calculates *userCounts* by using the *cardinality* protocol. Provider B obtains the number of common users for each sensitive value s of each record of T_B^*. The inputs parameters are the set of the existing users' *UID*s from Provider A and the set

Table 2. Internal anonymized table T_A^*, T_B^* and joined-anonymized table T^*

(a) Provider A's internal anonymized table (T_A^*)

(b) Provider B's internal anonymized table (T_B^*)

(c) Joined-anonymized table (T^*)

Initial division

GID	userIDs	income
1	User 1-15	*

GID	userIDs	time	userCounts
1	User 1-15	0:00-23:59	-

First division

GID	userIDs	income
2	User 1-10	*
3	User 11-15	*

GID	userIDs	time	userCounts
2	User 1-10	0:00-18:59	-
3	User 11-15	19:00-24:00	-

Second division

GID	userIDs	income
4	User 1-5	<40K
5	User 6-10	40K<=
3	User 11-15	*

GID	userIDs	time	userCounts
4	User 1-5	0:00-18:59	X movie: 1 / Y sports: 1
5	User 6-10	0:00-18:59	X movie: 1 / Y sports: 1
3	User 11-15	19:00-23:59	X movie: 1 / Y sports: 1

(c) Joined-anonymized table (T^*)

income	time	title
<40K	0:00-18:59	X movie
<40K	0:00-18:59	Y sports
40K<=	0:00-18:59	X movie
40K<=	0:00-18:59	Y sports
*	19:00-23:59	X movie
*	19:00-23:59	Y sports

of *UIDs* of the users who exist in and have *s* from Provider B. As an example of the records of Users 1-5 in Table 2(b)(Second division), the number of common users who watched *X movie* is one.

Joining Protocol. Finally, Provider C, who wants to obtain a T^*, requests Providers n to obtain cleaned T_n^*. Before sending T_n^*, Provider n delete *userIDs* of T_n^*. Also, to prevent inference from sequential *GIDs*, Provider A shuffles the *GIDs* and sends the interaction of shuffled *GIDs* to Provide B. Provide B updates the *GIDs* in accordance with the interaction. After that, the providers send cleaned T_n^* to Provide C. Provider C joins the cleaned T_n^* by *GID* to acquire a joined-anonymized table T^* (Table 2(c)).

5.2   Heuristic Function for Dummy User Protocol

This section proposes a new heuristic function to decide a division point for Dummy user protocol. The heuristic function of the Mondrian algorithm [8] selects the attribute that has the longest normalized range and selects the median of the selected attribute. In order that Dummy user protocol satisfies δ-*max-site-presence* additionally, we take an approach to extend the heuristic function of Mondrian. We consider that it is effective that the divided groups uniformly have dummy users. As an example of Table 1(c), which is the case of dividing on *50K* of *income*, the dummy users are not uniformly allocated across the divided groups. In contrast, in the case of dividing on *60K* of *income* (Table 1(d)), the dummy users are uniformly allocated.

Therefore, we introduce an entropy of dummy users (Dummy Entropy,*DE*) in Provider n across the divided groups:

$$DE(c,n) = - \sum_{U_i \in U_{hi}, U_{low}} \frac{|dummy(n, U_i)|}{|U_i|} log(\frac{|dummy(n, U_i)|}{|U_i|}) \qquad (3)$$

where c is a candidate of dividing points that divide a group U_p into the upper group U_{hi} and the lower group U_{low}, and $dummy(n, U_i)$ is the set of UIDs of the dummy users of Provider n in the group U_i.

By using this DE, we define the heuristic function for Dummy user protocol. First, the same as Mondrian, the function selects the attribute that has the longest normalized range. Then, for each candidate c_i of the selected attribute, the function calculates the following score S:

$$S(c_i) = \alpha \left(\frac{-L(c_i)}{\max_{x_j \in X}(L(x_j))} \right) + (1 - \alpha) \frac{1}{2} \sum_{n \in \{A,B\}} \left(\frac{DE(c_i, n)}{\max_{x_j \in X}(DE(x_j, n))} \right) \quad (4)$$

$$L(c_i) = \sum_{x_j \in X} |x_j - c_i| \quad (5)$$

where $\alpha(0 \leq \alpha \leq 1)$ is the weight parameter to adjust the effect of DE, and L is the sum of the distance between the value of c_i and each x_i, which is a value of the selected attribute. Note that if we set $\alpha=1$, then S will be maximized when c_i is median because the median is the value that minimizes the L. This means that the function selects the median the same way as Mondrian when $\alpha=1$.

Then the function selects the c_i whose score S is the maximum as the division point. The division point is expected to divide a group into two groups across which the dummy users are allocated uniformly. As a result, the T_n^* may be divided many times.

Secure Computation. The providers use three kinds of secure computations at some intermediate calculations in the function(Fig. 2(c)). First, the providers compare the normalized ranges that the providers calculate locally by *secure comparison*[15] protocol and decide a dividing provider.

Next, the providers calculate DE locally. However, nondividing provider cannot calculate $|dummy(n, U_i)|$ (the number of the dummy users in the group after division) because the provider does not know the groups after division of the candidate. Thus, the nondividing provider use the *cardinality* protocol of the *secure set intersection* to obtain $|dummy(n, U_i)|$. Also, $|U_i|$ (the number of the UIDs in the group after division) is necessary to calculate DE too, thus the dividing provider sends $|U_i|$ to the nondividing provider. As a result of these processes, the providers can calculate DE locally. Note that neither $|dummy(n, U_i)|$ nor $|U_i|$ contain the attribute value of the candidate of division point or UIDs. For example, when Provider A is the provider of the candidate c_i, Provider B can obtain the number of the UIDs of the group after dividing on candidate c_i but cannot obtain the attribute value or UIDs of the c_i.

Finally, the providers use *secure k-nearest neighbor* protocol[16] in order to decide the c_i that maximizes the score S. The dividing provider obtains the attribute value of the division point. As mentioned above, the providers can decide the division point without disclosing the attribute values or the existence of a user.

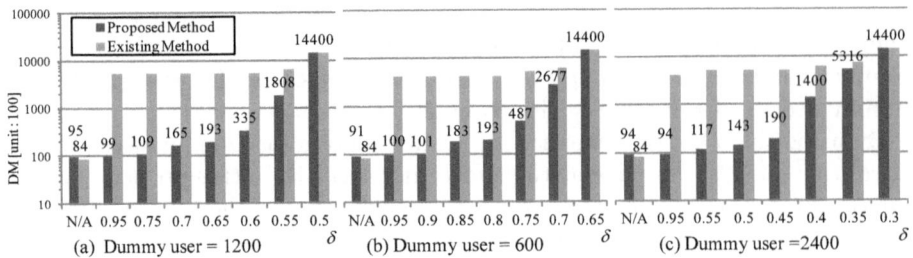

Fig. 4. Dummy user protocol vs Mondrian

6   Experimental Evaluation

We implemented a prototype of Dummy user protocol and evaluated it. The prototype was implemented in Java 1.6. It works in test architecture in which the providers are distributed virtually and communicate with each other virtually. We use an "Adult" data set in UCI Repository [2]. Furthermore, we divide the "Adult" data set the same way as Mohammed et al. [10].

Adult is almost 30K lines of data with 14 types of attributes and one class of data, income class (50K< or not). Also, we treat all records of "Adult" as a user set U and select the groups from the top: the users who exist in both $(U_A \cap U_B)$, the users who exist only in Providers A or B $((U_A - U_A \cap U_B), (U_B - U_A \cap U_B))$, and the remaining users who exist in neither (U_O). In our experiment, we fixed the number of $U_A \cap U_B$ at 1200 and changed the number of $U_A - U_A \cap U_B$ and $U_B - U_A \cap U_B$ from 600 to 12000. The evaluation results are average values of the results of 20 experiments.

The same as the evaluation of δ-presence[11], we use the Discernibility Metric (DM)[1] as the indicator of our evaluation. DM measures information loss. Thus, the lower the loss, the better. By letting $qids$ be the set of values of the *quasi-identifier* of T^*, DM can be calculated as follows:

$$DM = \sum_{q_i \in qids} |T^*[q_i]|^2 \tag{6}$$

For example, if 1200 records are divided into 150 8-record groups, $DM = 8^2 \times 150 = 9600$. Because data mining is the process of discovering rough patterns from data, we consider that even if a table is divided into 8-record groups, the table is useful for data mining.

6.1   Comparison with Mondrian Algorithm

δ-max-site-presence. In order to evaluate Dummy user protocol, we compare Dummy user protocol and extended Mondrian, which is a simple distributed anonymization protocol extended from Mondrian, in order to compare fairly.

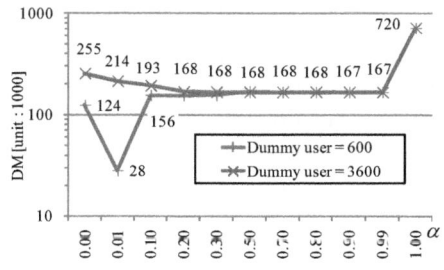

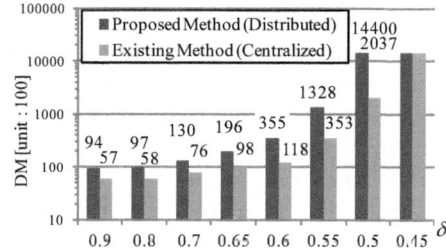

Fig. 5. DM in several α **Fig. 6.** Distributed vs. centralized

The extended Mondrian divides when satisfying not only k-*anonymity* but also δ-*max-site-presence* and outputs the table that contains the record of the common users.

First, we evaluated DM when the number of the dummy users existing only in Provider A or B is 2400, $k = 2$ and $\delta = \{0.95, \cdots, 0.5\}$. Fig. 4(a) shows DMs of the proposed method(Dummy user protocol) and existing method(extended Mondrian). Moreover, the weight α is set at 0.5 in order to make the effect of DE half.

These results show that when we make the method ignore δ-max-site-presence, the information loss of the existing method has slightly lower than that of the proposed method. On the other hand, when we set $\delta = \{0.95, \cdots, 0.5\}$ to hide the existence of a user, the information loss of the proposed method is lower than that of the existing method. When $\delta = \{0.95, \cdots, 0.75\}$, the results of proposed method are especially reasonable because the DM is almost $10,000$. This is because adding the dummy entropy into the heuristic function and the update of the dummy values enable the selection of a suitable division point that can hide the existence. As a result, the proposed method can reduce the information loss without disclosing the existence of users. However, when the δ is set at near 0.6, DM gets worse rapidly. This is because the proposed method cannot find a more effective division point.

Number of Dummy Users. Next, we changed the number of dummy users and evaluated DM. Figure 4(b) and (c) show the results of the evaluation when the number of dummy users is decreased to 600 and increased to 2400.

The results show that when the number of the dummy users is increased to 2400, DM can be kept at a low value from $\delta = 0.95$ to $\delta = 0.5$. In contrast, when it is decreased to 600, DM gets worse at nearby $\delta = 0.75$. This is because the selected division point cannot satisfy δ-*max-site-presence* and the dividing is stopped. Note that, if the number of the dummy users increases to 12000, DM can be kept at a low value even if $\delta = 0.3$. According to the results above, increasing the number of the dummy users effectively hides user presence.

Weight α. In order to find the best weight α, we changed α and evaluated DM. Fig. 5 shows the results of DM when the numbers of the dummy users are 600 and 3600. We set the δ at 0.75 and 0.3 along with the number of the dummy users. These results show that when there are many dummy users, it is better to set α at a larger value in order to weaken the effect of the DE. On the other hand, when there are few dummy users, it is better to set α at a smaller value. This is because when there are few dummy users, the density of the dummy users is likely to not be uniform and it's hard to satisfy $\delta\text{-}max\text{-}site\text{-}presence$. Therefore, it is better to strengthen the effect of the DE. According to the above results, α should be set on the basis of the number of the dummy users.

6.2   Comparison with Centralized Algorithm

Also, we compared the proposed method for a distributed environment and the existing method for a centralized environment, named MPALM [11]. MPALM is an algorithm that is also extended from Mondrian to hide the existence of a user in an anonymized table. The big difference is that the existing method hides only one side of the existence, but the proposed method hides both sides. Thus, to evaluate them fairly, we create data in which the numbers of the dummy users who exist only in Provider {A,B} is {1200, 0} respectively.

Fig. 6 shows DM of the proposed method and the existing method in the case of varying δ. These results show that DM of the proposed method gets worse rapidly at nearby $\delta = 0.6$. On the other hand, DM of the existing method is kept lower. This is because the algorithm of the proposed method selects the division point by calculating the heuristic function. On the other hand, the algorithm of the existing method selects it by trying to divide the anonymized table and check the indicators of the users' existence. This means that existing methods can retry division many times, so the existing methods can make the information loss lower than the proposed method. If the existing algorithm is used in a distributed environment, the user's existence will be disclosed by knowing whether the table in the existing algorithm can be divided or not. However, when $\delta \geq 0.7$, the DM of the proposed method is almost the same as that of the existing method. According to the results mentioned above, the proposed method can obtain the same utility as the existing centralized method when δ is not very small.

Finally, according to all results mentioned above, we can conclude that the proposed method can hide the existence of a user and anonymize personal information in distributed environments with little information loss.

7   Security Evaluation

In this section, we evaluate the security of the proposed protocol. If the providers cannot know more information than that we expect to be leaked, then we say that the protocol is secure. First, we will prove that Provider n ($n \in \{A, B\}$) cannot learn any more information than T_n^* and two types of the intermediate information. Next, we will show that even if these types of the information of a provider are known by the other provider, the privacy risk is low.

Our proof will show that, given T_n, T_n^* and the two types of the intermediate information, the simulator S can simulate all messages that Provider n receives during the execution of Dividing protocol in Dummy user protocol. Since the simulated messages do not contain any more information than that given, clearly Provider n cannot learn any more than the given information[7,5]. Also, our proof will use *composition theorem*[5], because the proposed protocol uses some secure computations such as a *secure set intersection*. When a protocol F is composed of smaller secure functionalities $f_1 \ldots f_n$, the composition theorem states that if the protocol F in *hybrid model* where the $f_1 \ldots f_n$ are replaced with protocols that use a trusted third party (TTP) is secure, then the protocol F is secure. In our proof, we show that the simulator S can simulate the messages of Dividing protocol in the hybrid model. After that, we show that Dividing protocol is secure by using composition theorem.

Theorem 1. *Provider $n \in \{A, B\}$ cannot learn any more information than T_n^* and Intermediates 1 and 2 from the messages of Dividing protocol of Dummy user protocol.*

- **Intermediate 1**: *For the candidates of the other provider of Provider n, the number of the UIDs and the dummy users in the group after division (Note that UIDs are not revealed)*
- **Intermediate 2**: *For the canceled division points, the provider of the division point, the attribute value of the division point (only if Provider n is the dividing provider), the number of the existing users in the group after division in the nondividing provider (only if Provider n is the nondividing provider), and the reason for the cancellation (k-anonymity or δ-max-site-presence)*

Proof. We will show that simulator S_n can simulate the message received by Provider $n \in \{A, B\}$ from T_n, T_n^* and Intermediates 1 and 2. First, we show the messages received by Provider A can be simulated from T_A, T_A^* and Intermediates 1 and 2. S_A analyzes the dividing sequence by using GID, which is assigned sequentially, in T_A^*. (e.g. in Table 2(a)(Second division) group $GID = 2$ is divided into the groups $GID = 4$ and $GID = 5$.) Also, by comparing the two divided records, S_A can infer the dividing provider and $UIDs$ of the group after division. In addition, if the dividing provider is Provider A, S_A can easily infer the attribute value of the division point. (e.g. in Table 2(a)(Second division) Users 1-10 are divided on *40K* of *income* in *Provider A* into Users 1-5 and Users 6-10.) We call the result of this analysis an *analyzed sequence*.

Then, S_A starts the simulation by using the analyzed sequence. In Dividing protocol, the communication is performed in Steps 2 and 3 (Fig. 2(a)). In Step 2, the communication is performed during the calculation of the heuristic function(Fig. 2(c)), the verification(Fig. 2(b)), and sending of the $UIDs$. First, S_A simulates the message at the first division. The messages of the calculation the heuristic function are three: (1) the provider of the dividing point (both providers received), (2) the number of the $UIDs$ and the dummy users in the group after division of the candidate (the nondividing provider received), and (3) the attribute value of the division point (the dividing provider received). S_A learns (1) and (3) from the analyzed sequence, and (2) from Intermediate 1. Thus, S_A can simulate the messages that Provider A receives. After that, S_A learns whether the division continues or not by checking the analyzed sequence. If the division continues, then S_A simulates the message of the verification as correct. Also, if the division is on Provider B, S_A simulates the sending of the *UIDs* by using the analyzed sequence. After that, S_A performs the above function recursively in terms of the divided groups.

If the division does not continue, then S_A simulates the message of the validation as correct and simulates the messages of the calculation of the heuristic function the same way as described above by using Intermediates 1 and 2. Intermediate 2 contains all intermediate information of the cancelled division. After that, S_A simulates the message of the validation as incorrect by using Intermediate 2. As mentioned above, S_A performs these processes recursively, and the messages can be simulated.

Next, we show the message received by Provider B can be simulated from T_B, T_B^*, and Intermediates 1 and 2. In Step 2, S_B can simulate the same way as described above. In Step 3, Provider B receives *userCount*. Because *userCount* is contained in T_B^*, S_B can simulate it.

As mentioned above, S_n can simulate all messages of Dividing protocol in the hybrid model. Also, by using composition theorem, if Dividing protocol is secure in the hybrid model, then the protocol is secure. Therefore, Provider $n \in \{A, B\}$ cannot learn any more information than T_n^* and Intermediates 1 and 2 from the message of Dividing protocol of Dummy user protocol.   □

Because T_n^* has no attribute value of the provider other than Provider n, T_n^* is not very sensitive. Also, Intermediates 1 and 2 are statistic information, which does not include *UID*. Thus, it is difficult to reveal the *sensitive attributes* or the existence of a user from the T_n^* and Intermediates 1 and 2. Furthermore, the T_n^* and Intermediates 1 and 2 are known by Providers A and B but not C. We assumed there is some kind of contract between Providers A and B in the real-world business. Therefore, we consider that the privacy risk is low.

8   Conclusion and Future Works

We showed that there is a problem of the existence of a user in the data of different service providers being revealed when the providers have different sets of users. To solve the problem, we proposed *δ-max-site-presence*, which indicates the probability of the existence of a user being revealed in distributed environment, and Dummy user protocol. Our evaluation results show that the proposed protocol can anonymize users in accordance with the policy of hiding users' existence and user anonymity with little information loss.

In the future, we plan to calculate computation and communication costs. Also, we hope to evaluate the proposed protocol by using the real data. In addition, we plan to improve the heuristic function to make it more efficient and extend the protocol to support multiple sites.

References

1. Bayardo, R.J., Agrawal, R.: Data privacy through optimal k-anonymization. In: Proc. ICDE 2005, pp. 217–228. IEEE (2005)
2. Blake, C.L., Merz, C.J.: Uci repository of machine learning databases (1998), http://archive.ics.uci.edu/ml/
3. Freedman, M.J., Nissim, K., Pinkas, B.: Efficient Private Matching and Set Intersection. In: Cachin, C., Camenisch, J.L. (eds.) EUROCRYPT 2004. LNCS, vol. 3027, pp. 1–19. Springer, Heidelberg (2004)
4. Fung, B., Wang, K., Fu, A., Yu, P.: Privacy-Preserving Data Publishing: Concepts and Techniques, ch. 11-12. CRC Press (2010)
5. Goldreich, O.: Foundations of Cryptography. Basic Applications, vol. 2. Cambridge University Press (2004)
6. Jiang, W., Clifton, C.: Privacy-Preserving Distributed k-Anonymity. In: Jajodia, S., Wijesekera, D. (eds.) Data and Applications Security 2005. LNCS, vol. 3654, pp. 166–177. Springer, Heidelberg (2005)
7. Jurczyk, P., Xiong, L.: Distributed Anonymization: Achieving Privacy for Both Data Subjects and Data Providers. In: Gudes, E., Vaidya, J. (eds.) Data and Applications Security 2009. LNCS, vol. 5645, pp. 191–207. Springer, Heidelberg (2009)
8. LeFevre, K., DeWitt, D.J., Ramakrishnan, R.: Mondrian multidimensional k-anonymity. In: Proc. ICDE 2006, p. 25. IEEE (2006)
9. Lindell, Y., Pinkas, B.: Secure multiparty computation for privacy-preserving data mining. Journal of Privacy and Confidentiality 1, 59–98 (2009)
10. Mohammed, N., Fung, B.C.M., Wang, K., Hung, P.C.K.: Privacy-preserving data mashup. In: Proc. EDBT 2009, pp. 228–239. ACM (2009)
11. Nergiz, M.E., Atzori, M., Clifton, C.: Hiding the presence of individuals from shared databases. In: Proc. SIGMOD 2007, pp. 665–676. ACM (2007)
12. Samarati, P.: Protecting respondents' identities in microdata release. IEEE Transactions on Knowledge and Data Engineering 13, 1010–1027 (2001)
13. Sweeney, L.: k-anonymity: a model for protecting privacy. Int. J. Uncertain. Fuzziness Knowl.-Based Syst. 10, 557–570 (2002)
14. Wang, K., Fung, B.C.M., Dong, G.: Integrating Private Databases for Data Analysis. In: Kantor, P., Muresan, G., Roberts, F., Zeng, D.D., Wang, F.-Y., Chen, H., Merkle, R.C. (eds.) ISI 2005. LNCS, vol. 3495, pp. 171–182. Springer, Heidelberg (2005)
15. Yao, A.C.: Protocols for secure computations. In: Proc. SFCS 1982, pp. 160–164. IEEE Computer Society (1982)
16. Zhan, J., Chang, L., Matwin, S.: Privacy preserving k-nearest neighbor classification. International Journal of Network Security (2005)

Improving Virtualization Security by Splitting Hypervisor into Smaller Components

Wuqiong Pan[1,2,*], Yulong Zhang[2], Meng Yu[2], and Jiwu Jing[1]

[1] State Key Laboratory of Information Security, Institute of Information Engineering,
Chinese Academy of Sciences, Beijing, China
{wqpan,jing}@lois.cn
[2] Department of Computer Science, Virginia Commonwealth University,
Richmond, VA, 23284 USA
{wpan,zhangy44,myu}@vcu.edu

Abstract. In cloud computing, the security of infrastructure is determined by hypervisor (or Virtual Machine Monitor, VMM) designs. Unfortunately, in recent years, many attacks have been developed to compromise the hypervisor, taking over all virtual machines running above the hypervisor. Due to the functions a hypervisor provides, it is very hard to reduce its size. Including a big hypervisor in the Trusted Computing Base (TCB) is not acceptable for a secure system design. Several secure, small, and innovative hypervisor designs, e.g., TrustVisor, CloudVisor, etc., have been proposed to solve the problem. However, these designs either have reduced functionalities or pose strong restrictions to the virtual machines. In this paper, we propose an innovative hypervisor design that splits hypervisor's functions into a small enough component in the TCB, and other components to provide full functionalities. Our design can significantly reduce the TCB size without sacrificing functionalities. Our experiments also show acceptable costs of our design.

Keywords: VMM, Hypervisor, Cloud computing, TCB.

1 Introduction

Virtualization techniques allow multiple operating systems (OSs) to run concurrently on a host computer. By sharing hardware, resource utilization can greatly be improved. Virtualization is also the key technology of cloud computing. Some software, such as Xen [1], can provide hardware virtualization by adding a new software layer called hypervisor beneath all Virtual Machines (VMs). A hypervisor emulates independent hardware resources for every VM. Both Intel and AMD have developed new extensions [2,3] for hardware based virtualization, which can simplify hypervisor designs.

In cloud computing, legal users and attackers may share the same physical server. Thus, it is important to isolate VMs from each other. In a virtualization

---

* This work was done while the first author worked at Virginia Commonwealth University.

N. Cuppens-Boulahia et al. (Eds.): DBSec 2012, LNCS 7371, pp. 298–313, 2012.
© IFIP International Federation for Information Processing 2012

architecture, the hypervisor is responsible for isolation. In current hypervisor designs, it also emulates the hardware and provides other security functions for VMs. Many researchers argued that the current hypervisor designs include too many functions [4,5]. For an example, Xen has about 270k lines of codes (LOC), which is difficulty to implement without bugs. Unfortunately, many vulnerabilities in hypervisor have already been discovered [6,7,8,9,10,11].

To address the problem, a lot of efforts [4,5,12,13,14,15,16] have been made to improve the isolation. Among these work, Overshadow [12] provides fine-grained isolation which protects applications from a compromised OS. SecureME [13] has the same function and can defend against hardware attacks by using a secure processor substrate. Bastion [15] defines a struct of module and protects it from hardware attacks. Overshadow, SecureME and Briston can provide fine-gained isolation for applications or modules. These work add new functions to the hypervisor. As a result, the size of the hypervisor is increased. Therefore, the added codes increase the size of TCB and they may introduce new vulnerabilities as well.

In contrast, some efforts try to reduce the number of functions provided by a hypervisor to make the hypervisor, thus TCB, smaller. For examples, SICE [14] provides hardware isolation for one workload on a core by using the System Management Mode (SMM) of x86 processors. NoHype [4,5] eliminates the hypervisor attack surface by removing the hypervisor after booting the guest VMs. NoHype pre-allocate resources for a VM before the boot procedure. CloudVisor [16] can protect VMs from a compromised hypervisor by adding an additional hypervisor layer. These new designs also introduce some restrictions. For examples, both SICE and Nohype have the limit of one protected workload, or VM per core on multi-core processors. Also, NoHype does not support dynamic resources allocation. CloudVisor does not allow hypervisor to access VM pages and it becomes a big burden for CloudVisor to determine which pages of the guest VMs can be accessed by the master hypervisor.

Moreover, the security problem of shared management domain (or hypervisor in CloudVisor) are not completely eliminated in afore mentioned designs. For examples, in NoHype design, the attackers can attack the management VM. Although the management OS may be well configured by cloud provider, an OS is usually considered more vulnerable to attacks than a hypervisor. In CloudVisor design, a new hypervisor is used to intercept all communications between the master hypervisor and VMs. Even though the attack surface is smaller, once the master hypervisor is compromised, all VMs can still be affected.

In this paper, we propose a split hypervisor architecture, called *SplitVisor*, which has a small TCB and does not limit the functions of a hypervisor. Our architecture leverages nested virtualization [17,18]. In our design, every VM has its own hypervisor, called *GuestVisor*. Users can customize the GuestVisor. A SplitVisor is underneath all GuestVisors and VMs. The SplitVisor is responsible for isolation, which is the only part in TCB.

Table 1 shows the comparison of the proposed architecture and existing ones. SplitVisor, CloudVisor and SICE are the only ones which have a small TCB. All of them sacrifice some of hypervisor functions except SplitVisor. SplitVisor is the only one which has a stable TCB, which means that the TCB need little or even none of change when adding new functions to hypervisor. All others need to modify the TCB if they add new functions to the hypervisor. A stable TCB can greatly reduce the verification costs when upgrading secure software. SplitVisor, CloudVisor, and SICE can defend against attacks from a compromised hypervisor. The security of SICE is ensured by the SMM mode of x86 processors while the security of SplitVisor and CloudVisor are ensured by a nested architecture. CloudVisor can block a hypervisor from accessing its VMs' data. The protected units of these work are also different. Only SplitVisor, CloudVisor, and Nohype have a VM as the unit of protection. Other work either have applications [12,13], or hardware [19,20,21,22], as the unit of protection, which are out of the scope of this paper. These protections are still compatible with SplitVisor design.

Table 1. Comparison of different designs

	TCB Stability[1]	Functions	Hypervisor[2]	VM[3]	Protection	Assertion	HD[4]
SplitVisor	small √	full	√	√	VM	√	
CloudVisor [16]	small	partial	√	√	VM	√	
SICE [14]	small	partial	√	√	region	√	
NoHype [4,5]	large	partial			VM		
Overshadow [12]	large	full			application		
Bastion [15]	large	full			module		√
SecureME [13]	large	full			application		√
XOM [19,20]	large	full			region		√
AEGIS [21]	large	full			region		√
AISE [22]	large	full			region		√

The main contributions of our work include:

- A small and stable TCB. Both GuestVisors and the management OS are not in the TCB. They cannot access other VMs. Most of functions can be added to GuestVisors without modifying the TCB.
- Supporting full-functions. Unlike other architectures which have a small TCB, SplitVisor does not eliminate any functions from hypervisors.
- Allowing users to verify the execution environment. Users can get the assertion of the environment.

The rest of this paper is organized as follows. Section 2 discusses our goals and shows the whole architecture of SplitVisor. Section 3 describes SplitVisor in booting, memory management, scheduling and some other details. Section 4 compares SplitVisor with other recent work. Section 5 shows the related work in isolation. Section 6 gives an conclusion of the paper.

---

[1] A TCB design is *stable* if the TCB needs little or even none of changes when adding new functions to the hypervisor.
[2] Attacks can be confined to the hypervisor when the hypervisor is compromised.
[3] VMs are protected when the hypervisor is compromised.
[4] Hardware level attacks, such as memory tapping [23], can be handled.

2   Overview

2.1   Design Principles

Cloud services are usually provided using virtualization techniques. The examples are Amazon EC2 [24], Eucalyptus [25], FlexiScale [26], Nimbus [27], and RackSpace [28]. In order to attract more cloud users, they continuously improve their products by adding more and more functions to the platform. As the result, the size of TCB, the hypervisor, also increases prominently. Table 2 shows the TCB size of Xen from version 2.0 to version 4.0 [1].

Table 2. Xen TCB size (by Lines of Code)

	Hypervisor	Kernel of Domain 0	Tools	TCB
Xen 2.0	45k	4,136k	26k	4,027k
Xen 3.0	121k	4,807k	143k	5,071k
Xen 4.0	270k	7,560k	647k	8,477k

In the table, the size of Xen 4.0 hypervisor is six times as large as that of Xen 2.0. Although Xen 2.0 is more secure in term of hypervisor size, Xen 4.0 is more attractive to users because of the new features, such as Xen access control, I/O optimization, and Memory page sharing. Domain 0 and tools are also parts of the TCB in Xen in addition to the hypervisor, because they can access the data of all VMs. For an example, the xm tool can dump the memory of a VM.

A possible way to reduce the size of TCB is to remove Domain 0 and Xen tools from the TCB, which requires disabling some functions, or encrypting data to prevent accesses from Domain 0. However, this will limit the functionalities of Xen. Cloud providers usually try to provide as many functions as them can. But for a particular user, he may only need a small set of the functions. In SplitVisor, every user can choose a *GuestVisor* which serves as the current hypervisor. The user can choose a hypervisor with the necessary functions, then he will not suffer vulnerabilities of the unneeded functions. Although the GuestVisor still can be attacked by its own VM, the attack will not affect other GuestVisors and VMs. The isolation is ensured by the SplitVisor. Thus, the SplitVisor should have a small code base to ensure its security.

2.2   Assumptions

Our assumptions are described as follows.

Adversaries: We assume that the attackers can easily control a VM. For an example, the attackers can buy a VM. The attackers can invade a management OS and a rich-functions hypervisor. They can send any instructions in the name of the hypervisor. The attackers can also sniff I/O and steal users' OS images. We assume that the attackers cannot physically access the machines. The data centers are usually protected by well-trained guards. The cloud providers have no motivation to use malicious hardware. Using malicious hardware easily leaves evidence, which definitely ruins cloud providers' reputation. The cloud providers are usually famous companies, while cloud clients are usually small companies. They are not competitors in most cases.

Security Guarantees: The main goal of SplitVisor is to provide isolated environments for VMs. SplitVisor directly provides CPU isolation and memory isolation for VMs. Other security attributes can be achieved based on these two attributes. SplitVisor allows users to verify its TCB. Users can get evidence that their OSs run in the expected environment.

Non-guaranteed Goals: SplitVisor cannot guarantee the availability of a particular VM. If a VM cannot get CPU time slices from the management software, the VM will be blocked, but this is easily discovered by outside.

SplitVisor is not designed to defend against side-channel attacks. Xen provides Chinese Wall (CHWALL) policy, which can control the set of VMs on the same machine. Well-defined CHWALL policy can reduce side-channel attacks [29]. Besides hypervisor level policies, some applications also have built-in mechanisms to reduce side-channel attacks [30].

2.3   Architecture

Both Intel and AMD have added new extensions to support the hypervisor layer, like Intel's Virtualization Technology for x86 (VT-x) and AMD's Secure Virtual Machine (SVM). With the new extensions, the Intel processors have two operation modes: virtual-machine extension (VMX) root mode and VMX non-root mode. In general, a hypervisor will run in VMX root mode and a guest OS will run in VMX non-root mode. The control information of VMX transition is stored in a data structure called virtual-machine structure (VMCS). A VMCS includes almost all environment parameters of a guest OS, such as registers.

SplitVisor has two types of components that belong to one hypervisor before: GuestVisors and a SplitVisor, as shown in Fig 1. The SplitVisor runs in VMX root mode, while both the GuestVisors and VMs run in VMX non-root mode. Both GuestVisors and VMs have VMCSs. A GuestVisor's VMCS is controlled by the SplitVisor, and a VM's VMCS is controlled by a GuestVisor. Most requests from a guest OS are handled by the GuestVisor under it. A GuestVisor is responsible for the execution environment of a guest OS. If a user wants to add some functions into the current hypervisor layer, he can add them into the GuestVisor. The isolation among all GuestVisors and VMs is ensured by the SplitVisor. All VMX transition instructions are executed in the SplitVisor. The SplitVisor is also responsible for memory isolation.

The features of the SplitVisor, a GuestVisor and a VM are shown in Table 3. The SplitVisor does not provide a complete virtualized running environment to a GuestVisor. A GuestVisor knows that it is not running on the bare hardware. A VM runs in a virtualized environment emulated by a GuestVisor. A VM can be para-virtualized or full-virtualized.

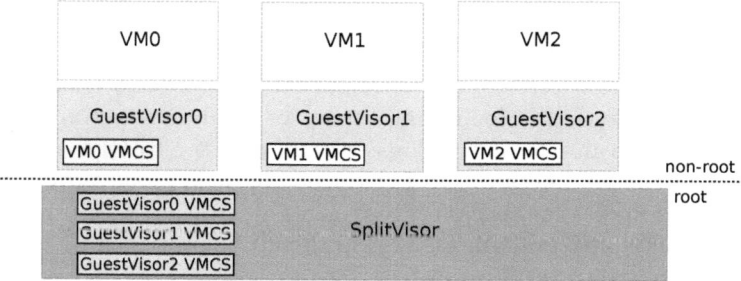

Fig. 1. SplitVisor architecture

Table 3. Units in SplitVisor

Unit	Functions	Transparency
SplitVisor	VMX transition, memory isolation	-
GuestVisor	OS execution environment, extra functions	Not support
VM	Running OS	Both para- and full- virtualization

Security Analysis: If a GuestVisor is compromised, the SplitVisor will ensure other GuestVisors and VMs are not affected. In other words, the isolation is ensured. But the VM of the compromised GuestVisor is not protected.

3   Design Details

3.1   Secure Boot

The boot process of a computer is the first step to set up a secure environment for guest OS. In our design, the only trusted component is the SplitVisor. The secure boot process is described as follows.

First, *verify the SplitVisor.* Users can verify the SplitVisor, which is supported by a Trusted Platform Module (TPM) [31]. A TPM is an secure chip which can help to protect the integrity of the boot process and the SplitVisor. Users can also obtain the SplitVisor's public key with the help of the TPM.

Second, *make a new image.* The binary codes of both user's VM and GuestVisor should be provided by users. Users must encrypt all or part of the codes with a symmetric key. The symmetric key should be encrypted by the SplitVisor's public key, so the SplitVisor can decrypt it. All the plain text should be signed to protect against unauthorized modification. The total data sent to cloud server includes the following: encrypted image, encrypted symmetric key, plain data, signature over everything, and the public key certificate of the user (signer).

Third, *prepare the environment of a GuestVisor.* When the SplitVisor receives the above data from users. It will decrypt all data and verify the signature. If the verification goes through, the SplitVisor will set up the running environment for a GuestVisor. The allocation of memory and CPU is decided by the management software. The boot process of VM is controlled by the GuestVisor.

Fourth, *authentication between the user and the guest OS*. In traditional login, only OS verifies users. In cloud computing, we need two-way authentication, because attackers can run a malicious system, and try to trick user to login. Users can put the key in guest OS before sending it to cloud providers. When guest OS boots up in remote server, they can authenticate each other based on the key. All communication after that also should be protected by the key.

Boot Process of the Management VM: The management VM is the one we should boot first. The boot process is a little different at the first step and the second step. Cloud providers should get the public key of the TPM through some public channels. The making of the management VM's image is the same as that of others. However, the image must be stored in a disk, or other storage devices which can be easily read by a SplitVisor. All the memory except the SplitVisor's, CPUs and devices are belonged to the management VM at first. It will assign the resources to other VMs in the future execution.

Security Analysis: Without the authentication key, a malicious OS cannot cheat users. Attackers cannot get the key which is encrypted in the second step. If attackers do not crack the key but modify the image, it will be detected when verifying signatures in the third step.

3.2   Memory Management

Both Intel and AMD have extended two layer address translation to three layer address translation, called Extended Page-Table (EPT) and nested paging. The page table (PT) specified by CR3 is still responsible for translating virtual frame number (VFN) to physical frame number (PFN). A new table called EPT for Intel is responsible for translating PFN to machine frame number (MFN). The address of EPT is specified by EPTP, a VMCS field. The three layer model is shown in Fig 2 to compare our SplitVisor design and the current hypervisor design.

The SplitVisor will set up a default EPT for a GuestVisor. The SplitVisor assigns all allowed memory page to the GuestVisor at the beginning. The range of PFN is fixed, for an example, it always starts from 0. For further processing, SplitVisor will allow GuestVisor to read its own EPT.

After a GuestVisor is booted, it will prepare the environment for a VM. Firstly, GuestVisor sets up an EPT for the VM. The GuestVisor can get the addresses of its MFNs by reading its own EPT. It can keep some MFNs for its own use, and assign others to VM. Finally, GuestVisor write the address of VM's EPT to VM's VMCS.

When a VM is booted, SplitVisor will check its EPT to see if all the MFNs are from the GuestVisor. So a user cannot access others' MFN by booting VM. After the checking, the SplitVisor marks the EPT page as read-only.

A GuestVisor has two means to manipulate the address translation of a VM. One is to modify the VM's EPT, another is to modify the VM's PT. Modifying the PT is the same as shadow paging. Modifying the EPT is the same as that we

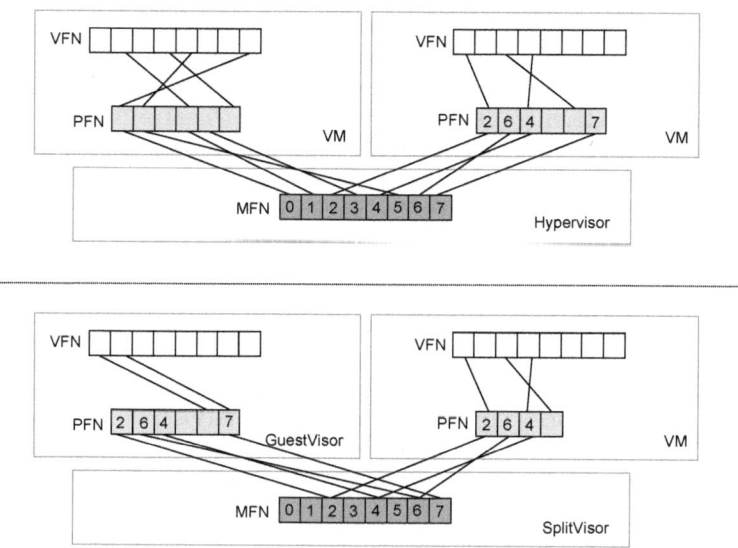

Fig. 2. Comparison of the current hypervisor design (top) and SplitVisor design (bottom) in terms of address translation

support EPT functions between a GuestVisor and a VM. The only difference is that a VM's EPT is marked as read-only by the SplitVisor. Any modification of the VM's EPT will cause a VM exit. The SplitVisor will check the modification and do it for the GuestVisor.

The SplitVisor allows a GuestVisor to transfer its memory to other GuestVisors, but the request must be originated by the GuestVisor who owns the memory. The SplitVisor un-maps the pages from the GuestVisor's and VM's EPT, and maps them to another GuestVisor. Then the SplitVisor adds all the pages to the EPT of the second GuestVisor, and the second GuestVisor can decide how to use it. Memory sharing is similar to memory transferring except some flags of the memory pages are different.

There is a trade off about whether the management software should be allowed to forcibly receive other VMs' memory. If it is allowed, the SplitVisor can reset the received pages to zero when they are remapped. But the management software can still observe the action of a GuestVisor when different pages are received. It will introduce a way to attack the GuestVisor and VMs. Therefore, our SplitVisor design does not allow it. In this situation, the GuestVisor may do not want to give up its memory even if it does not need it. We can charge more to the GuestVisor's owner, it is fair since the user pays more when using more resources. A problem of this strategy is that we cannot receive GuestVisor's memory when it is crashed. It is difficulty to know the crash for the SplitVisor. The SplitVisor can try to migrate the GuestVisor. If the GuestVisor does not respond , it is crashed. Then SplitVisor can reallocate its memory.

Table 4 shows the comparison of memory management in different designs. In CloudVisor design, VM management is implemented in a modified hypervisor above the bottom layer called CloudVisor. In SplitVisor design, VM management, memory transferring and memory sharing are controlled by GuestVisors. Thus, the GuestVisors have corresponding privileges to manage VMs. The only work of the SplitVisor is to check the page access permissions. Xen implements all functions in the hypervisor. If Xen wants to add new functions, it has to modify the hypervisor.

Table 4. Comparison of Memory Designs - Where The Functions Are Implemented

Function	SplitVisor	CloudVisor	Xen
Memory isolation	SplitVisor	CloudVisor	Hypervisor
VM management	GuestVisor	Hypervisor	Hypervisor
Memory transferring, sharing	GuestVisor	CloudVisor	Hypervisor

3.3   Scheduling and VMCS

The transition from the root mode to the non-root mode is called *VM entry*, while the opposite is called *VM exit*. The transitions are controlled by a VMCS. The SplitVisor creates a default VMCS for a GuestVisor and controls the scheduling of GuestVisors. A GuestVisor controls the scheduling of the VM running above it. For an example, A GuestVisor may provide many VCPUs to the VM for special purpose. All the VCPUs are managed by the GuestVisor.

Every time a GuestVisor gets a time slice, the SplitVisor will switch into the GuestVisor first. The GuestVisor decides on which VCPU to start and switches back to the SplitVisor, then the SplitVisor switches into the VM. Before switching into the VM, the SplitVisor must modify the VM' VMCS. Some fields in the VMCS should be rewritten by the SplitVisor, such as host registers, which determine the CPU status when returning from the VM. The rewriting should be done every time entering the VM, because the GuestVisor may modify the VMCS.

The SplitVisor does not handle VM exits of a VM. It transfers all VM exits to the GuestVisor except timer and some I/O interrupts. Operations causing VM exits are specified by a VMCS. A GuestVisor can configure VM exits by modifying a VM's VMCS. For an example, if a GuestVisor wants to intercept VM's system calls, it just modifies control filed of the 80h interrupt.

Table 5 shows the comparison of different scheduling designs. VCPU management is implemented in the hypervisor of CloudVisor, which gives the hypervisor opportunities to attack VMs. Xen implements all functions in the hypervisor. A problem of scheduling is who can get the next time slice. If a scheduling algorithm is implemented in the management software, the attackers can deny the service of a VM if they controls the management software. If the scheduling algorithm is implemented in the SplitVisor, its parameters cannot be changed dynamically. Xen implements the algorithm in the hypervisor, but the parameters can be modified in the management software. The security level is the same as the one implemented in the management software. The implementation in SplitVisor can be determined case by case. For an example, in Amazon EC2 the time sharing of VMs is predefined. For such situations, the scheduling algorithm can be implemented in the SplitVisor. Otherwise, it can be implemented in management software.

Table 5. Comparison of Scheduling Designs - Where The Functions Are Implemented

Function	SplitVisor	CloudVisor	Xen
VCPUs management	GuestVisor	Hypervisor	Hypervisor
(GuestVisor) hypervisor's VMCS	SplitVisor	CloudVisor	-
VM's VMCS	GuestVisor	Hypervisor	Hypervisor

3.4   Interrupt and Device Management

In order to meet the needs of cloud computing, some device manufactures added virtualization support to their products, as mentioned in [4]. If a device can support virtualization by itself, every VM can get its own devices. For such situations, the SplitVisor can simply distribute the interrupts and I/O ports to each GuestVisor.

   If devices do not support virtualization, all VMs have to share the devices. It can be implemented by front-backend drivers. The SplitVisor assign the devices to one GuestVisor. All request are handled by the special GuestVisor. From the SplitVisor's perspective, some GuestVisors have shared memory with the special GuestVisor, and the SplitVisor does not need to emulate the devices.

3.5   Functions of a GuestVisor

Most functions of the current hypervisors are implemented in the GuestVisor. The GuestVisor is the key to reduce the size of TCB. We list the possible functions of GuestVisor in this section. Different versions of GuestVisors, from light weight ones to the ones with full functionalities, can be provided to the user as options.

I/O Encryption. The GuestVisor that controls devices may be controlled by attackers. The attackers can sniff the I/O data which pass through the devices. Some applications already have built-in mechanisms to protected I/O, such as Bitlocker and SSL. It is hard to decide whether I/O protection should be implemented in the hypervisor. It is more secure to be implemented in the hypervisor, while more flexible in the application. The SplitVisor leaves the decisions to users. If users want to implement I/O protection in the hypervisor, they can add the new function in the GuestVisor.

VM Life-Cycle Management. Some VM life-cycle functions, such as shutting down a VM, should be implemented in the GuestVisor. When a VM is shutting down, the GuestVisor needs to follow the boot preparation procedure as described in Section 3.1. Also, the GuestVisor writes all data to disk or send them to the management software. The same thing should be done when taking a snapshot of a VM. The management of snapshots is also the GuestVisor's responsibility.

Another important work of the GuestVisor is VM migration. The GuestVisor verifies the new server and creates a new VM image on the target server. All the steps described in secure boot in Section 3.1 should be done again.

Privileged Instructions. Privileged instructions cannot be executed in a VM. In SplitVisor, they are handled by the GuestVisor. If an instruction, e.g., a hyper call, has some arguments in the VM's memory, the GuestVisor can directly read it. In CloudVisor, things are more complex. CloudVisor must fetches page table entries and the arguments for the hypervisor, in the meanwhile the CloudVisor must make sure that no sensitive information is leaked to the hypervisor. The CloudVisor must know the exactly meaning of the instructions. Some work [13] may add new instructions, so it is a big burden to the CloudVisor.

Fine-Gained Isolation. Some architectures [12,13] provide fine-gained isolation. They can provide a secure environment for applications even in a malicious OS. The function is implemented in the hypervisor, which intercepts the communication between applications and OS. In SplitVisor, it can be implemented in the GuestVisor.

Monitoring and Virus Detection. It is much more secure to implement monitoring and virus detection in the hypervisor. When an OS is under the attackers' control, all traditional virus detections are useless. The virus detection must be reliable, because it can access VMs' memory. But virus detection software are usually large, it is not suitable to be added to the TCB. In SplitVisor, it can be implemented in the GuestVisor. It is not possible to be added in other architecture without increasing the size of TCB. The monitoring software is the same as virus detection software.

Virtual TPM. A GuestVisor can emulate some devices for VMs, such as a virtual TPM. The TPM is an important device for software protection.

Table 6 summarizes the functions of the GuestVisors. All the functions are implemented in the hypervisor of Xen, as a part of TCB. It is more secure that SplitVisor can implement them out of the TCB. Other functions can be implemented in the GuestVisor as long as they do not violate the isolation restriction.

Table 6. Comparison of different designs - Where The Functions Are Implemented

Functions	SplitVisor CloudVisor		Xen
I/O encryption	GuestVisor	CloudVisor	Hypervisor
VM life cycle	GuestVisor	CloudVisor	Hypervisor
Privileged instructions	GuestVisor	CloudVisor, hypervisor	Hypervisor
Fine-gained Isolation	GuestVisor	None	Hypervisor
Monitor and virus detection	GuestVisor	None	Hypervisor
Virtual TPM	GuestVisor	None	Hypervisor

4 Performance Evaluation

4.1 Memory Overhead

The memory overhead of SplitVisor is mainly caused by the GuestVisor. In general, users only need a small part of all functions. The GuestVisor is in users' memory space. If a users wants to save memory space, the user can choose a simple GuestVisor with less functions.

Xen occupies 64 MB memory. The GuestVisor has similar size. The memory of a typical VM is shown in Table 7, which is from Amazon EC2 [24]. The memory overhead is from 0.4% - 3.8%.

Table 7. Memory Overhead

VM type	VM	GuestVisor	Overhead
Small Instance	1.7 GB	64 MB	3.8%
Large Instance	7.5 GB	64 MB	0.9%
Extra Large Instance	15 GB	64 MB	0.4%

4.2 CPU and I/O Overhead

When running a VM, some privileged instructions and interrupts will cause VM exits, which introduces the major CPU and I/O overhead of SplitVisor when comparing with other approaches. In current hypervisor design, VM exits are delivered to and handled by the hypervisor. In SplitVisor, VM exits are transferred twice, from the SplitVisor to a GuestVisor. We firstly compare SplitVisor with other two-layer-hypervisor architectures: CloudVisor and nested hypervisor [17]. The upper part of Fig. 3 shows the process of VM exits in SplitVisor and CloudVisor. The labels in the figure indicate what messages are delivered. In CloudVisor, the hypervisor cannot access the memory of a VM. CloudVisor encrypts all the data from a VM to the hypervisor. In SplitVisor, all the memory of a VM are mapped in a GuestVisor, so the GuestVisor can easily deal with the VM's memory. The nested hypervisor provides full virtual environment for every level, where the L1 hypervisor does not know that it is in a virtual machine. So the L1 hypervisor may execute some privileged instructions which traps to the L0 hypervisor. The L0 hypervisor emulates hardware for L1 the hypervisor. The SplitVisor does not aim at providing a virtual environment for GuestVisors. A GuestVisor is aware of virtualization. The lower part of Fig. 3 shows the process of VM exits in SplitVisor and nested hypervisor. When a VM exit occurs in a VM (L2), it is delivered to a GuestVisor (L1). A GuestVisor usually handles the VM exit by itself, while an L1 hypervisor may cause many new VM exits.

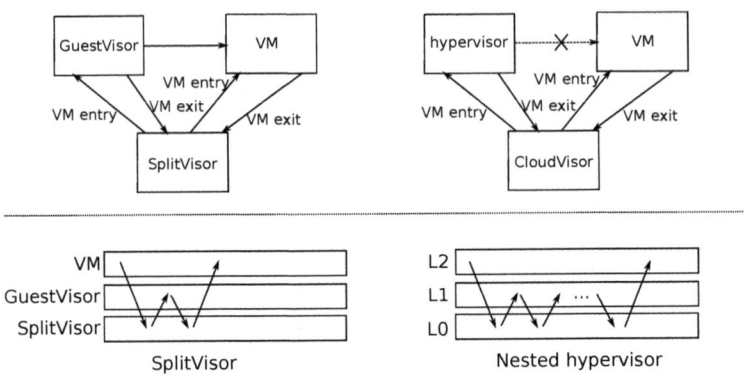

Fig. 3. Process of VM exit

A GuestVisor can control what can be intercepted by modifying a VM's VMCS. In our experiments, we assume that a GuestVisor intercepts the same VM exits as Xen. Then SplitVisor needs twice as many VM exits as Xen does. We choose SPECjbb2005 [32] as the benchmark program. We run SPECjbb2005 in Xen HVM with different numbers of JVMs. Then we count the number of VM exits and estimate the running time in SplitVisor. The results are shown in Fig 4. The average overhead is about 4.3%.

5   Related Work

Research similar to ours can be classified into several categories.

Hardware Level Protection: Some work, such as XOM [19,20], AEGIS [21] and AISE [22], can defend against hardware attacks. They use a specially designed CPU to defend against memory tampering. In these architecture, CPU encrypts all data that goes out of the CPU, and decrypts what are loaded into CPU. CPU maintains a hash tree that ensures data integrity. These work can protect all software from hardware based attacks, but do not address security problems inside the software.

Application Protection: Many vulnerabilities have been discovered in OSs. Many techniques, such as Overshadow [12,33] and SecureME [13], protect applications from a malicious OS. These work leverage the hypervisor to prevent

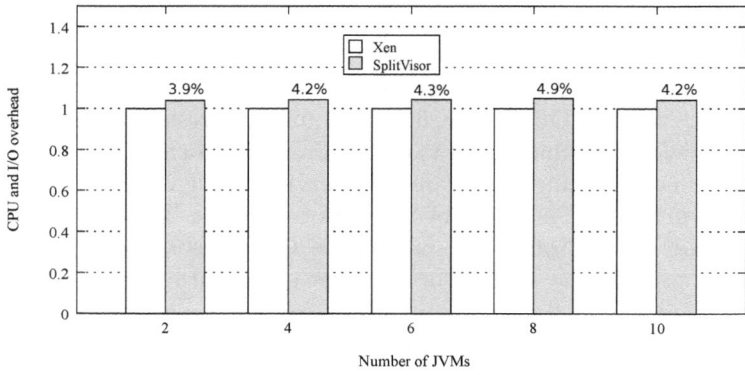

Fig. 4. CPU and I/O Overhead

the OS from directly accessing applications' memory and intercept the communication between the OS and applications. Bastion [15] provides module-level protection. Similar to application-level protection, entering into a module and going out of a module are intercepted by the hypervisor.

Hypervisor Protection: Hypervisor is the most important part in a virtualization architecture. TPM and other hardware protections [19,20,21,22] can verify the integrity of software when a hypervisor is loaded. HyperSentry [34] can verify the integrity dynamically. HyperSafe [35] can verify the integrity of control-flow of hypervisor execution. These work mainly focus on detecting attacks, instead of building a secure hypervisor. NoHype [4,5] cuts off the communication between hypervisor and VM after VM is booted. However, it cannot defend against the attacks from the management VM. So NoHype still has a large TCB.

VM Protection: The best way to protect a VM is to protect the hypervisor. If a hypervisor works correctly, the VM can be well protected by the hypervisor. Some work can provide protection without the help of a hypervisor. SICE [14] implements the protection mechanism in the SMM. Even if a hypervisor is malicious, a VM can be protected by SICE. Currently, SICE has a limit of protecting at most one VM on each core.

The turtles project [17] shows the architecture of nested hypervisors, which allows running hypervisors above a hypervisor. CloudVisor [16] leverages this architecture to protect VMs. In these designs, the orignial hypervisor is not at the highest level. All communications between a hypervisor and VMs must be verified by the CloudVisor, and the hypervisor cannot directly access VMs' memory. CloudVisor does not provide the protection of the hypervisor, because all requests from VM are forwarded to the hypervisor.

6   Conclusion

Existing virtualization architectures usually have either rich functions or a small TCB, but not both. In this paper, we propose an innovative virtualization architecture, SplitVisor, to support both. SplitVisor has a two-layer-virtualization structure: a SplitVisor and GuestVisors. The SplitVisor is responsible for isolation between different users' VMs. A GuestVisor is responsible for emulating hardware for VMs. A GuestVisor is not required to reside in the TCB, so the TCB of SplitVisor is small. SplitVisor allows users to choose their own hypervisors. It cannot be achieved in single hypervisor designs where all VMs share the same hypervisor. The TCB of SplitVisor is stable because we do not add new functions to the SplitVisor, but to the GuestVisors. This also offers the opportunity to store the TCB in firmware or to optimize it with hardware.

Acknowledgement. This work was supported in part by NSF Grants CNS-1100221 and CNS-0905153.

References

1. "Xen hypervisor project", http://www.xen.org/products/xenhyp.html
2. Neiger, G., Santoni, A., Leung, F., Rodgers, D., Uhlig, R.: Intel virtualization technology: Hardware support for efficient processor virtualization. Intel Technology Journal 10(3), 167–177 (2006)
3. AMD. Secure virtual machine architecture reference manual
4. Keller, E., Szefer, J., Rexford, J., Lee, R.: Nohype: virtualized cloud infrastructure without the virtualization. In: Proceedings of the 37th Annual International Symposium on Computer Architecture, pp. 350–361. ACM (2010)
5. Szefer, J., Keller, E., Lee, R., Rexford, J.: Eliminating the hypervisor attack surface for a more secure cloud. In: Proceedings of the 18th ACM Conference on Computer and Communications Security, pp. 401–412. ACM (2011)
6. Kortchinsky, K.: Hacking 3d (and breaking out of vmware). BlackHat USA (2009)
7. Cve-2007-4993: Xen guest root can escape to domain 0 through pygrub (2007), http://cve.mitre.org/cgi-bin/cvename.cgi?name=CVE-2007-4993
8. Cve-2007-5497: Vulnerability in xenserver could result in privilege escalation and arbitrary code executionr (2007), http://support.citrix.com/article/CTX118766
9. Wojtczuk, R.: Subverting the xen hypervisor. BlackHat USA (2008)
10. Cve-2008-2100: Vmware buffer overflows in vix api let local users execute arbitrary code in host os (2008), http://cve.mitre.org/cgi-bin/cvename.cgi?name=CVE-2008-2100
11. Ristenpart, T., Tromer, E., Shacham, H., Savage, S.: Hey, you, get off of my cloud! exploring information leakage in third-party compute clouds. Computer and Communications Security (2009)
12. Chen, X., Garfinkel, T., Lewis, E., Subrahmanyam, P., Waldspurger, C., Boneh, D., Dwoskin, J., Ports, D.: Overshadow: a virtualization-based approach to retrofitting protection in commodity operating systems. In: ACM SIGARCH Computer Architecture News, vol. 36, pp. 2–13. ACM (2008)
13. Chhabra, S., Rogers, B., Solihin, Y., Prvulovic, X., Chen, M., Garfinkel, T., Lewis, E., Subrahmanyam, P., Waldspurger, C., Boneh, D., Dwoskin, J., Ports, D.: Secureme: a hardware-software approach to full system security. In: Proceedings of the International Conference on Supercomputing, pp. 108–119. ACM (2011)

14. Zhang, X., Azab, A., Ning, P.: Sice: A hardware-level strongly isolated computing environment for x86 multi-core platforms. In: 18th ACM Conference on Computer and Communications Security (2011)
15. Champagne, D., Lee, R.: Scalable architectural support for trusted software. In: 2010 IEEE 16th International Symposium on High Performance Computer Architecture (HPCA), pp. 1–12. IEEE (2010)
16. Zhang, F., Chen, J., Chen, H., Zang, B.: Cloudvisor: retrofitting protection of virtual machines in multi-tenant cloud with nested virtualization. In: Proceedings of the Twenty-Third ACM Symposium on Operating Systems Principles, pp. 203–216. ACM (2011)
17. Ben-Yehuda, M., Day, M., Dubitzky, Z., Factor, M., Har'El, N., Gordon, A., Liguori, A., Wasserman, O., Yassour, B.: The turtles project: Design and implementation of nested virtualization. In: 9th USENIX Symposium on Operating Systems Design and Implementation (OSDI), Vancouver, British Columbia, Canada, pp. 423–436 (October 2010)
18. Goldberg, R.: Architecture of virtual machines. In: Proceedings of the Workshop on Virtual Computer Systems, pp. 74–112. ACM (1973)
19. Lie, D., Thekkath, C., Mitchell, M., Lincoln, P., Boneh, D., Mitchell, J., Horowitz, M.: Architectural support for copy and tamper resistant software. ACM SIGPLAN Notices 35(11), 168–177 (2000)
20. Lie, D., Thekkath, C., Horowitz, M.: Implementing an untrusted operating system on trusted hardware. ACM SIGOPS Operating Systems Review 37(5), 178–192 (2003)
21. Suh, G., Clarke, D., Gassend, B., Van Dijk, M., Devadas, S.: Aegis: architecture for tamper-evident and tamper-resistant processing. In: Proceedings of the 17th Annual International Conference on Supercomputing, pp. 160–171. ACM (2003)
22. Chhabra, S., Rogers, B., Solihin, Y., Prvulovic, M.: Making secure processors os- and performance-friendly. ACM Transactions on Architecture and Code Optimization (TACO) 5(4), 16 (2009)
23. Huang, A.: Hacking the Xbox: an introduction to reverse engineering. No Starch Pr. (2003)
24. Amazon elastic compute cloud, http://aws.amazon.com/
25. Eucalyptus cloud computing software, http://www.eucalyptus.com/
26. Flexiscale cloud computing services, http://www.flexiscale.com/
27. Nimbus platform, http://www.nimbusproject.org/
28. Rackspace hosting, http://www.rackspace.com/
29. Xen users' manual v3.3, http://www.xen.org/products/xenhyp.html
30. Witteman, M., Oostdijk, M.: Secure application programming in the presence of side channel attacks. In: RSA Conference, vol. 2008 (2008)
31. Tpm main specification, http://www.trustedcomputinggroup.org/
32. Specjbb2005 (java server benchmark), http://www.spec.org/jbb2005/
33. Yang, J., Shin, K.: Using hypervisor to provide data secrecy for user applications on a per-page basis. In: Proceedings of the Fourth ACM SIGPLAN/SIGOPS International Conference on Virtual Execution Environments, pp. 71–80. ACM (2008)
34. Azab, A., Ning, P., Wang, Z., Jiang, X., Zhang, X., Skalsky, N.: Hypersentry: Enabling stealthy in-context measurement of hypervisor integrity. In: Proceedings of the 17th ACM Conference on Computer and Communications Security, pp. 38–49. ACM (2010)
35. Wang, Z., Jiang, X.: Hypersafe: A lightweight approach to provide lifetime hypervisor control-flow integrity. In: 2010 IEEE Symposium on Security and Privacy, pp. 380–395. IEEE (2010)

Enforcing Subscription-Based Authorization Policies in Cloud Scenarios

Sabrina De Capitani di Vimercati[1], Sara Foresti[1],
Sushil Jajodia[2], and Giovanni Livraga[1]

[1] Università degli Studi di Milano, 26013 Crema, Italy
`firstname.lastname@unimi.it`
[2] George Mason University, Fairfax, VA 22030-4444, USA
`jajodia@gmu.edu`

Abstract. The rapid advances in the Information and Communication Technologies have brought to the development of on-demand high quality applications and services allowing users to easily access resources anywhere anytime. Users can pay for a service and access the resources made available during their subscriptions until the subscribed periods expire. Users are then forced to download such resources if they want to access them also after the subscribed periods. To avoid this burden to the users, we propose the adoption of a subscription-based access control policy that combines a flexible key derivation structure with selective encryption. The publication of new resources as well as the management of subscriptions are accommodated by adapting the key derivation structure in a transparent way for the users.

Keywords: access control, subscription-based policies, data outsourcing.

1   Introduction

The advances in the Information and Communication Technologies (ICTs) have driven the users into the Globalization era, where the techniques for processing, storing, and accessing information have radically changed. New emerging computing paradigms (e.g., data outsourcing and cloud computing) offer enormous advantages to both users and organizations. Users can now subscribe to a variety of services, and access them anywhere anytime: at home from their laptop, on the train from their tablet, or while waiting in a queue from their smartphone. Organizations are more and more resorting to external elastic storage and computational services for creating and running business over the Internet in new ways. Organizations can then provide large-scale cloud data services widely accessible to a variety of users, possibly restricting access to resources on the basis of users' subscriptions. These services can be offered at affordable prices, thanks to the use of external cloud storage servers for the management of data. As a side effect of this trend, security requirements are becoming more complex since cloud storage servers are typically trusted neither to access the resources content nor to restrict access to the services according to users' subscriptions.

N. Cuppens-Boulahia et al. (Eds.): DBSec 2012, LNCS 7371, pp. 314–329, 2012.

Emerging approaches in the data outsourcing scenario regulate access to resources through selective encryption, meaning that they translate the privilege to access a resource into the knowledge of the key used to encrypt the resource itself (e.g., [10]). These approaches, however, while representing important steps towards the support of flexible access control solutions in data outsourcing, are still in their infancy. In fact, they cannot easily support a scenario where both the set of users who can access a resource and the set of resources change frequently over time, due to new subscriptions and the publication of new resources. Also access control solutions developed for publish/subscribe systems (e.g., [11,20]), which may seem to have some similarities with the publication scenario we consider, are not suitable since they have been developed for a different problem. We take into account scenarios where users pay for a service and then can freely access the resources made available during their subscriptions. In this context, to access resources also after the expiration of their subscriptions, users can download the resources for which they are authorized to their local machine. Our proposal aims at avoiding this burden to the users allowing them to maintain the right to access such resources without the worry that they will lose this right after the expiration of their subscriptions. For instance, users who have purchased an annual subscription for 2012 for a magazine should be able to access all and only the issues of the magazine published in 2012, and should be able to access them even after December 31, 2012. We therefore propose an approach that takes advantage of selective encryption to guarantee that users who subscribe for a service can access all and only the resources published during their subscriptions, while allowing the resources to self-enforce the subscription-based restrictions. Before being stored on the cloud storage server, the resources are encrypted, and a key derivation structure is built to guarantee that they can be decrypted only by authorized users. The key derivation structure is updated whenever new resources are published, new subscriptions are received, or users withdraw from their subscriptions.

The remainder of this paper is organized as follows. Section 2 describes the considered scenario and the protection requirements that the access control system should satisfy. Section 3 formalizes the concept of subscription-based policy. Section 4 presents our techniques for enforcing a subscription-based policy. Section 5 illustrates how the system publishes resources and manages subscriptions. Section 6 discusses related work. Finally, Section 7 reports our conclusions.

2 Scenario, Protection Requirements, and Motivation

We consider a scenario where a *resource provider* uses an external *cloud storage server* for storing its resources, thus taking advantage of the cost savings that the cloud storage server can provide. The resource provider periodically publishes new resources that should be able to self-enforce restrictions on who can access them and should not be accessible to the cloud storage server. *Users* can subscribe to the services offered by the resource provider for different periods of time, and can withdraw from a subscription at any time. We assume that users are trusted, that is, they do not redistribute resources they can access to unauthorized users.

In the considered scenario, accesses to resources should be regulated by a *subscription-based access control policy* according to which users are authorized to access all and only the resources that have been published by the resource provider during their subscribed periods. A peculiarity of our scenario is that user authorizations remain valid also after the expiration of their subscriptions. The subscription-based access control policy takes then into consideration both the subscriptions of the users and the time when resources have been published. Existing solutions result limited for our scenario. We can classify such existing solutions in two main categories.

- *Account-based.* Traditional access control solutions (e.g., [17]), including those emerging in the data outsourcing scenario (e.g., [10]), are based on the assumption that when users leave the system their authorizations terminate and they cannot access the resources anymore. Furthermore, access control solutions for data outsourcing cannot easily support a dynamic scenario where resources are continuously created, and new users can join the system and old users can leave the system at any time.
- *Time-based.* Temporal-based access control solutions (e.g., [4]) enforce time restrictions in a way that is different from what we need. In fact, these solutions consider a scenario where resources are stored and managed by the party who creates them, and assume that authorizations apply only to specific time intervals and/or that authorizations can be applied following a periodic pattern (e.g., a user can access a file only during the working days from 8:00 a.m. to 5:00 p.m.).

We then put forward the idea of using the same principles at the basis of the access control solutions developed for the data outsourcing scenario (which encrypt resources for protecting their confidentiality from the storage server and adopt key derivation techniques for efficiently combining authorization-based access control and cryptographic protection) to enforce a subscription-based access control policy without delegating it to the cloud storage server. Our solution should guarantee the correct enforcement of the subscription-based access control policy (i.e., users should be able to access the resources made available during their subscribed periods also after the expiration of their subscriptions) and the *forward* and *backward* protection requirements. Forward protection means that users cannot access resources published before the beginning of their subscriptions (e.g, users who subscribe to a magazine for 2012 cannot access the issues of the magazine published before January 1, 2012). Backward protection means that users cannot access resources published after the expiration of their subscriptions (e.g., users who subscribe to a magazine for 2012 cannot access the issues of the magazine published during 2013). Like for the data outsourcing scenario, with our solution the published resources are encrypted so that they self-enforce the subscription-based access restrictions. In addition to the correct enforcement of the subscription-based policy and the satisfaction of the forward and backward protection requirements mentioned above, our solution should avoid re-encryption of resources and re-distribution of keys whenever users subscribe to services or withdraw from their subscriptions.

3   Subscription-Based Policy

A resource provider offers a set $\mathcal{P}$ of services to which users can subscribe. Each service $p \in \mathcal{P}$ consists in a period of publication of resources, and each user subscribing to service p can access all the resources published for p during her subscription. We denote with $\mathcal{U}$ and $\mathcal{R}$ the set of users subscribed to service p and the set of published resources for p, respectively. For simplicity, but without loss of generality, we focus on the management of accesses to a single service. We also note that, although in this paper we consider time-based subscriptions, our approach can be easily adapted to other scenarios where subscriptions to a service can be defined on the basis of different criteria (e.g., topic of interest, geographical region).

Given a *time domain* $(\mathcal{TS}, \leq)$, with $\mathcal{TS}$ a set of time instants and $\leq$ a total order relationship on $\mathcal{TS}$ [5], the resource provider assigns to each resource $r \in \mathcal{R}$ a timestamp $r.t$ in $\mathcal{TS}$ that represents the time when the resource has been published. The resource provider may combine contiguous time instants into time windows, defined on arbitrary granularities, forming a *time hierarchy*. Intuitively, these time windows represent the periods of time for which the resource provider allows users to subscribe to the service offered. Formally, a time hierarchy $\mathcal{H_T}$ is a pair $(\mathcal{T}, \succeq)$, where $\mathcal{T}$ is a set of time windows, and $\succeq$ is a partial order relationship over $\mathcal{T}$. A time window T_i in $\mathcal{T}$ is a pair $[t_i^s, t_i^e]$ of time instants and represents the set of time instants $t \in \mathcal{TS}$ such that $t_i^s \leq t \leq t_i^e$. Given two time windows T_i and T_j in $\mathcal{T}$, T_i *dominates* T_j, denoted $T_i \succeq T_j$, if $t_i^s \leq t_j^s$ and $t_j^e \leq t_i^e$ (i.e., the time instants in T_j represent a subset of the time instants in T_i). The leaves of the time hierarchy correspond to time instants in $\mathcal{TS}$, which can be seen as time windows with $t^s = t^e$. The time hierarchy can be graphically represented as a directed acyclic graph with vertices representing time windows in $\mathcal{T}$ and edges representing direct dominance relationships. For simplicity, but without loss of generality, in this paper we assume $\mathcal{H_T}$ to be a tree. As an example, consider resource provider Condé Nast, monthly publishing magazine *Glamour* and offering the possibility to buy subscriptions for a month (single issue), a trimester, a semester, or a year. Figure 1 illustrates the time hierarchy defined by the resource provider. For the sake of readability, in the figure we denote leaves with the time instant they represent. Each user $u \in \mathcal{U}$ can subscribe to the service offered by the resource provider for an arbitrary set, denoted $u.\mathcal{S}$, of time windows in $\mathcal{H_T}$ (i.e., $u.\mathcal{S} \subseteq \mathcal{T}$).

The timestamps assigned to resources along with the user subscriptions establish the set of resources that each user can access: user $u \in \mathcal{U}$ can access resource $r \in \mathcal{R}$ if she subscribed for a time window including $r.t$. Formally, the subscription-based policy regulating access to the resources is defined as follows.

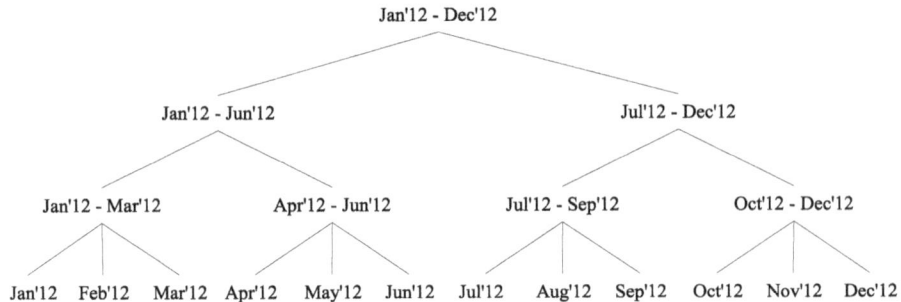

Fig. 1. An example of time hierarchy

Definition 1 (Subscription-based policy). *Let $\mathcal{H}_{\mathcal{T}}(\mathcal{T},\succeq)$ be a time hierarchy defined on time domain $(\mathcal{TS},\leq)$, $\mathcal{U}$ be a set of users with $u.\mathcal{S}\subseteq\mathcal{T}$ for all $u\in\mathcal{U}$, and $\mathcal{R}$ be a set of resources with $r.t\in\mathcal{TS}$ for all $r\in\mathcal{R}$. The* subscription-based policy $\mathcal{A}$ *on $\mathcal{U}$ and $\mathcal{R}$ grants $u\in\mathcal{U}$ access to $r\in\mathcal{R}$ iff $\exists[t^s,t^e]\in u.\mathcal{S}$ s.t. $t^s\leq r.t\leq t^e$.*

Example 1. Suppose that three issues of magazine *Glamour* have been published with timestamp Jan'12, Feb'12, and Mar'12, respectively (i.e., $\mathcal{R}=\{$ *Glam-01,Glam-02,Glam-03*$\}$). Assume now that two users $\mathcal{U}=\{$*Alice, Barbara*$\}$ subscribe to the magazine for the first trimester of 2012 ([Jan'12,Mar'12]), and for the first issue of the year ([Jan'12,Jan'12]), respectively. The subscription-based policy grants *Alice* access to all the issues of the magazine in $\mathcal{R}$, while it grants *Barbara* access only to the first issue *Glam-01*.

4 Graph Modeling of the Subscription-Based Policy

We propose to enforce the subscription-based policy by combining *selective encryption* [10] with a key derivation technique that uses a key derivation structure based on public *tokens* [1]. Given two keys k_i and k_j in a set $\mathcal{K}$ of keys, token $d_{i,j}=k_j\oplus h(k_i,l_j)$, with l_j a public label associated with k_j, $\oplus$ the bitwise xor operator, and h a deterministic cryptographic function, permits to derive k_j from the knowledge of k_i and label l_j. The derivation relationship between keys can be either *direct*, via a single token, or *indirect*, via a chain of tokens. Our idea consists in defining a key derivation structure so that each resource is encrypted only once with a single key, and each user receives only one key from which she can derive all and only the keys used for encrypting the resources that she can access according to the subscription-based policy. To fix ideas and make the discussion clear, we consider the system at a specific point in time when some resources have been published and some users have subscribed to the service offered by the resource provider. We first discuss how resources are encrypted and then describe how to model users' subscriptions.

   The techniques developed for enforcing an access control policy in the data outsourcing scenario build a key derivation structure on the basis of the sets of users that can access resources. In our scenario, such sets of users vary frequently over time, and therefore it is not convenient to exploit them for building the key derivation structure. We then use the time hierarchy $\mathcal{H}_\mathcal{T}$ defined by the resource provider as a key derivation structure where each time window is associated with a key, and each edge corresponds to a token. The timestamp associated with a published resource, therefore, identifies the time window in the time hierarchy representing the key used to encrypt the resource itself. The keys associated with time windows including more than a time instant (i.e., internal vertices) are not used for encrypting resources, but only for derivation purposes. Clearly, not all the time windows in the time hierarchy are necessary for enforcing the subscription-based policy, but only those corresponding to the timestamps of published resources along with all the time windows dominating them. For instance, with respect to Example 1, the time windows that must be represented in the key derivation structure are Jan'12, Feb'12, and Mar'12, which are the timestamps of the three published resources, and all the time windows dominating them in the time hierarchy in Figure 1, that is, [Jan'12,Mar'12], [Jan'12,Jun'12], and [Jan'12,Dec'12]. In this way, from the knowledge, for example, of the key associated with [Jan'12,Mar'12] we can derive the keys used for encrypting all the resources published during the first trimester of 2012.

For each user in the system, the resource provider generates a new key and communicates it to the user. With this unique key, the user should be able to access all and only the resources for which she is authorized according to her subscriptions. The idea is to "hook the user" through a token on each time window T for which she subscribed. In this way, the user can adopt her key to directly derive the key associated with time window T. From this key she can directly or indirectly derive the keys used to encrypt all and only the resources whose timestamp is included in T. For instance, according to the subscriptions in Example 1, *Alice* can access all the resources published in the first trimester of 2012. The resource provider then creates a token from *Alice*'s key to the key associated with [Jan'12,Mar'12]. By construction, all resources published in Jan'12, Feb'12, and Mar'12 will be encrypted with a key derivable from the key associated with [Jan'12,Mar'12], which *Alice* can derive. Note that it may happen that a user subscribes for a time window for which no resource has been published (e.g., a user subscribes to a magazine for April'12 and the issue of April has not been published yet). The key derivation structure must then include also the time windows representing users' subscriptions, along with their ancestors in $\mathcal{H}_\mathcal{T}$. The resulting key derivation structure, which we call *user and resource graph*, can be formally defined as follows.

Definition 2 (User and resource graph). *Let $\mathcal{H}_\mathcal{T}(\mathcal{T},\succeq)$ be a time hierarchy on time domain $(\mathcal{TS},\leq)$, $\mathcal{U}$ be a set of users with $u.\mathcal{S}\subseteq\mathcal{T}$ for all $u\in\mathcal{U}$, and $\mathcal{R}$ be a set of resources with $r.t\in\mathcal{TS}$ for all $r\in\mathcal{R}$. A user and resource graph over $\mathcal{U}$, $\mathcal{R}$, and $\mathcal{H}_\mathcal{T}$ is a graph $G(V,E)$, with:*

- *$V = \mathcal{T}_r \cup \mathcal{T}_s \cup \mathcal{T}_p \cup \mathcal{U}$, with $\mathcal{T}_r = \bigcup_{r\in\mathcal{R}}[r.t,r.t]$, $\mathcal{T}_s = \bigcup_{u\in\mathcal{U}}u.\mathcal{S}$, and $\mathcal{T}_p = \{T \in \mathcal{T} \mid \exists T' \in \mathcal{T}_s \cup \mathcal{T}_r$ such that $T\succeq T'\}$*

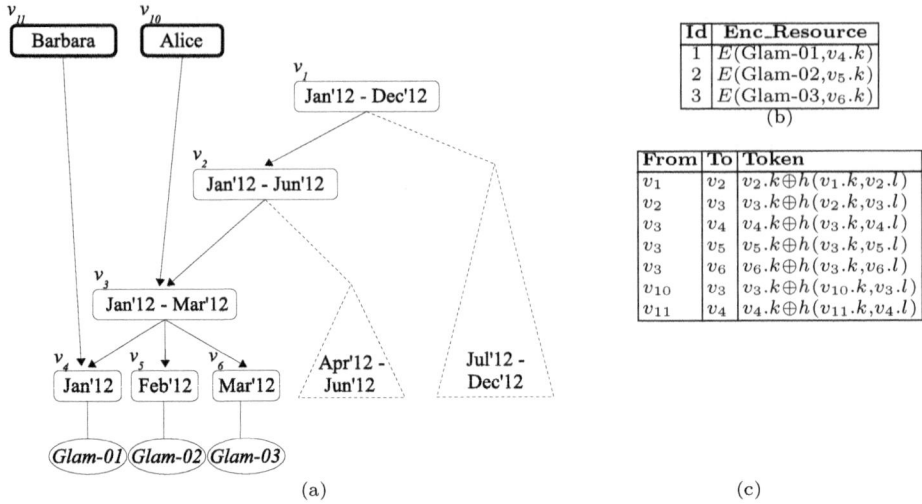

Fig. 2. An example of user and resource graph (a), published resources (b), and token catalog (c)

$$- \; E = \{(u,T) \mid u \in \mathcal{U} \wedge T \in V \backslash \mathcal{U} \wedge T \in u.\mathcal{S}\} \cup$$
$$\{(T_i,T_j) \mid T_i,T_j \in V \backslash \mathcal{U} \wedge T_i \succeq T_j \wedge (\nexists T_z \in V \backslash \mathcal{U}, \; T_i \succeq T_z \succeq T_j \wedge T_z \neq T_i \neq T_j)\}$$

The vertices in the user and resource graph represent the keys of the system, while the edges represent the tokens in the *token catalog* $\mathcal{D}$ stored at the external cloud storage server together with the encrypted resources.

Example 2. Consider the time hierarchy in Figure 1 and the subscription-based policy in Example 1. Figure 2(a) shows the corresponding user and resource graph, where dotted triangles represent subtrees of the time hierarchy that are not associated with a vertex in the graph. For the sake of clarity, the figure also reports the published resources, represented as ovals connected with the vertices in the graph representing their timestamp and whose keys are used to encrypt them. Figure 2(b) shows the encrypted resources stored at the external cloud storage server, with **Id** the resource identifier and **Enc_Resource** the encrypted resource ($E(r, k)$ denotes the encryption of r with k), and Figure 2(c) illustrates the token catalog resulting from the user and resource graph in Figure 2(a).

The user and resource graph in Definition 2 guarantees the correct enforcement of the subscription-based policy since each user can decrypt all and only the resources with a timestamp included in at least one of the time windows in the user's subscriptions. This is formalized by the following theorem, whose proof is omitted from the paper for space constraints.

Theorem 1 (Correct enforcement of subscription-based policy). *Let $\mathcal{H}_T(\mathcal{T}, \succeq)$ be a time hierarchy on time domain $(\mathcal{TS}, \leq)$, $\mathcal{U}$ be a set of users with*

PUBLISH_RESOURCE(r)
1: $\mathcal{R} := \mathcal{R} \cup \{r\}$
2: $v := \textbf{Get\_Vertex}([r.t,r.t])$ /* retrieve the vertex representing the timestamp of the resource */
3: **Encrypt**$(r,v.k)$
4: publish the encrypted resource

GET_VERTEX(T)
5: **if** $T{\in}V$ **then** /* T already belongs to G */
6:    let $v{\in}V$ be the vertex with $v{=}T$
7:    **return**(v)
8: generate vertex $v := T$
9: generate encryption key $v.k$
10: generate public label $v.l$
11: $V := V \cup \{v\}$ /* insert the vertex into the user and resource graph */
12: let $T_i{\in}\mathcal{T}$: $T_i{\succeq}T \wedge \nexists T_j$: $T_i{\succeq}T_j{\succeq}T$, $T_j{\neq}T_i{\neq}T$ /* determine the direct ancestor of T in $\mathcal{H}_\mathcal{T}$ */
13: **if** $T_i{\neq}$NULL **then**
14:   $v_i := \textbf{Get\_Vertex}(T_i)$ /* retrieve the vertex in G that represents T_i */
15:   $E := E \cup \{(v_i,v)\}$ /* insert the edge connecting T_i to T in G */
16:   $\mathcal{D} := \mathcal{D} \cup \{v.k{\oplus}h(v_i.k,v.l)\}$ /* publish the corresponding token */
17: **return**(v)

Fig. 3. Pseudocodes of procedure **Publish_Resource** and function **Get_Vertex**

$u.\mathcal{S}{\subseteq}\mathcal{T}$ *for all* $u{\in}\mathcal{U}$, *and* $\mathcal{R}$ *be a set of resources with* $r.t{\in}\mathcal{TS}$ *for all* $r{\in}\mathcal{R}$. *The user and resource graph* $G(V,E)$ *correctly enforces a subscription-based policy* $\mathcal{A}$ *on* $\mathcal{U}$ *and* $\mathcal{R}$ *when* $\forall u{\in}\mathcal{U}$, $\forall r{\in}\mathcal{R}$:

$$\exists [t^s,t^e] \in u.\mathcal{S} \ s.t. \ t^s{\leq}r.t{\leq}t^e \iff \langle u,[r.t,r.t]\rangle \ is \ a \ path \ in \ G.$$

5   Management of Resources and Subscriptions

Whenever there is a change in the subscription-based policy (e.g., a new resource is published, a user subscribes to a service for a specific time window, or a user decides to withdraw from a subscription), the user and resource graph has to be updated accordingly. In the following, we discuss how changes to the policy can be managed in a transparent way for the users.

5.1   Resource Publishing

At initialization time, the user and resource graph is empty (no key is necessary for resource encryption) and it is dynamically built as resources are published. Figure 3 illustrates the pseudocode of procedure **Publish_Resource** that the resource provider calls whenever it needs to publish a resource. The procedure takes a resource r as input and publishes its encrypted representation. The procedure first calls function **Get_Vertex** on time window $T{=}[r.t,r.t]$ (line 2). This function checks whether the vertex representing $[r.t,r.t]$ is in the user and resource graph, since its key has to be used for encrypting r. If such a vertex exists, the function returns it (lines 5–7). Otherwise, the function first creates a vertex v representing T, along with the corresponding encryption key $v.k$ and public label $v.l$, and inserts v into the set V of vertices of the user and resource

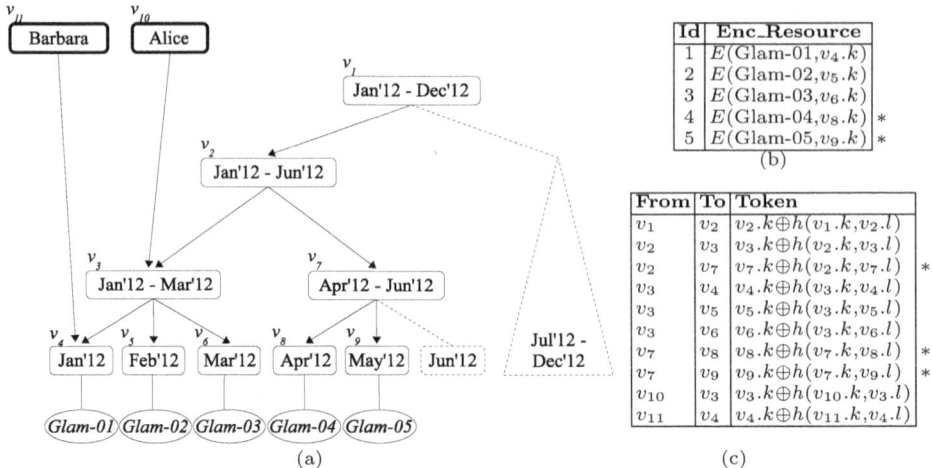

Fig. 4. User and resource graph (a), published resources (b), and token catalog after *Glam-04* and *Glam-05* are published (c)

graph (lines 8-11). To guarantee that the time window T_i directly dominating T in the time hierarchy is represented in the user and resource graph, function **Get_Vertex** recursively calls itself on T_i, obtaining the vertex v_i representing T_i in the graph (lines 12-14). The function inserts into G edge (v_i,v) and publishes the corresponding token (lines 15-16). We note that the recursive nature of function **Get_Vertex** guarantees that all the ancestors of T in $\mathcal{H}_\mathcal{T}$ are represented by a vertex in the user and resource graph, and that each vertex is connected to all its direct descendants represented in the graph. The function then returns vertex v representing $[r.t,r.t]$ (line 17). Finally, procedure **Publish_Resource** encrypts r with $v.k$ and publishes the resulting encrypted resource (lines 3-4).

Example 3. Consider the user and resource graph, published resources, and token catalog in Figure 2 and assume that Condé Nast publishes the fourth issue of *Glamour* in April'12. The resource provider calls procedure **Publish_Resource** on resource *Glam-04* that in turn calls function **Get_Vertex** on [Apr'12,Apr'12]. The function inserts vertex v_8 representing [Apr'12,Apr'12] and its direct ancestor v_7 representing [Apr'12,Jun'12]. Procedure **Publish_Resource** then encrypts *Glam-04* with the key of vertex v_8. Assume now that Condé Nast publishes the fifth issue of *Glamour* in May'12, calling procedure **Publish_Resource** on resource *Glam-05*. Function **Get_Vertex** inserts vertex v_9 representing [May'12,May'12] and directly connects it to [Apr'12,Jun'12], since it is already included in the graph. Resource *Glam-05* is encrypted with the key of vertex v_9. Figure 4 illustrates the resulting user and resource graph, published resources, and token catalog, where new resources and tokens are denoted with a *.

SUBSCRIBE(u,T)
1: **if** $u \notin \mathcal{U}$ **then** /* u is a new user in the system */
2:   $\mathcal{U} := \mathcal{U} \cup \{u\}$
3:     generate vertex $v_u := u$
4:     generate encryption key $v_u.k$
5:     generate public label $v_u.l$
6:   $V := V \cup \{v_u\}$
7: **else** let $v_u \in V$ be the vertex with $v_u = u$
8:   $u.\mathcal{S} := u.\mathcal{S} \cup \{T\}$
9:   $v_T := $ **Get_Vertex**(T)
10:   $E := E \cup \{(v_u,v_T)\}$
11:   $\mathcal{D} := \mathcal{D} \cup \{v_T.k \oplus h(v_u.k,v_T.l)\}$
12:   let $T_i \in \mathcal{T}$: $T_i \succeq T \wedge (\nexists T_j: T_i \succeq T_j \succeq T, T_j \neq T_i \neq T)$ /* determine the direct ancestor of T in $\mathcal{H_T}$ */
13:   $\mathcal{T}' := \{T_j \in u.\mathcal{S} \mid T_i \succeq T_j \wedge (\nexists T_z \in \mathcal{T}: T_i \succeq T_z \succeq T_j, T_i \neq T_z \neq T_j)\}$
14:   **if** $\bigcup_{T_j \in \mathcal{T}'} T_j = T_i$ **then**
15:     $u.\mathcal{S} := u.\mathcal{S} \setminus \mathcal{T}'$
16:     $E := E \setminus \{(v_i,v_j) \mid v_i = u \wedge v_j = T_j, T_j \in \mathcal{T}'\}$
17:     $\mathcal{D} := \mathcal{D} \setminus \{v_j.k \oplus h(v_i.k,v_j.l) \mid v_i = u \wedge v_j = T_j, T_j \in \mathcal{T}'\}$
18:     **Subscribe**(u,T_i)

Fig. 5. Pseudocode of procedure **Subscribe**

5.2 New Subscription

Both new and existing users can subscribe to a service for a time window at any point in time (i.e., before the beginning, during, or even after the expiration of the window). Figure 5 illustrates procedure **Subscribe** that manages new subscriptions. The procedure takes a user u and a time window T as input and works as follows. If u is a new user, the procedure creates a vertex v_u representing u, her encryption key $v_u.k$, and public label $v_u.l$ (lines 1-6). Otherwise, the procedure identifies the vertex v_u representing the user in G (line 7). The procedure then inserts T into $u.\mathcal{S}$, calls function **Get_Vertex** on T so that the vertex v_T representing T and its ancestors are possibly added to the graph, and inserts edge (v_u,v_T) in the user and resource graph, publishing the corresponding token (lines 8-11). Through this token, the user can directly derive from her key the key of the time window to which she is subscribing.

To keep the number of tokens under control, the procedure verifies whether the set $u.\mathcal{S}$ of subscriptions includes all the time windows directly dominated by T_i that in turn directly dominates T in $\mathcal{H_T}$ (e.g., a user may be subscribed for three issues of a magazine that correspond to a trimester). In this case, instead of maintaining a token from u to all the direct descendants of T_i, it is possible to replace them with a single token from vertex u to T_i. To this purpose, procedure **Subscribe** identifies the direct ancestor T_i of the time window T to which u is subscribing and checks if $u.\mathcal{S}$ includes all the descendants $T_j,\ldots,T_l$ of T_i (lines 12-14). In this case, it removes $T_j,\ldots,T_l$ from $u.\mathcal{S}$, the edges connecting v_u to the vertices representing them, and the corresponding tokens (lines 15-17). The procedure then recursively calls itself to subscribe u to T_i to possibly propagate up in the graph this factorization (line 18).

Example 4. Consider the user and resource graph, published resources, and token catalog in Figure 4, and assume that *Alice* renews her subscription to

324   S. De Capitani di Vimercati et al.

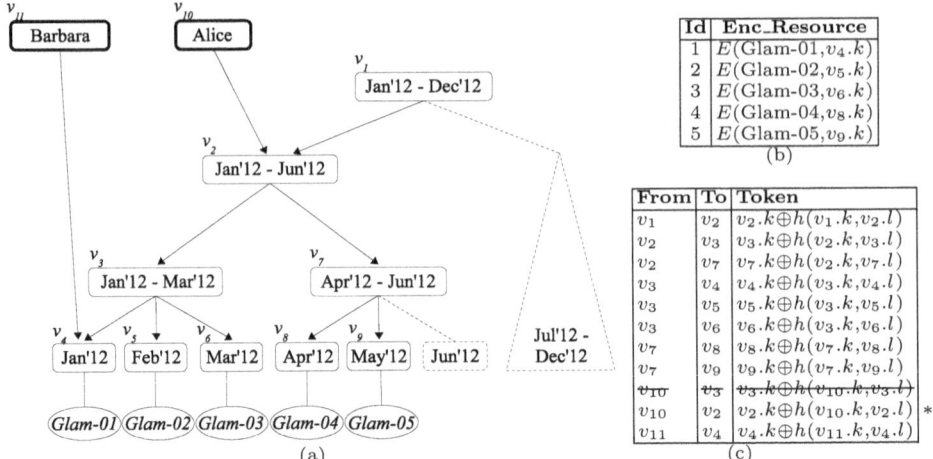

Fig. 6. User and resource graph (a), published resources (b), and token catalog after *Alice* subscribes for [Apr'12,Jun'12] (c)

Glamour for trimester [Apr'12,Jun'12]. Since both *Alice* and [Apr'12,Jun'12] are already in the graph (vertices v_{10} and v_7, respectively), procedure **Subscribe** only inserts edge (v_{10},v_7) and publishes the corresponding token. Renewing her subscription, *Alice* is now subscribed for the first semester of year 2012. Procedure **Subscribe** factorizes the two subscriptions for [Jan'12,Mar'12] and [Apr'12,Jun'12] in a unique subscription for [Jan'12,Jun'12]. Figure 6 illustrates the resulting user and resource graph, published resources, and token catalog (removed tokens are ~~crossed out~~). Assume now that *Carol* joins the system and subscribes for [Apr'12,Jun'12]. Procedure **Subscribe** first inserts vertex v_{12} representing *Carol* in the graph, and communicates her the corresponding key. It then inserts edge (v_{12},v_7) in the graph. Figure 7 illustrates the resulting user and resource graph, published resources, and token catalog.

5.3   Withdrawal from a Subscription

As our system provides high flexibility in defining the time windows available for subscription, withdrawal from a subscription represents an exception in the working of the system and must be managed as a special case. In fact, no action is needed when a subscription naturally expires. When a user withdraws from a subscription for time window $[t^s,t^e]$, starting from time instant t, the resource provider must guarantee that: *i)* she cannot access the resources with timestamp in $(t,t^e]$ (backward protection), and *ii)* she continues to access the resources with timestamp in $[t^s,t]$. For instance, consider Example 4. In May'12 *Alice* could decide to withdraw from her subscription for the first semester of year 2012. In this case, she should not be able to decrypt the issue of June of the magazine, while she will continue to access the issues of January, February, March, April, and May. Clearly, a user can withdraw from her subscription at time t only

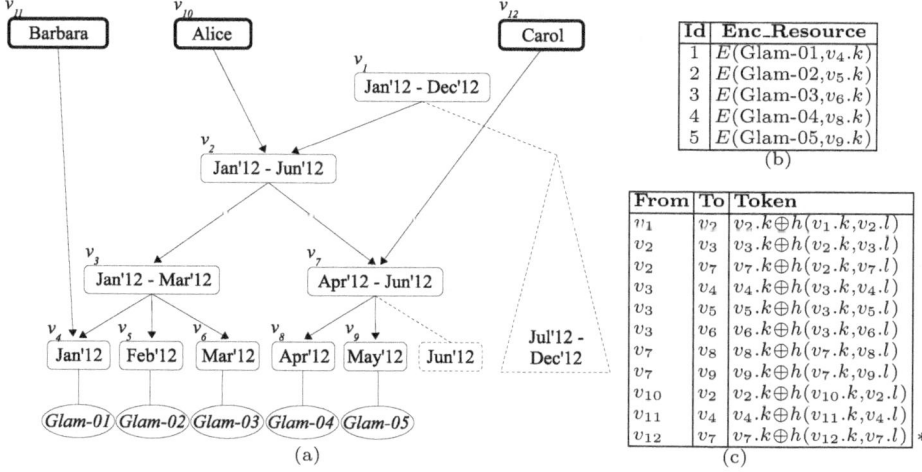

Fig. 7. User and resource graph (a), published resources (b), and token catalog after *Carol* subscribes for [Apr'12,Jun'12] (c)

if no resource with timestamp in $(t,t^e]$ has been published yet, since otherwise she could have accessed it before withdrawal. To guarantee that withdrawals are transparent for all the users and cause a limited overhead to the resource provider, our approach avoids re-keying and re-encryption operations.

Figure 8 illustrates procedure **Withdraw_Subscription**, which takes a user u and a time instant t as input, and updates the user and resource graph. The procedure first identifies the vertex v_u representing the user in G and the time window $[t^s,t^e]$ in $u.\mathcal{S}$ that includes t (lines 1-2). If such a time window does not exist or if at least a resource with timestamp in $(t,t^e]$ has been published, the procedure terminates notifying the problem to the resource provider (line 3). Otherwise, procedure **Withdraw_Subscription** removes the subscription by first substituting $[t^s,t^e]$ with $[t^s,t]$ in $u.\mathcal{S}$ (line 4). Since user u already knows the keys of the vertices along the path from vertex $[t^s,t^e]$ to t if they are represented in the user and resource graph, the resource provider must guarantee that all the resources with a timestamp following t will be encrypted with a key that is not derivable from the keys along this path. To this purpose, the procedure updates the time window $[t_i^s,t_i^e]$ that each of these vertices represents by setting t_i^e to t, creates a new set of vertices representing the time windows that has been changed, and connects them in a path of the user and resource graph. Also, the procedure inserts an edge between each new vertex $[t_i^s,t_i^e]$ to vertex $[t_i^s,t]$ since $[t_i^s,t_i^e]$ clearly dominates $[t_i^s,t]$. Finally, for each user u such that $[t_i^s,t_i^e]\in u.\mathcal{S}$, the procedure substitutes the token (and corresponding edge) between u and $[t_i^s,t]$ (i.e., the vertex that represented $[t_i^s,t_i^e]$ before the change performed by procedure **Withdraw_Subscription**) with the token (and corresponding edge) between u and the new vertex representing $[t_i^s,t_i^e]$, to preserve her ability to derive all the keys of the time windows dominated by $[t_i^s,t_i^e]$.

WITHDRAW_SUBSCRIPTION(u,t)
1: let $v_u \in V$ be the vertex with $v_u = u$
2: let $T=[t^s,t^e] \in u.\mathcal{S}$ s.t. $t^s \leq t \leq t^e$
3: **if** T=NULL $\vee$ ($\exists r \in \mathcal{R}$ s.t. $t < r.t \leq t^e$) **then exit**
4: $u.\mathcal{S} := u.\mathcal{S} \backslash \{T\} \cup \{[t^s,t]\}$ /* update the time window in user subscriptions */
5: let $v_T \in V$ be the vertex with $v_T = T$
6: **while** $t^e \neq t \wedge t^s \neq t^e \wedge T \in V$ **do** /* visit the path from T to $[t,t]$ */
7:    $T_{new} := [t^s,t^e]$
8:    $v_T := [t^s,t]$ /* update the label of the vertex */
9:    $v_{new} := $ **Get_Vertex**(T_{new}) /* create a vertex representing T_{new} */
10:    $E := E \cup \{(v_{new},v_T)\}$ /* $[t^s,t^e]$ dominates $[t^s,t]$ */
11:    $\mathcal{D} := \mathcal{D} \cup \{v_T.k \oplus h(v_{new}.k,v_T.l)\}$
12:    **for each** (v_u,v_T) s.t. $v_u \in \mathcal{U} \backslash \{u\}$ **do** /* update users' subscriptions */
13:       $E := E \cup \{(v_u,v_{new})\} \backslash \{(v_u,v_T)\}$
14:       $\mathcal{D} := \mathcal{D} \cup \{v_{new}.k \oplus h(v_u.k,v_{new}.l)\} \backslash \{v_T.k \oplus h(v_u.k,v_T.l)\}$
15:    let $T=[t^s,t^e] \in \mathcal{T}$ s.t. $T_{new} \succeq T \wedge t^s \leq t \leq t^e \wedge \nexists T_j : T_{new} \succeq T_j \succeq T, T_j \neq T_{new} \neq T$
16:    let $v_T \in V$ be the vertex with $v_T = T$

Fig. 8. Pseudocode of procedure **Withdraw_Subscription**

Note that the keys along the path from T to t, whose time windows have been updated by procedure **Withdraw_Subscription**, are not affected. Therefore, users who have already computed these keys can still use their local copy. The number of additional vertices and edges in the user and resource graph is limited and is at most h-1 and 2(h-1), respectively, where h is the height of the time hierarchy. The number of updated edges is $|\mathcal{U}|$-1 in the worst case.

Example 5. Consider the user and resource graph, published resources, and token catalog in Figure 7, and assume that *Alice* withdraws from her subscription in May'12. Procedure **Withdraw_Subscription** updates her subscription for [Jan'12,Jun'12] to [Jan'12,May'12], and visits the path from vertex v_2 (representing [Jan'12,Jun'12]) to the vertex representing [May'12,May'12]. First, it visits vertex v_2, updates its time window to [Jan'12,May'12], creates a new vertex v_2' for time window [Jan'12,Jun'12], and inserts edge (v_2',v_2) in the user and resource graph. The procedure executes the same operations when visiting v_7. Since *Carol* should still be able to access all the issues of *Glamour* published in [Apr'12,Jun'12], the procedure substitutes edge (v_{12},v_7) with edge (v_{12},v_7'). From her key *Alice* can derive, after this update, the keys used to encrypt the issues published in [Jan'12,May'12], while *Carol* can still derive keys used to encrypt issues published in [Apr'12,Jun'12]. Figure 9 illustrates the user and resource graph, published resources, and token catalog after *Alice*'s withdrawal.

5.4   Correctness

The procedures described in this section correctly enforce changes to the subscription-based policy. This is formally stated by the following theorem, whose proof is omitted from the paper for space constraints.

Theorem 2 (Correct enforcement of policy updates). *Let $\mathcal{H}_T(\mathcal{T},\succeq)$ be a time hierarchy on time domain $(\mathcal{TS},\leq)$, $\mathcal{U}$ be a set of users with $u.\mathcal{S} \subseteq \mathcal{T}$ for all*

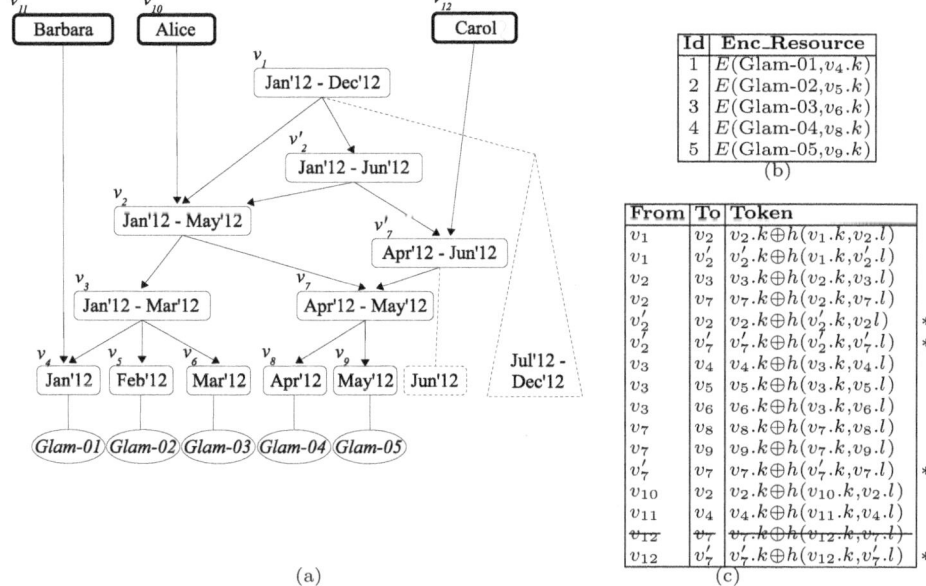

Id	Enc_Resource
1	$E(\text{Glam-01},v_4.k)$
2	$E(\text{Glam-02},v_5.k)$
3	$E(\text{Glam-03},v_6.k)$
4	$E(\text{Glam-04},v_8.k)$
5	$E(\text{Glam-05},v_9.k)$

(b)

From	To	Token	
v_1	v_2	$v_2.k\oplus h(v_1.k,v_2.l)$	
v_1	v_2'	$v_2'.k\oplus h(v_1.k,v_2'.l)$	
v_2	v_3	$v_3.k\oplus h(v_2.k,v_3.l)$	
v_2	v_7	$v_7.k\oplus h(v_2.k,v_7.l)$	
v_2'	v_2	$v_2.k\oplus h(v_2'.k,v_2 l)$	*
v_2'	v_7'	$v_7'.k\oplus h(v_2'.k,v_7'.l)$	*
v_3	v_4	$v_4.k\oplus h(v_3.k,v_4.l)$	
v_3	v_5	$v_5.k\oplus h(v_3.k,v_5.l)$	
v_3	v_6	$v_6.k\oplus h(v_3.k,v_6.l)$	
v_7	v_8	$v_8.k\oplus h(v_7.k,v_8.l)$	
v_7	v_9	$v_9.k\oplus h(v_7.k,v_9.l)$	
v_7'	v_7	$v_7.k\oplus h(v_7'.k,v_7.l)$	*
v_{10}	v_2	$v_2.k\oplus h(v_{10}.k,v_2.l)$	
v_{11}	v_4	$v_4.k\oplus h(v_{11}.k,v_4.l)$	
$\sout{v_{12}}$	$\sout{v_7}$	$\sout{v_7.k\oplus h(v_{12}.k,v_7.l)}$	
v_{12}	v_7'	$v_7'.k\oplus h(v_{12}.k,v_7'.l)$	*

(c)

(a)

Fig. 9. User and resource graph (a), published resources (b), and token catalog after *Alice* withdraws from her subscription in May'12 (c)

$u\in\mathcal{U}$, $\mathcal{R}$ be a set of resources with $r.t\in\mathcal{TS}$ for all $r\in\mathcal{R}$, and $G(V,E)$ be the user and resource graph over $\mathcal{U}$, $\mathcal{R}$, and $\mathcal{H_T}$.

1. Procedure **Publish_Resource**(r) generates a user and resource graph that correctly enforces the subscription-based policy on $\mathcal{U}$ and $\mathcal{R}\cup\{r\}$.
2. Procedure **Subscribe**(u,T) generates a user and resource graph that correctly enforces the subscription-based policy on $\mathcal{U}\cup\{u\}$ and $\mathcal{R}$, with $u.\mathcal{S}\cup\{T\}$.
3. Procedure **Withdraw_Subscription**(u,t) generates a user and resource graph that correctly enforces the subscription-based policy on $\mathcal{U}$ and $\mathcal{R}$, with $u.\mathcal{S}\setminus\{[t^s,t^e]\}\cup\{[t^s,t]\}$.

6   Related Work

Previous work close to ours is in the area of data outsourcing [18], where many approaches focused on efficient query evaluation at the external server (e.g., [8,12,21]), and on guaranteeing data integrity and authenticity (e.g., [15]). Recent works have also addressed access control enforcement (e.g., [10,14,23]), but these approaches are not suited for the scenario considered in this paper, as they assume the sets of users, resources, and authorizations not to change frequently.

The problem of enforcing access control policies with time-based restrictions has been widely studied (e.g., [4,19]). However, these works restrict access to resources depending on the time when the access is requested. Recently, time-based access control restrictions have been enforced also in the data outsourcing scenario, by integrating them in the key derivation process (e.g., [2,3,7]). The solutions in [2,3] allow users to derive encryption keys only within the time windows for which they are authorized. The approach in [7] proposes instead a more general model for enforcing any interval-based restriction (e.g., time and space). These solutions mainly focus on the security of key derivation and on minimizing the number of edges in the key derivation graph. Our proposal is instead aimed at correctly enforcing a subscription-based policy and at guaranteeing transparency for users in subscription management and resource publishing.

Our work may bring some resemblance with access control in publish/subscribe systems, characterized by a set of users who publish events, a set of users who subscribe to the system declaring their interests, and a service responsible to deliver published events to the users whose interests match with the event attributes [11,20]. However, in publish/subscribe systems the access control policy depends on some properties related to the events. Also, publish/subscribe systems typically rely on a trusted party that can access events and enforce access restrictions.

Another related but different line of work addresses the problem of enforcing time-based restrictions to users when accessing broadcasting services (e.g., [6,22]). These approaches are not applicable in our scenario where we assume to publish persistent resources as opposed to data streams.

7   Conclusions

We proposed an approach for effectively restricting access to published resources based on the subscriptions of the users to a service. Our solution is based on selective encryption so that the encrypted resources self-enforce the subscription-based restrictions. A key derivation structure is also used for easily enforcing changes in the subscription-based policy due to the addition of new users and resources, and to the withdrawal of users from their subscriptions.

Acknowledgements. We would like to thank Pierangela Samarati for discussions, suggestions, and comments. This work was partially supported by the Italian Ministry of Research within the PRIN 2008 project "PEPPER" (2008SY2PH4). The work of Sushil Jajodia was partially supported by the National Science Foundation under grants CCF-1037987 and CT-20013A.

References

1. Atallah, M.J., Blanton, M., Fazio, N., Frikken, K.B.: Dynamic and efficient key management for access hierarchies. ACM TISSEC 12(3), 18:1–18:43 (2009)
2. Atallah, M.J., Blanton, M., Frikken, K.B.: Incorporating Temporal Capabilities in Existing Key Management Schemes. In: Biskup, J., López, J. (eds.) ESORICS 2007. LNCS, vol. 4734, pp. 515–530. Springer, Heidelberg (2007)
3. Ateniese, G., De Santis, A., Ferrara, A.L., Masucci, B.: Provably-secure time-bound hierarchical key assignment schemes. Journal of Cryptology 25(2), 243–270 (2012)

 4. Bertino, E., Bettini, C., Ferrari, E., Samarati, P.: An access control model support-
    ing periodicity constraints and temporal reasoning. ACM TODS 23(3), 231–285
    (1998)
 5. Bettini, C., Dyreson, C.E., Evans, W.S., Snodgrass, R.T., Wang, X.S.: A Glos-
    sary of Time Granularity Concepts. In: Etzion, O., Jajodia, S., Sripada, S. (eds.)
    Dagstuhl Seminar 1997. LNCS, vol. 1399, pp. 406–413. Springer, Heidelberg (1998)
 6. Blanton, M., Frikken, K.B.: Efficient Multi-dimensional Key Management in
    Broadcast Services. In: Gritzalis, D., Preneel, B., Theoharidou, M. (eds.) ESORICS
    2010. LNCS, vol. 6345, pp. 424–440. Springer, Heidelberg (2010)
 7. Crampton, J.: Practical and efficient cryptographic enforcement of interval-based
    access control policies. ACM TISSEC 14(1), 14:1–14:30 (2011)
 8. Damiani, E., De Capitani di Vimercati, S., Jajodia, S., Paraboschi, S., Samarati, P.:
    Balancing confidentiality and efficiency in untrusted relational DBMSs. In: Proc.
    of CCS 2003, Washington, DC, USA (October 2003)
 9. De Capitani di Vimercati, S., Foresti, S., Jajodia, S., Paraboschi, S., Samarati, P.:
    A data outsourcing architecture combining cryptography and access control. In:
    Proc. of CSAW 2007, Fairfax, VA, USA (November 2007)
10. De Capitani di Vimercati, S., Foresti, S., Jajodia, S., Paraboschi, S., Samarati, P.:
    Encryption policies for regulating access to outsourced data. ACM TODS 35(2),
    12:1–12:46 (2010)
11. Eugster, P.T., Felber, P.A., Guerraoui, R., Kermarrec, A.: The many faces of pub-
    lish/subscribe. ACM CSUR 35(2), 114–131 (2003)
12. Hacigümüs, H., Iyer, B., Mehrotra, S., Li, C.: Executing SQL over encrypted data
    in the database-service-provider model. In: Proc. of the SIGMOD 2002, Madison,
    WI, USA (June 2002)
13. Jhawar, R., Piuri, V., Santambrogio, M.D.: A comprehensive conceptual system-
    level approach to fault tolerance in cloud computing. In: Proc. of IEEE SysCon
    2012, Vancouver, BC, Canada (March 2012)
14. Miklau, G., Suciu, D.: Controlling access to published data using cryptography. In:
    Proc. of VLDB 2003, Berlin, Germany (September 2003)
15. Mykletun, E., Narasimha, M., Tsudik, G.: Authentication and integrity in out-
    sourced databases. ACM TOS 2(2), 107–138 (2006)
16. Preda, S., Cuppens-Boulahia, N., Cuppens, F., Toutain, L.: Architecture-aware
    adaptive deployment of contextual security policies. In: Proc. of ARES 2010,
    Krakow, Poland (2010)
17. Samarati, P., De Capitani di Vimercati, S.: Access Control: Policies, Models, and
    Mechanisms. In: Focardi, R., Gorrieri, R. (eds.) FOSAD 2000. LNCS, vol. 2171,
    pp. 137–196. Springer, Heidelberg (2001)
18. Samarati, P., De Capitani di Vimercati, S.: Data protection in outsourcing scenarios:
    Issues and directions. In: Proc. of ASIACCS 2010, Beijing, China (April 2010)
19. Toahchoodee, M., Ray, I.: On the formalization and analysis of a spatio-temporal
    role-based access control model. JCS 19(3), 399–452 (2011)
20. Wang, C., Carzaniga, A., Evans, D., Wolf, A.: Security issues and requirements for
    internet-scale publish-subscribe systems. In: Proc. of HICSS 2002, Big Island, HI,
    USA (January 2002)
21. Wang, H., Lakshmanan, L.V.S.: Efficient secure query evaluation over encrypted
    XML databases. In: Proc. of VLDB 2006, Seoul, Korea (September 2006)
22. Wong, C.K., Gouda, M., Lam, S.S.: Secure group communications using key graphs.
    IEEE/ACM TON 8(1), 16–30 (2000)
23. Yu, S., Wang, C., Ren, K., Lou, W.: Achieving secure, scalable, and fine-grained
    data access control in cloud computing. In: Proc. of INFOCOM 2010, San Diego,
    CA, USA (March 2010)

Author Index

GPSR Compliance

The European Union's (EU) General Product Safety Regulation (GPSR) is a set of rules that requires consumer products to be safe and our obligations to ensure this.

If you have any concerns about our products, you can contact us on ProductSafety@springernature.com

In case Publisher is established outside the EU, the EU authorized representative is:

Springer Nature Customer Service Center GmbH
Europaplatz 3
69115 Heidelberg, Germany

Batch number: 09474011

Printed by Printforce, the Netherlands